Calcium-Binding Protein Protocols
Volume I

METHODS IN MOLECULAR BIOLOGY™

John M. Walker, Series Editor

204. **Molecular Cytogenetics:** *Methods and Protocols,* edited by *Yao-Shan Fan, 2002*
203. **In Situ Detection of DNA Damage:** *Methods and Protocols,* edited by *Vladimir V. Didenko, 2002*
202. **Thyroid Hormone Receptors:** *Methods and Protocols,* edited by *Aria Baniahmad, 2002*
201. **Combinatorial Library Methods and Protocols,** edited by *Lisa B. English, 2002*
200. **DNA Methylation Protocols,** edited by *Ken I. Mills and Bernie H, Ramsahoye, 2002*
199. **Liposome Methods and Protocols,** edited by *Subhash C. Basu and Manju Basu, 2002*
198. **Neural Stem Cells:** *Methods and Protocols,* edited by *Tanja Zigova, Juan R. Sanchez-Ramos, and Paul R. Sanberg, 2002*
197. **Mitochondrial DNA:** *Methods and Protocols,* edited by *William C. Copeland, 2002*
196. **Oxidants and Antioxidants:** *Ultrastructural and Molecular Biology Protocols,* edited by *Donald Armstrong, 2002*
195. **Quantitative Trait Loci:** *Methods and Protocols,* edited by *Nicola J. Camp and Angela Cox, 2002*
194. **Post-translational Modification Reactions,** edited by *Christoph Kannicht, 2002*
193. **RT-PCR Protocols,** edited by *Joseph O'Connell, 2002*
192. **PCR Cloning Protocols, 2nd ed.,** edited by *Bing-Yuan Chen and Harry W. Janes, 2002*
191. **Telomeres and Telomerase:** *Methods and Protocols,* edited by *John A. Double and Michael J. Thompson, 2002*
190. **High Throughput Screening:** *Methods and Protocols,* edited by *William P. Janzen, 2002*
189. **GTPase Protocols:** *The RAS Superfamily,* edited by *Edward J. Manser and Thomas Leung, 2002*
188. **Epithelial Cell Culture Protocols,** edited by *Clare Wise, 2002*
187. **PCR Mutation Detection Protocols,** edited by *Bimal D. M. Theophilus and Ralph Rapley, 2002*
186. **Oxidative Stress and Antioxidant Protocols,** edited by *Donald Armstrong, 2002*
185. **Embryonic Stem Cells:** *Methods and Protocols,* edited by *Kursad Turksen, 2002*
184. **Biostatistical Methods,** edited by *Stephen W. Looney, 2002*
183. **Green Fluorescent Protein:** *Applications and Protocols,* edited by *Barry W. Hicks, 2002*
182. **In Vitro Mutagenesis Protocols, 2nd ed.,** edited by *Jeff Braman, 2002*
181. **Genomic Imprinting:** *Methods and Protocols,* edited by *Andrew Ward, 2002*
180. **Transgenesis Techniques, 2nd ed.:** *Principles and Protocols,* edited by *Alan R. Clarke, 2002*
179. **Gene Probes:** *Principles and Protocols,* edited by *Marilena Aquino de Muro and Ralph Rapley, 2002*
178. **Antibody Phage Display:** *Methods and Protocols,* edited by *Philippa M. O'Brien and Robert Aitken, 2001*
177. **Two-Hybrid Systems:** *Methods and Protocols,* edited by *Paul N. MacDonald, 2001*
176. **Steroid Receptor Methods:** *Protocols and Assays,* edited by *Benjamin A. Lieberman, 2001*
175. **Genomics Protocols,** edited by *Michael P. Starkey and Ramnath Elaswarapu, 2001*
174. **Epstein-Barr Virus Protocols,** edited by *Joanna B. Wilson and Gerhard H. W. May, 2001*

173. **Calcium-Binding Protein Protocols, Volume 2:** *Methods and Techniques,* edited by *Hans J. Vogel, 2001*
172. **Calcium-Binding Protein Protocols, Volume 1:** *Reviews and Case Histories,* edited by *Hans J. Vogel, 2001*
171. **Proteoglycan Protocols,** edited by *Renato V. Iozzo, 2001*
170. **DNA Arrays:** *Methods and Protocols,* edited by *Jang B. Rampal, 2001*
169. **Neurotrophin Protocols,** edited by *Robert A. Rush, 2001*
168. **Protein Structure, Stability, and Folding,** edited by *Kenneth P. Murphy, 2001*
167. **DNA Sequencing Protocols,** *Second Edition,* edited by *Colin A. Graham and Alison J. M. Hill, 2001*
166. **Immunotoxin Methods and Protocols,** edited by *Walter A. Hall, 2001*
165. **SV40 Protocols,** edited by *Leda Raptis, 2001*
164. **Kinesin Protocols,** edited by *Isabelle Vernos, 2001*
163. **Capillary Electrophoresis of Nucleic Acids, Volume 2:** *Practical Applications of Capillary Electrophoresis,* edited by *Keith R. Mitchelson and Jing Cheng, 2001*
162. **Capillary Electrophoresis of Nucleic Acids, Volume 1:** *Introduction to the Capillary Electrophoresis of Nucleic Acids,* edited by *Keith R. Mitchelson and Jing Cheng, 2001*
161. **Cytoskeleton Methods and Protocols,** edited by *Ray H. Gavin, 2001*
160. **Nuclease Methods and Protocols,** edited by *Catherine H. Schein, 2001*
159. **Amino Acid Analysis Protocols,** edited by *Catherine Cooper, Nicole Packer, and Keith Williams, 2001*
158. **Gene Knockoout Protocols,** edited by *Martin J. Tymms and Ismail Kola, 2001*
157. **Mycotoxin Protocols,** edited by *Mary W. Trucksess and Albert E. Pohland, 2001*
156. **Antigen Processing and Presentation Protocols,** edited by *Joyce C. Solheim, 2001*
155. **Adipose Tissue Protocols,** edited by *Gérard Ailhaud, 2000*
154. **Connexin Methods and Protocols,** edited by *Roberto Bruzzone and Christian Giaume, 2001*
153. **Neuropeptide Y Protocols,** edited by *Ambikaipakan Balasubramaniam, 2000*
152. **DNA Repair Protocols:** *Prokaryotic Systems,* edited by *Patrick Vaughan, 2000*
151. **Matrix Metalloproteinase Protocols,** edited by *Ian M. Clark, 2001*
150. **Complement Methods and Protocols,** edited by *B. Paul Morgan, 2000*
149. **The ELISA Guidebook,** edited by *John R. Crowther, 2000*
148. **DNA–Protein Interactions:** *Principles and Protocols* **(2nd ed.),** edited by *Tom Moss, 2001*
147. **Affinity Chromatography:** *Methods and Protocols,* edited by *Pascal Bailon, George K. Ehrlich, Wen-Jian Fung, and Wolfgang Berthold, 2000*
146. **Mass Spectrometry of Proteins and Peptides,** edited by *John R. Chapman, 2000*
145. **Bacterial Toxins:** *Methods and Protocols,* edited by *Otto Holst, 2000*
144. **Calpain Methods and Protocols,** edited by *John S. Elce, 2000*
143. **Protein Structure Prediction:** *Methods and Protocols,* edited by *David Webster, 2000*
142. **Transforming Growth Factor-Beta Protocols,** edited by *Philip H. Howe, 2000*

METHODS IN MOLECULAR BIOLOGY™

Calcium-Binding Protein Protocols

Volume 1: Reviews and Case Studies

Edited by

Hans J. Vogel

*Department of Biological Sciences, University of Calgary
Calgary, AB, Canada*

Humana Press ❋ Totowa, New Jersey

© 2002 Humana Press Inc.
999 Riverview Drive, Suite 208
Totowa, New Jersey 07512

www.humanapress.com

All rights reserved. No part of this book may be reproduced, stored in a retrieval system, or transmitted in any form or by any means, electronic, mechanical, photocopying, microfilming, recording, or otherwise without written permission from the Publisher. Methods in Molecular Biology™ is a trademark of The Humana Press Inc.

The content and opinions expressed in this book are the sole work of the authors and editors, who have warranted due diligence in the creation and issuance of their work. The publisher, editors, and authors are not responsible for errors or omissions or for any consequences arising from the information or opinions presented in this book and make no warranty, express or implied, with respect to its contents.

This publication is printed on acid-free paper. ∞
ANSI Z39.48-1984 (American Standards Institute) Permanence of Paper for Printed Library Materials.

Cover design by Patricia F. Cleary.
Cover illustration: From Fig. 1A in Chapter 3, Vol. 1 "Crystal Structure of Calpain and Insights into Ca^{2+}-Dependent Activation" by Zongchao Jia, Christopher M. Hosfield, Peter L. Davies, and John S. Elce.

Production Editor: Kim Hoather-Potter.

For additional copies, pricing for bulk purchases, and/or information about other Humana titles, contact Humana at the above address or at any of the following numbers: Tel: 973-256-1699; Fax: 973-256-8341; E-mail: humana@humanapr.com, or visit our Website at www.humanapress.com

Photocopy Authorization Policy:
Authorization to photocopy items for internal or personal use, or the internal or personal use of specific clients, is granted by Humana Press Inc., provided that the base fee of US $10.00 per copy, plus US $00.25 per page, is paid directly to the Copyright Clearance Center at 222 Rosewood Drive, Danvers, MA 01923. For those organizations that have been granted a photocopy license from the CCC, a separate system of payment has been arranged and is acceptable to Humana Press Inc. The fee code for users of the Transactional Reporting Service is: [0-89603-688-X/02 $10.00 + $00.25].

Printed in the United States of America. 10 9 8 7 6 5 4 3 2 1

Library of Congress Cataloging in Publication Data

Main entry under title: Methods in molecular biology™.

Calcium-binding protein protocols / edited by Hans J. Vogel
    p. cm. -- (Methods in molecular biology; v. v. 172-)
  Includes bibliographical references and index.
  Contents: v. 1. Reviews and case studies.
  ISBN 0-89603-688-X (alk. paper)
  1. Calcium-binding proteins--Research--Methodology I. Vogel, Hans J. II. Methods in molecular biology (Clifton, N.J.) ; . v. 172, etc.

QP552.C24 C33 2001
572'.69--dc21
                                                                                01-063354

Dedication

This book is dedicated to the memory of Dr. J. David Johnson (Columbus, OH) whose untimely death on January 21, 2000 has deeply shocked all his colleagues and friends. David has made numerous excellent contributions to our understanding of calcium-binding proteins. His insight and enthusiasm will be sadly missed.

Hans J. Vogel, PhD

Preface

Calcium plays an important role in a wide variety of biological processes. This divalent metal ion can bind to a large number of proteins; by doing so it modifies their biological activity or their stability. Because of its distinct chemical properties calcium is uniquely suited to act as an on–off switch or as a light dimmer of biological activities. The two books entitled *Calcium-Binding Protein Protocols* (Volumes I and II) focus on modern experimental analyses and methodologies for the study of calcium-binding proteins. Both extracellular and intracellular calcium-binding proteins are discussed in detail. However, proteins involved in calcium handling (e.g., calcium pumps and calcium channels), fall outside of the scope of these two volumes. Also, calcium-binding proteins involved in bone deposition will not be discussed, as this specific topic has been addressed previously. The focus of these two books is on studies of the calcium-binding proteins and their behavior in vitro and in vivo. The primary emphasis is on protein chemistry and biophysical methods. Many of the methods described will also be applicable to proteins that do not bind calcium.

Calcium-Binding Protein Protocols is divided into three main sections. The section entitled *Introduction and Reviews* provides information on the role of calcium in intracellular secondary messenger activation mechanisms. Moreover, unique aspects of calcium chemistry and the utilization of calcium in dairy proteins, as well as calcium-binding proteins involved in blood clotting, are addressed. The second section entitled *Calcium-Binding Proteins: Case Studies* provides a wealth of information about protein purification and characterization strategies, X-ray crystallography and other studies that are focused on specific calcium-binding proteins. Together, these two sections comprise Volume I of this series. By introducing the various classes of intra- and extracellular calcium-binding proteins and their modes of action, these two sections set the stage and provide the necessary background for the third section. The final section entitled *Methods and Techniques to Study Calcium-Binding Proteins* makes up Volume II of *Calcium-Binding Protein Protocols*. Here the focus is on the use of a range of modern experimental techniques that can be employed to study the solution structure, stability, dynamics, calcium-binding properties, and biological activity of calcium-binding proteins in general. As well, studies of their ligand-binding properties and their distribution in cells are included. In addition to enzymatic assays and more routine spectroscopic and protein chemistry techniques, particular attention has been paid in the second volume to modern NMR approaches, thermodynamic analyses, kinetic measurements such as surface plasmon resonance, strategies for amino acid sequence alignments, as well as fluorescence methods to study the distribution of calcium and calcium-binding proteins in cells. In preparing their chapters, all the authors have attempted to share the little secrets that are required to successfully apply these methods to related proteins. Together the two volumes of *Calcium-Binding Protein Protocols* provide the reader with a host of experimental methods that can be applied either to uncover new aspects of earlier characterized calcium-binding proteins or to study newly discovered proteins.

As more and more calcium-binding proteins are being uncovered through genome sequencing efforts and protein interaction studies (e.g., affinity chromatography, crosslinking or yeast two-hybrid systems) the time seemed right to collect all the methods used to characterize these proteins in a book. The methods detailed here should provide the reader with the essential tools for their analysis in terms of structure, dynamics, and function. The hope is that these two volumes will contribute to our understanding of the part of the proteome, which relies on interactions with calcium to carry out its functions.

In closing, I would like to thank Margaret Tew for her invaluable assistance with the editing and organization of these two books. Finally, I would like to thank the authors of the individual chapters, who are all experts in this field, for their cooperation in producing these two volumes in a timely fashion.

Hans J. Vogel, PhD

Contents

Dedication v
Preface vii
Contents of Companion Volume xi
Contributors xiii

PART I. INTRODUCTION AND REVIEWS

1. Calcium-Binding Proteins
   Hans J. Vogel, Richard D. Brokx, and Hui Ouyang 3
2. Calcium
   Robert J. P. Williams 21
3. Crystal Structure of Calpain and Insights into Ca^{2+}-Dependent Activation
   Zongchao Jia, Christopher M. Hosfield, Peter L. Davies, and John S. Elce 51
4. The Multifunctional S100 Protein Family
   Claus W. Heizmann 69
5. Ca^{2+} Binding to Proteins Containing γ-Carboxyglutamic Acid Residues
   Egon Persson 81
6. The Caseins of Milk as Calcium-Binding Proteins
   Harold M. Farrell, Jr., Thomas F. Kumosinski, Edyth L. Malin, and Eleanor M. Brown 97

PART II. CALCIUM-BINDING PROTEINS: CASE STUDIES

7. Preparation of Recombinant Plant Calmodulin Isoforms
   Raymond E. Zielinski 143
8. Isolation of Recombinant Cardiac Troponin C
   John A. Putkey and Wen Liu 151
9. Skeletal Muscle Troponin C: *Expression and Purification of the Recombinant Intact Protein and Its Isolated N- and C-Domain Fragments*
   Joyce R. Pearlstone and Lawrence B. Smillie 161
10. Purification of Recombinant Calbindin D_{9k}
    Eva Thulin 175

11	S100 Proteins: *From Purification to Functions* **Jean Christophe Deloulme, Gaëlh Ouengue Mbele, and Jacques Baudier** 185
12	Cadherins **Jean-René Alattia, Kit I. Tong, Masatoshi Takeichi, and Mitsuhiko Ikura** 199
13	α-Lactalbumin and (Calcium-Binding) Lysozyme **Katsutoshi Nitta** 211
14	Recombinant Annexin II Tetramer **Hyoung-Min Kang, Nolan R. Filipenko, Geetha Kassam, and David M. Waisman** 225
15	Purification and Characterization of ALG-2: *A Novel Apoptosis-Linked Ca^{2+}-Binding Protein* **Mingjie Zhang and Kevin W.-H. Lo** 235
16	Crystallization and Structural Details of Ca^{2+}-Induced Conformational Changes in the EF-Hand Domain VI of Calpain **Miroslaw Cygler, Pawel Grochulski, and Helen Blanchard** 243
17	Neurocalcin: *Role in Neuronal Signaling* **Senadhi Vijay-Kumar and Vinod D. Kumar** 261
18	Crystallization and Structure–Function of Calsequestrin **ChulHee Kang, William R. Trumble, and A. Keith Dunker** 281
19	Use of Fluorescence Resonance Energy Transfer to Monitor Ca^{2+}-Triggered Membrane Docking of C2 Domains **Eric A. Nalefski and Joseph J. Falke** 295
20	Ca^{2+}-Binding Mode of the C_2A-Domain of Synaptotagmin **Josep Rizo, Josep Ubach, and Jesús García** 305
21	Study of Calcineurin Structure by Limited Proteolysis **Seun-Ah Yang and Claude Klee** 317
Index	 335

CONTENTS OF THE COMPANION VOLUME

Calcium-Binding Protein Protocols

Volume II: Methods and Techniques

PART III. METHODS AND TECHNIQUES TO STUDY CALCIUM-BINDING PROTEINS

1. Quantitative Analysis of Ca^{2+}-Binding by Flow Dialysis
   Michio Yazawa
2. Calcium Binding to Proteins Studied via Competition with Chromophoric Chelators
   Sara Linse
3. Deconvolution of Calcium-Binding Curves: *Facts and Fantasies*
   Jacques Haiech and Marie-Claude Kilhoffer
4. Absorption and Circular Dichroism Spectroscopy
   Stephen R. Martin and Peter M. Bayley
5. Fourier Transform Infrared Spectroscopy of Calcium-Binding Proteins
   Heinz Fabian and Hans J. Vogel
6. Steady-State Fluorescence Spectroscopy
   Aalim M. Weljie and Hans J. Vogel
7. Fluorescence Methods for Measuring Calcium Affinity and Calcium Exchange with Proteins
   J. David Johnson and Svetlana B. Tikunova
8. Surface Plasmon Resonance of Calcium-Binding Proteins
   Karin Julenius
9. Differential Scanning Calorimetry
   Maria M. Lopez and George I. Makhatadze
10. Isothermal Titration Calorimetry
    Maria M. Lopez and George I. Makhatadze
11. Multiangle Laser Light Scattering and Sedimentation Equilibrium
    Leslie D. Hicks, Jean-René Alattia, Mitsuhiko Ikuru, and Cyril M. Kay
12. Small-Angle Solution Scattering Reveals Information on Conformational Dynamics in Calcium-Binding Proteins and in their Interactions with Regulatory Targets
    Jill Trewhella and Joanna K. Krueger
13. Investigation of Calcium-Binding Proteins Using Electrospray Ionization Mass Spectrometry
    Amanda L. Doherty-Kirby and Gilles A. Lajoie
14. Synthetic Calcium-Binding Peptides
    Gary S. Shaw

15 Proteolytic Fragments of Calcium-Binding Proteins
    Richard D. Brokx and Hans J. Vogel
16 Electron Magnetic Resonance Studies of Calcium-Binding Proteins
    Lawrence J. Berliner
17 Cadmium-113 and Lead-207 NMR Spectroscopic Studies of Calcium-Binding Proteins
    Teresa E. Clarke and Hans J. Vogel
18 Calcium-43 NMR of Calcium-Binding Proteins
    Torbjörn Drakenberg
19 Exploring Familial Relationships Using Multiple Sequence Alignment
    Aalim M. Weljie and Jaap Heringa
20 Structure Determination by NMR: *Isotope Labeling*
    Monica X. Li, David C. Corson, and Brian D. Sykes
21 Protein Structure Calculation from NMR Data
    Tapas K. Mal, Stefan Bagby, and Mitsuhiko Ikura
22 Shape and Dynamics of a Calcium-Binding Protein Investigated by Nitrogen-15 NMR Relaxation
    Jörn M. Werner, Iain D. Campbell, and A. Kristina Downing
23 The Use of Dipolar Couplings for the Structure Refinement of a Pair of Calcium Binding EGF Domains
    Jonathan Boyd, Iain D. Campbell, and A. Kristina Downing
24 Vector Geometry Mapping: *A Method to Characterize the Conformation of Helix-Loop-Helix Calcium Binding Proteins*
    Kyoko L. Yap, James B. Ames, Mark B. Swindells, and Mitsuhiko Ikura
25 Use of Calmodulin Antagonists and S-100 Protein Interacting Drugs for Affinity Chromatography
    Ryoji Kobayashi
26 Enzymatic Assays to Compare Calmodulin Isoforms, Mutants, and Chimeras
    Michael P. Walsh, Jacquelyn E. Van Lierop, Cindy Sutherland, Ritsu Kondo, and J. David Johnson
27 Gene Expression in Transfected Cells
    Kate Hughes, Juha Saarikettu, and Thomas Grundström
28 Monitoring the Intracellular Free Ca^{2+}-Calmodulin Concentration with Genetically-Encoded Fluorescent Indicator Proteins
    Anthony Persechini
29 Studying the Spatial Distribution of Ca^{2+}-Binding Proteins: *How Does it Work for Calmodulin?*
    Katalin Török, Richard Thorogate, and Steven Howell

Contributors

JEAN-RENÉ ALATTIA • *Division of Molecular and Structural Biology, Ontario Cancer Institute, Department of Medical Biophysics, University of Toronto, Toronto, ON, Canada*

JACQUES BAUDIER • *Departement de Biologie Moleculaire et Structurale du CEA, EMI-INSERM104, Grenoble Cedex, France*

HELEN BLANCHARD • *Biochemisches Institüt, Universität Zürich, Zürich, Switzerland*

RICHARD D. BROKX • *Department of Biological Sciences, University of Calgary, Calgary, AB, Canada*

ELEANOR M. BROWN • *Eastern Regional Research Center, United States Department of Agriculture, ARS, Wyndmoor, PA*

MIROSLAW CYGLER • *National Research Council, Montreal, Biotechnology Research Institute, Quebec, Canada*

PETER L. DAVIES • *Department of Biochemistry, Queen's University and the Protein Engineering Network of Centres of Excellence, Kingston, ON, Canada*

JEAN CHRISTOPHE DELOULME • *Departement de Biologie Moleculaire et Structurale du CEA, EMI-INSERM104, Grenoble Cedex, France*

A. KEITH DUNKER • *Department of Biochemistry and Biophysics, Washington State University, Pullman, WA*

JOHN S. ELCE • *Department of Biochemistry, Queen's University and the Protein Engineering Network of Centres of Excellence, Kingston, ON, Canada*

JOSEPH J. FALKE • *Department of Chemistry and Biochemistry, University of Colorado, Boulder, CO*

HAROLD M. FARRELL, JR. • *Eastern Regional Research Center, United States Department of Agriculture, ARS, Wyndmoor, PA*

NOLAN R. FILIPENKO • *Department of Biochemistry and Molecular Biology, University of Calgary, Calgary, AB, Canada*

JESÚS GARCÍA • *Departments of Biochemistry and Pharmacology, University of Texas Southwestern Medical Centre, Dallas, TX*

PAWEL GROCHULSKI • *BioMep Inc., Department of Biochemistry, University of Montreal, Quebec, Canada*

CLAUS W. HEIZMANN • *Department of Pediatrics, Division of Clinical Chemistry and Biochemistry, University of Zurich, Zurich, Switzerland*

CHRISTOPHER M. HOSFIELD • *Department of Biochemistry, Queen's University and the Protein Engineering Network of Centres of Excellence, Kingston, ON, Canada*

MITSUHIKO IKURA • *Division of Medical and Structural Biology, Department of Medical Biophysics, Ontario Cancer Institute, University of Toronto, Toronto, ON, Canada*

ZONGCHAO JIA • *Department of Biochemistry, Queen's University and the Protein Engineering Network of Centres of Excellence, Kingston, ON, Canada*

CHULHEE KANG • *Department of Biochemistry and Biophysics, Washington State University, Pullman, Washington*

HYOUNG-MIN KANG • *Department of Laboratory Medicine, University of Washington, Seattle, WA*

GEETHA KASSAM • *Department of Biochemistry and Molecular Biology, University of Calgary, Calgary, AB, Canada*

CLAUDE KLEE • *Laboratory of Biochemistry, National Cancer Institute, Bethesda, MD*

VINOD D. KUMAR • *Department of Microbiology and Immunology, Kimmel Cancer Institute, Thomas Jefferson University, Philadelphia, PA*

THOMAS F. KUMOSINSKI • *Eastern Regional Research Center, United States Department of Agriculture, ARS, Wyndmoor, PA*

WEN LIU • *Department of Biochemistry and Molecular Biology, University of Texas Medical School, Houston, TX*

KEVIN W.-H. LO • *Department of Biochemistry, Hong Kong University of Science and Technology, Hong Kong, PR China*

EDYTH L. MALIN • *Eastern Regional Research Center, United States Department of Agriculture, ARS, Wyndmoor, PA*

GAËLH OUENGUE MBELE • *Departement de Biologie Moleculaire et Structurale du CEA, EMI-INSERM104, Grenoble Cedex, France*

ERIC A. NALEFSKI • *Department of Chemistry and Biochemistry, University of Colorado, Boulder, CO*

KATSUTOSHI NITTA • *Division of Biological Sciences, Graduate School of Science, Hokkaido University, Kitaku, Sapporo, Japan*

HUI OUYANG • *Department of Biological Sciences, University of Calgary, Calgary, AB, Canada*

JOYCE R. PEARLSTONE • *Department of Biochemistry, University of Alberta, Edmonton, AB, Canada*

EGON PERSSON • *Vascular Biochemistry, Novo Nordisk A/S, Novo Nordisk Park, Malov, Denmark*

JOHN A. PUTKEY • *Department of Biochemistry and Molecular Biology, University of Texas Medical Center, Houston, TX*

Contributors

JOSEP RIZO • *Departments of Biochemistry and Pharmacology, University of Texas Southwestern Medical Centre, Dallas, TX*

LAWRENCE B. SMILLIE • *Department of Biochemistry, University of Alberta, Edmonton, AB, Canada*

MASATOSHI TAKEICHI • *Department of Biophysics, Faculty of Science, Kyoto University, Kitashirakawa, Sakyo-ku, Kyoto 606, Japan*

EVA THULIN • *Department of Physical Chemistry 2, Lund University, Lund, Sweden*

KIT I. TONG • *Division of Molecular and Structural Biology, Ontario Cancer Institute, Department of Medical Biophysics, University of Toronto, Toronto, ON, Canada*

WILLIAM R. TRUMBLE • *Agricultural Experiment Station, University of New Hampshire, Durham, New Hampshire*

JOSEP UBACH • *Departments of Biochemistry and Pharmacology, University of Texas Southwestern Medical Centre, Dallas, TX*

SENADHI VIJAY-KUMAR • *Department of Biochemistry, Fels Institute for Cancer Research and Molecular Biology, Temple University School of Medicine, Philadelphia, PA*

HANS J. VOGEL • *Department of Biological Sciences, University of Calgary, Calgary, AB, Canada*

DAVID M. WAISMAN • *Department of Biochemistry and Molecular Biology, University of Calgary, Calgary, AB, Canada*

ROBERT J. P. WILLIAMS • *Inorganic Chemistry Laboratory, University of Oxford, Oxford, UK*

SEUN-AH YANG • *Laboratory of Biochemistry, National Cancer Institute, Bethesda, MD*

MINGJIE ZHANG • *Department of Biochemistry, Hong Kong University of Science and Technology, Hong Kong, PR China*

RAYMOND E. ZIELINSKI • *Department of Plant Biology, University of Illinois, Urbana, IL*

I

INTRODUCTION AND REVIEWS

1

Calcium-Binding Proteins

Hans J. Vogel, Richard D. Brokx, and Hui Ouyang

1. Introduction

Calcium plays an important role in many biological processes. At first, it might seem that the most important role of calcium in biology is a structural one, and indeed it may be argued that this is true. Hydroxyapatite, a co-crystal of calcium, phosphate, and hydroxide ions, forms the matrix of tooth enamel, the hardest substance in the human body. Calcium phosphate is also responsible for the rigidity of bone, and the deposition of this matrix in bone is a very tightly controlled biological process (1). In fact, the vast majority of calcium (more than 99%) is immobilized in bones and teeth in humans. Poor diet or improper regulation of calcium deposition can lead to diseases such as childhood rickets or osteoporosis in older adults. Moreover, calcium is also important structurally in other organisms; for example, calcium carbonate is the major component of egg shells and also of the exoskeleton of animals such as mollusks and barnacles. Nutritionally, calcium is found in many foods, but of course the major source in most human diets is dairy products. In milk, a predominant class of proteins is the caseins, which function to solubulize calcium phosphate microgranules by surrounding them in a micellar structure (2), providing an important mineral nutrient in liquid form.

In this book, our attention is focused primarily on the role of the Ca^{2+} ion in solution, rather than on the solid deposits aforementioned. Numerous calcium-binding proteins interact with calcium in solution, mediating a wide range of physiological processes. For example, calcium triggers the binding of many proteins to biological membranes (3), or it plays a role as an intracellular messenger. In this introductory chapter, we will briefly review the role of calcium-binding proteins in the extracellular and intracellular environment. Because the calcium concentrations in these environments are quite different, these proteins have characteristically different calcium-binding motifs. Finally, we will discuss the ubiquitous regulatory protein calmodulin, for which the calcium activation mechanism has been clarified through numerous studies.

2. Extracellular Calcium-Binding Proteins

Calcium plays a key role in the blood-clotting process, where many enzymes involved in the blood-clotting cascade have several post-translationally modified γ-carboxyglutamate (Gla) residues in an N-terminal Gla domain, that specifically binds calcium. It was originally thought that the Ca^{2+} ion, by binding to the Gla residues, forms a bridge that allows these proteins to bind directly to the phospholipid membranes. It appears now that this is likely not the case; Ca^{2+}-dependent membrane binding may be brought about through the exposure of hydrophobic residues on the protein *(4)*, although this is still open to debate *(3)*. The absence of Ca^{2+} severely limits blood clotting, and, in fact, the removal of free Ca^{2+} by the addition of chelators is a popular method of preventing blood samples from clotting in clinical laboratories. In addition to the Gla domains, many blood-clotting factors, as well as other extracellular proteins, have epidermal growth factor-like (EGF) domains, some of which bind calcium *(5–8)*. Many other extracellular proteins also bind Ca^{2+} ions, in which the Ca^{2+} ion usually plays a structural role *(6)*. These include the C-type lysozymes and the related protein α-lactalbumin *(9–11)*, a protein involved in lactose biosynthesis, the pancreatic enzyme deoxyribonuclease I *(12)* and the bacterial protease subtilisin *(13)*. Interestingly, for subtilisin and the related subtilases, the number of Ca^{2+} ions bound and the tightness of binding correlate well with the thermostability of the enzyme *(13,* and references therein). Intracellular proteolytic enzymes bind Ca^{2+} as well, but here the role of the Ca^{2+} ion is often more complex. Calpains are cytosolic thiol proteases with substrates involved in many different cell-signaling processes. In the absence of Ca^{2+}, the active site of these enzymes is not functional *(14,15)*, requiring Ca^{2+} as the activator of the protease rather than a more conventional mechanism of zymogen activation, such as limited cleavage.

Many actin filament-severing proteins are also Ca^{2+}-binding proteins, including gelsolin *(16)*, villin *(17)*, and adseverin *(18)*. These proteins sever actin filaments and cap the high-affinity end of these filaments upon binding Ca^{2+}. Gelsolin is the first protein identified to have actin filament severing activity, and most proteins in this family are homologs of gelsolin *(19)*. Gelsolin contains six structurally related domains, which are probably the result of gene triplication followed by gene duplication. X-ray crystallography *(20,21)* reveals that the binding of Ca^{2+} by gelsolin induces a dramatic conformational change in the protein that exposes the actin-binding sites, which are not sufficiently exposed in Ca^{2+}-free gelsolin.

3. Calcium Metabolism

The relative abundance and physiological function of the Ca^{2+} ion inside the cell is quite different from that in extracellular matrices such as blood plasma; furthermore, intracellular Ca^{2+} concentrations are tightly controlled. This leads to perhaps the most important biological role of Ca^{2+}: that of a secondary messenger in eukaryotic cells. Usually, a eukaryotic cell exists at a so-called resting state, where the cytoplasmic level of the Ca^{2+} ion is quite low, typically 10^{-8}–10^{-7} M *(22)*, compared to an extracellular concentration of ~10^{-3} M. This low cytoplasmic concentration of Ca^{2+} is maintained by an extensive array of ATP-dependent Ca^{2+} pumps (Ca^{2+}-ATPases), which pump Ca^{2+} out of the cell or into specialized organelles such as the sarcoplasmic reticulum in muscle cells. However, external stimuli, such as hormonal signals or neural impulses, cause an increase in cytoplasmic Ca^{2+} to an activated state of ~10^{-6} M *(22,23)*. This activation is short-lived as the cell usually quickly returns to resting Ca^{2+} levels through the action of the aforementioned Ca^{2+}-ATPases, as well as Ca^{2+} "buffering" proteins, such as calsequestrin and parvalbumin. Parvalbumin is a muscle protein with a high Ca^{2+}-binding affinity, whose function is to remove free Ca^{2+} in the cytosol much faster than the outward Ca^{2+} pumps can operate. This buffering property is particularly important for cells requiring very fast relaxation, like for fast muscle cells *(24)*. The "spike" in cytoplasmic Ca^{2+} is sometimes termed a "calcium transient." This calcium transient is a very important cellular signal, and thus Ca^{2+} is an essential secondary messenger.

It is curious why calcium, and not some other cation, plays a pivotal role in signal transduction. It is believed that this has to do with the chemistry of calcium. Ca^{2+} belongs to the "hard" class of metal ions (for a discussion on the (bio)chemistry of metal ions *see* Chapter 2 in this volume, and **refs. 25** and **26**). It favors oxygen ligands such as the carbonyl and carboxyl groups of proteins. The most obvious alternative to play the role of Ca^{2+} would be Mg^{2+}, another readily available, soluble, divalent metal cation of the alkaline earth metal group. Like Ca^{2+}, it favors oxygen ligands but an important difference is that Mg^{2+} has a much slower rate for loss of water molecules from its hydration shell *(25)*, and its complexes are generally only six coordinate, with water molecules occupying at least two sites. Mg^{2+}, then, is not favorable for the high-coordinate irregular binding sites of Ca^{2+}-binding proteins. It is these types of sites that produce the greatest structural change upon metal binding, which enables Ca^{2+} binding to act as a conformational trigger.

Other ions may also be considered for this type of signaling role. Monovalent ions such as Na^+, K^+, or Cl^-, are readily bioavailable, but they generally act only as bulk ions and are not considered to be important protein ligands. Other di- or trivalent metals such as Zn^{2+}, Mn^{2+}, Cu^{2+}, or Fe^{3+}, can certainly bind to proteins, but their off-rate is usually too slow to be effective as a reversible conformational switch. These ions generally play the role of structural stabilization or act as Lewis acid or redox catalysts in enzymatic reactions. Moreover, the solubility of these larger metal ions is often low, making them difficult to obtain in sufficient amounts to play a signaling role.

As an aside, Ca^{2+} insolubility may be problematic in the cell as well. Although Ca^{2+} exists at an extracellular concentration of 10^{-3} M, inorganic phosphate and phosphate-containing ligands such as phosphoproteins, phosphosugars, and nucleic acids, would form insoluble complexes with Ca^{2+} at high concentration *(22)*. Consequently, even during a Ca^{2+} transient the cytosolic level of Ca^{2+} generally only raises 10-fold to a concentration of ~10^{-6} M and never approaches the levels outside the cell. Although this is only a modest increase in intracellular Ca^{2+} levels, many very important events occur in the cell during a Ca^{2+} transient. Many of these changes are mediated through regulatory Ca^{2+}-binding proteins, which become saturated with Ca^{2+} during the activated state of the cell.

4. Intracellular Calcium-Binding Proteins

Like the vitamin K-dependent blood-clotting factors found in blood plasma, there are numerous Ca^{2+}-dependent membrane-binding proteins in the cytoplasm of cells, including the pentraxins *(3)*, proteins of unknown function, and the annexins *(4,27)*, which play many roles in cellular signaling. In the crystal structure of these soluble proteins, there are often water molecules found in the Ca^{2+}-binding sites, which are thought to be displaced by the phospholipid head groups as the proteins bind to membranes. The annexins, in turn, are substrates for another class of Ca^{2+}-binding proteins, the protein kinases C *(27)*. Protein kinases C bind Ca^{2+} through their C2 domain calcium-binding motifs; it is now known that C2 domains exist in nearly 100 known proteins of several different classes *(28,29)*. The basic C2 domain consists of an eight-stranded antiparallel β-sandwich of approx 130 amino acid residues; up to three Ca^{2+} ions are bound by connecting loops on one side of the sandwich. Like the pentraxins and annexins, C2 domains bind phospholipids in a Ca^{2+}-dependent manner; this enables localization to the membrane. Many lipid-modifying enzymes such as the phospholipases C also have C2 domains for the same reason. In the case of other proteins, Ca^{2+} is responsible for localization of the proteins to the cell membrane where signaling events take place. Moreover, many C2 domains are also responsible for binding other substrates, including other proteins, making these domains quite versatile and complex.

Another very important class of Ca^{2+}-binding proteins is the "EF-hand" family *(30–32)*, a name used to describe the structural characteristics of the calcium-binding sites of these proteins. This term was co,ined by Kretsinger and

Nockolds *(33)* when they determined the structure of carp parvalbumin, and it refers to the conserved helix-loop-helix Ca^{2+}-binding motifs found in these proteins. The EF-hand family includes "structural" Ca^{2+}-binding proteins (also called Ca^{2+} "buffers") such as parvalbumin, the sarcoplasmic Ca^{2+}-binding proteins *(34)*, and the vitamin D-dependent intestinal Ca^{2+}-binding proteins (calbindins). Other EF-hand helix-loop-helix Ca^{2+}-binding proteins include the aforementioned calpains *(14,15)*. Another interesting EF-hand protein is the retinal protein recoverin *(35)*, in which Ca^{2+} binding causes the exposure of an N-terminal myristoyl group (which is buried in the Ca^{2+}-free state of the protein), thereby enabling targeting to membranes.

Probably the largest group of regulatory helix-loop-helix Ca^{2+}-binding proteins bind to other target proteins in response to the transient increases in Ca^{2+} concentrations. These include the S100 proteins *(36–38)*, some of which can bind other metals like Zn^{2+} and Cu^{2+}, a host of neuronal Ca^{2+}-binding proteins *(39*, and references therein), and a newly discovered calmodulin-like protein from human epidermis *(40)*. Two of the more important regulatory Ca^{2+}-binding proteins of this family include troponin-C, the calcium-binding component of the troponin complex in cardiac and skeletal muscle cells *(41–43)*, and the structurally analogous protein calmodulin.

5. Calmodulin

Calmodulin (CaM) is a ubiquitous, acidic protein found in almost all eukaryotic cells, in organisms ranging from yeast to plants to humans *(22,31,44)*. In response to elevated intracellular Ca^{2+} levels, CaM can bind four Ca^{2+} ions, one in each of its EF-hand calcium-binding motifs. In turn, CaM binds and activates a range of different target proteins *(22,23,45)*, enabling it to affect numerous cellular processes, such as smooth muscle contraction, biosynthesis of other messenger molecules, protein phosphorylation and dephosphorylation, gene expression, and cell-cycle control. Thus, CaM has a pivotal role in cellular metabolism (*see* **Table 1**).

The X-ray crystal structure of Ca^{2+}-saturated CaM *(46,47)* reveals a largely α-helical structure with an elongated dumbbell shape with two lobes, each containing a pair of EF-hand motifs, separated by a long, central α-helical linker (*see* **Fig. 1A**). Through the results of solution structural studies of CaM, such as small-angle X-ray scattering *(48)* and NMR *(49,50)*, it has been proven that this central linker of CaM is actually quite flexible.

CaM belongs to the EF-hand superfamily of calcium-binding proteins. There is considerable homology between calcium-binding motifs, both internally among the motifs in a single protein, and externally among various proteins in the EF-hand family. Sequences have been compared and examined extensively in the literature *(51,52)*. A typical Ca^{2+}-binding loop from an EF-hand helix-

Table 1
List of the Major CaM-Binding Proteins[a]

Function	Protein
Muscle contraction	Myosin light-chain kinase (smooth and skeletal)
	Caldesmon
	Calponin
Protein phosphorylation/ dephosphorylation	CaM-dependent protein kinases
	Calcineurin (protein phosphataseIIb)
	Phosphorylase kinase
Cell messengers	3':5'-Cyclic nucleotide phosphodiesterase
	Adenylate cyclase
	Inositol trisphosphate kinase
	Constitutive nitric oxide synthase (endothelial and neuronal)
	Inducible (macrophage) nitric oxide synthase
	Calcium ATPase
	Calcium and other ion channels
Nuclear proteins	CaM-dependent protein kinases
	Basic helix-loop-helix transcription factors
	CaM-dependent endonuclease
	RNA helicase
	Transcription elongation factor 1α
Other activities	Plant glutamate decarboxylase
	MARCKS and MRPs
	Dystrophin
	Multidrug resistance P-glycoprotein
	HIV/SIV glycoprotein
Apo-CaM binding proteins	Neuromodulin (GAP43/P–57)
	Neurogranin
	Inducible (macrophage) nitric oxide synthase
	3':5'-Cyclic nucleotide phosphodiesterase
	Phosphorylase kinase

[a]This list is by no means complete, and many more CaM-binding proteins are surely still to be discovered. Adapted from **refs. 22** and **59**. Abbreviations: MARCKS: myristoylated alanine-rich C-kinase substrate; MRP: MARCKS-related protein; HIV: human immunodeficiency virus; SIV: simian immunodeficiency virus.

Fig. 1. Ribbon representations of representative structures of the four basic states of calmodulin. **(A)** NMR structure of Ca^{2+}-free CaM (**ref. 58**, PDB accession code 1dmo). **(B)** Crystal structure of calcium-saturated calmodulin (**ref. 47**, PDB accession code 1cll). **(C)** NMR structure of calmodulin with a target peptide bound at the C-terminal lobe (**ref. 65**, PDB accession code 1cff); the α-helical peptide is seen with its axis perpendicular to the page at the bottom of the structure. **(D)** NMR

Calcium-Binding Proteins

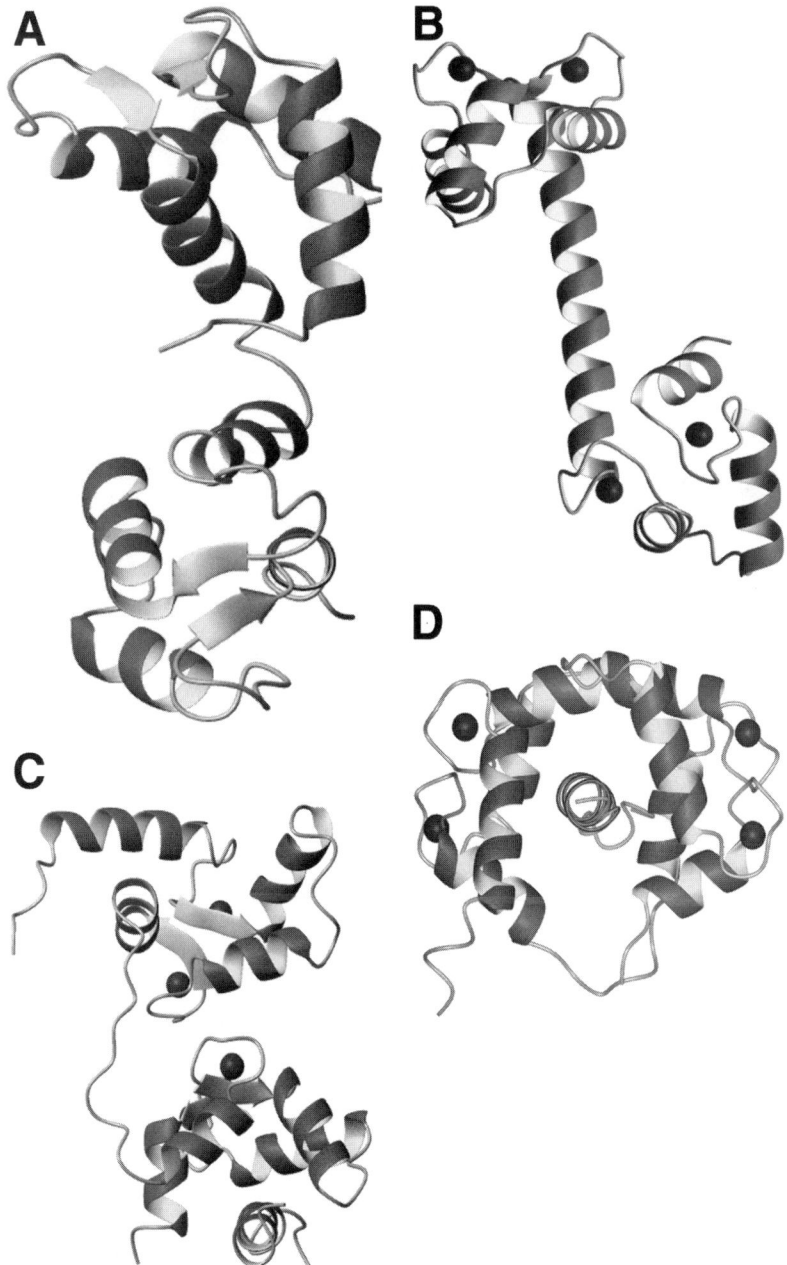

(Fig. 1. *continued from opposite page*) structure of Ca^{2+}-CaM complexed with a peptide from the CaM-binding domain of skeletal muscle light-chain kinase (**ref. 62**, PDB accession code 2bbm). The peptide is seen as an α-helix with its axis perpendicular to the page in the center of the structure. These figures were generated with the MOLMOL molecular representation program (Koradi et al., 1996) *(70)*.

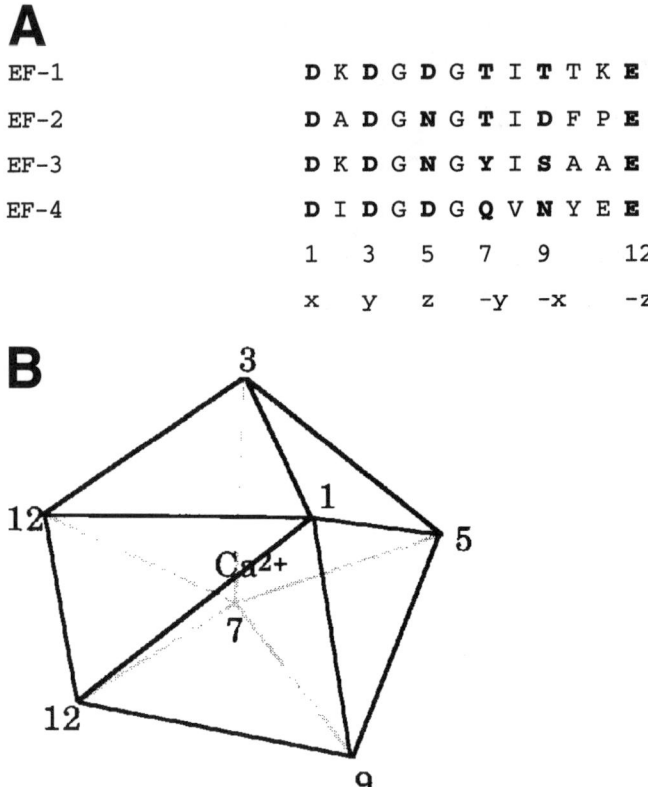

Fig. 2. **(A)** Sequence alignment of the four Ca^{2+}-binding loops of CaM. Ca^{2+}-liganding residues are indicated in bold type, whereas the positions of these residues in the Ca^{2+} coordination sphere is shown below. **(B)** Schematic diagram of the pentagonal bipyramidal Ca^{2+} coordination sphere in an EF-hand.

loop-helix motif has 12 residues. Crystallographic evidence now indicates that the last three residues in this 12-residue sequence actually comprise the N-terminal part of the helix following the Ca^{2+}-binding loop *(51)*. There are six residues involved in Ca^{2+} liganding, located at positions 1, 3, 5, 7, 9, and 12 of the loop. Originally, it was thought that these ligands formed an octahedral arrangement around the Ca^{2+} ion, but it is now known that there are, in fact, seven liganding interactions with the Ca^{2+} ion, and that the site resembles a distorted pentagonal bipyramid (*see* **Fig. 2**). All of the ligand atoms are oxygens, contributed either by sidechains of Asp, Asn, Glu, Gln, and Ser residues, main chain carbonyls, or in some cases by water oxygens.

Sequence comparisons among calcium-binding loops reveal a high degree of identity at various positions *(52)*. Position 1 is almost invariably an aspartate residue, and position 3 is most often an Asp or an Asn. The residue at position 12 is almost always a glutamate; it binds the Ca^{2+} ion through both of its sidechain carboxyl oxygens in a "bidentate" fashion, making it responsible for donating the extra, seventh oxygen atom in the liganding sphere. Moreover, there are also conserved nonliganding residues within the calcium-binding loop. Position 6 is very often occupied by a glycine residue because of the fact that only Gly, with its absence of a sidechain, has the conformational freedom to form the proper EF-hand geometry. Position 8 is also important; it is most often isoleucine, or perhaps some other hydrophobic amino acid (*Val* or *Leu*). This residue forms part of the small, but important, two-stranded β-sheet that exists between opposite Ca^{2+}-binding loops in a pair of EF-hands.

The pairwise existence of these motifs is found across all classes of EF-hand proteins and is very important for their function. Interactions between the individual motifs in the pair allows for tight, cooperative binding of Ca^{2+} ions. If only a single helix-loop-helix domain were found in a protein, it would bind Ca^{2+} very poorly. This has been shown through studies of synthetic peptides of helix-loop-helix domains *(53,54)*, and proteolytic fragments of EF-hand proteins *(55,56)*.

With all the homology among various EF-hand calcium-binding proteins, it is difficult to understand why there is such diversity in their Ca^{2+}-binding properties and physiological functions. Some EF-hand proteins, such as the parvalbumins and the calbindins (vitamin D-dependent intestinal calcium-binding proteins), seem to act only as Ca^{2+} "buffers." All of their EF-hand sites bind Ca^{2+} with similar affinity, and little structural change occurs upon Ca^{2+} binding by the protein. They exist only to aid in absorption of the important Ca^{2+} ion from the intestinal lumen, or to prevent potentially harmful free Ca^{2+} ions from forming precipitates with DNA, phosphoproteins, or other compounds.

By contrast, the regulatory EF-hand proteins, including CaM, troponin-C, and the S100 proteins, all exhibit important conformational changes upon Ca^{2+} binding, which causes the exposure of hydrophobic patches on the surface of these proteins. It is through these patches that these proteins, in turn, interact with their respective target proteins. This exposure of a hydrophobic surface is the result of the movement of helices within a pair of EF-hand motifs, a conformational change that was first revealed by determining the structure of apo- (Ca^{2+}-free) CaM (*57,58*; and *see* **Fig. 1A**).

Interhelical angles in the four EF-hand motifs of CaM are altered by Ca^{2+} binding (*58*; and *see* **Fig. 1A,B**), causing a change from a "closed" to an "open" conformation. This results in the exposure of two hydrophobic clefts on the surface of CaM, one on each of the N- and C-terminal lobes. It is through these hydrophobic patches that CaM binds it target proteins. Although CaM targets are numerous and varied, they share a general feature of a small, contiguous amino acid sequence of 17–25 amino acids in length, usually at or near the C terminus, which comprises the CaM-binding domain of the protein. These CaM-binding domains share very little sequence homology, although they all do contain a large number of basic residues and have the propensity to form an amphiphilic α-helix, with the positively charged residues on one face and hydrophobic residues on the other *(59)*. Also, there seems to be the important feature of two major hydrophobic "anchor" residues, usually either *Trp*, *Phe*, or a bulky aliphatic residue such as *Leu*, spaced either 9 or 13 residues apart.

A structure of an example of a calmodulin-activated protein, calmodulin-dependent protein kinase I (CaMKI), is shown in **Fig. 3**. *Trp*303, the major hydrophobic anchor residue in CaMKI, points away from the rest of the protein into the solvent, ready to be bound by CaM. The binding of this *Trp* residue by CaM is thought to be the primary event in activation of CaMKI by CaM.

Chemically synthesized peptides comprising the CaM-binding domains of target proteins can also independently bind CaM. CaM-binding peptides are usually 20–30 amino acids long and bind simultaneously to both domains of the protein, forming 1:1 complexes. Shorter peptides with partial CaM-binding domains can often form 2:1 complexes with CaM, with one peptide bound to each of the two lobes of the protein *(60)*. Another distinct CaM-binding peptide is the CaM-binding domain from plant glutamate decarboxylase (PGD). Although PGD is a full-length peptide, it too binds with 2:1 stoichiometry, again via both domains of CaM *(61)*. A few structures of complexes of these full-length peptides with CaM have been solved, including a complex of CaM with a synthetic skeletal muscle myosin light-chain kinase (skMLCK) peptide solved by NMR *(62)*, which is shown in **Fig. 1D**. The structures of CaM complexes with a chicken smooth-muscle MLCK (smMLCK) peptide *(63)*, and a brain CaM-dependent kinase IIα (CaMKIIα) peptide *(64)* have been solved by X-ray crystallography. They are both quite similar to the skMLCK complex structure. The CaM-peptide structures reveal that the peptides, which are normally non structured in solution, adopt a well-defined α-helical conformation in the complex. The central helix of CaM unwinds as it engulfs its target molecule, allowing for more intimate contact between CaM and the target peptide. There is a great deal of interaction between the hydrophobic face of the amphiphilic target peptide and the hydrophobic patches on the CaM surface. The anchor residues of the peptide are very important, as they interact with the extensively with the hydrophobic patches of CaM. For example, *Trp*303 of CaM kinase I (*see* **Fig. 3**) is a crucial anchoring residue for CaM. In fact, the hydrophobic surfaces on CaM are more similar to a "pocket" into which these anchor residues insert, than a "patch," but the term hydrophobic patch has been commonly used to describe CaM. More recently, the structures of two other complexes have been solved by NMR, namely those of CaM with a peptide

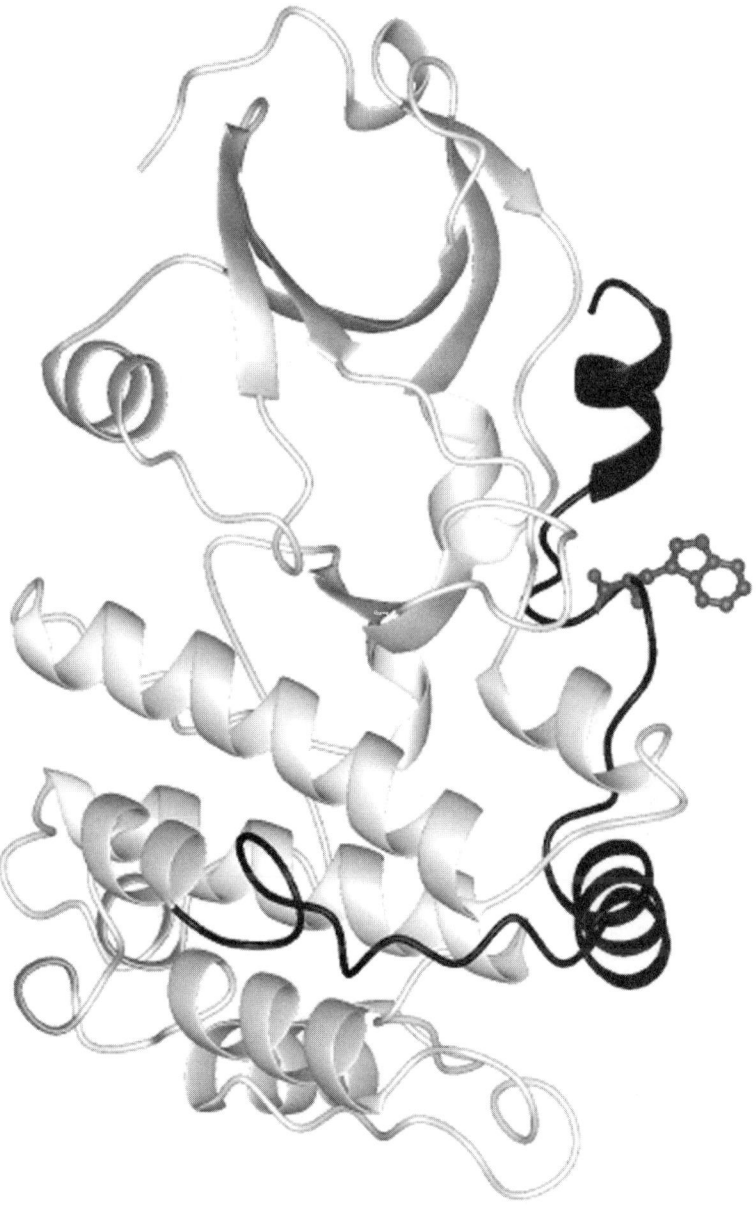

Fig. 3. The X-ray crystal structure of recombinant truncated rat brain calcium/calmodulin-dependent kinase I (CaMKI; **ref. 69**, PDB accession code 1AO6), in ribbon format. The activation domain of the protein is shown in black, with the CaM-binding domain shown at the top-right of the figure. *Trp*303, the important CaM-binding tryptophan residue, is shown in ball-and-stick format. This figure was generated with MOLMOL *(70)*.

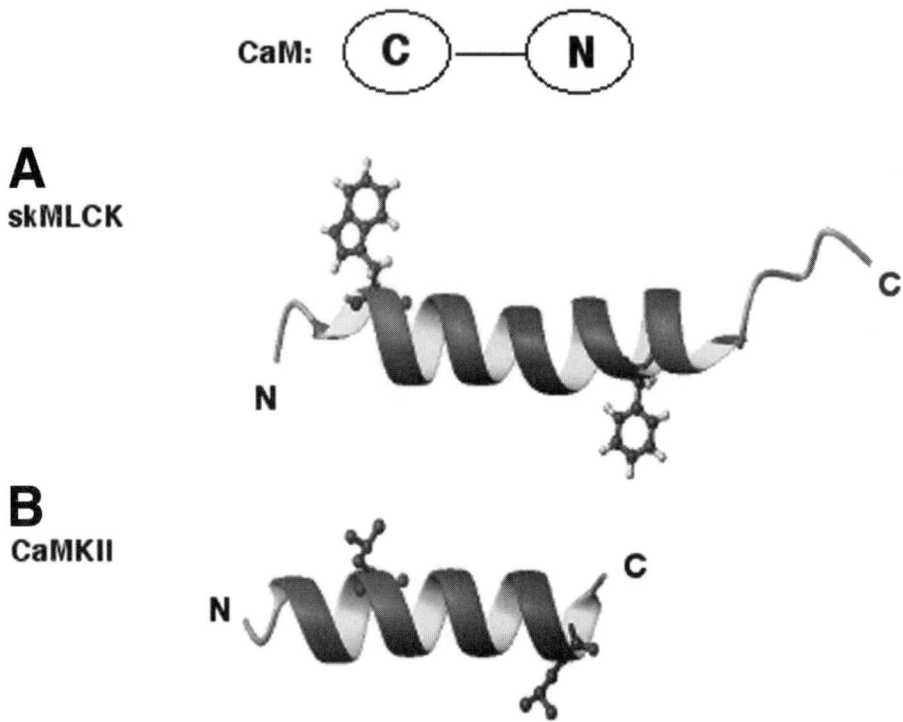

Fig. 4. Ribbon diagrams of the Ca^{2+}-CaM-bound structures of CaM-binding target peptides from (**A**) skeletal muscle myosin light-chain kinase (Ikura et al., 1992); (**B**) CaM-dependent protein kinase II (Meador et al., 1993). (*Continued on next page.*)

from the Ca^{2+} pump (*65*; and *see* **Fig. 1C**) and a peptide from CaM-dependent protein kinase kinase (CaMKK; *66*); the CaM-bound structures of these two peptides, as well as skMLCK and CaMKIIα, are shown in **Fig. 4**, with the hydrophobic anchor residues highlighted. The pump peptide is atypical in that it only binds to the C-terminal lobe of Ca^{2+}-CaM; this is not surprising given the fact that the second hydrophobic anchor residue is missing, although it could have been conceivable that the pump peptide bound with a 2:1 stoichiometry. Because it only binds to one lobe, the pump peptide is also unable to compact CaM into a globular structure; this structure is also seen as being representative of CaM-binding by more canonical target peptides, where binding to the C-terminal lobe is thought to be the initial step in the interaction. The CaMKK peptide is unique in that the CaM-bound conformation has a hairpin-type structure. What remains the same, though, is that the two hydrophobic anchor residues are still in the correct spatial orientation, and the overall shape of the complex is globular, similar to most other CaM-peptide complexes.

Calcium-Binding Proteins

C CaMPump

D CaMKK

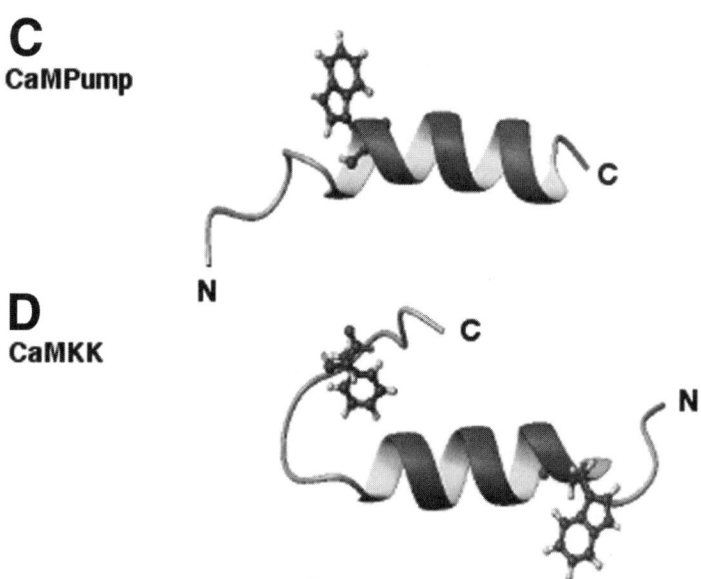

Fig. 4. (*continued*) (**D**) CaM-dependent protein kinase kinase (Osawa et al., 1999), with the two major hydrophobic "anchor" residues shown in ball-and-stick format. The first two peptides (skMLCK, CaMKII) bind CaM (shown schematically above the structures) in an antiparallel fashion, with the N-terminal anchor residue burying in the C-terminal hydrophobic patch of CaM. The pump peptide, which only binds to the C-terminal lobe of CaM, only has one hydrophobic anchor residue. The CaMKK peptide binds to CaM in a parallel fashion, with its tryptophan residue buried in the N-terminal lobe of CaM. This figure was generated with the MOLMOL molecular rendering program *(70)*.

By comparing **Fig. 1B** with **Fig. 1C,D**, it is apparent that the long, α-helical central linker of CaM, which predominates the crystal structure of the protein (**Fig. 1B**), is not seen in the structures of CaM:peptide complexes (**Fig. 1C,D**). In fact, the central linker is actually nonhelical and quite flexible in solution *(48–50)*; this is important because it can unwind to varying degrees, allowing CaM to bind a range of different target proteins *(64)*. The interactions between CaM and target peptides are solely through amino acid side chains, which is unique for a protein:protein complex. This demonstrates the importance of another feature of CaM that enables it to bind so many different peptides. This is related to the unusual composition of the two hydrophobic patches on the protein, which are exposed only in the Ca^{2+}-bound state. These patches are rich in methionine residues of which CaM has an unusually high number (9 out of 148 residues), and the polarizeable character of the *Met* side chains has been shown to be important for binding of target peptides by CaM *(67,68)*.

In conclusion, we are starting to understand at a molecular level how the ubiquitous intracellular Ca^{2+}-regulatory protein CaM exerts its action. Many other important Ca^{2+}-binding proteins, have been discovered, and continue to be uncovered in genome sequencing projects, but their mode of action still remains uncertain. Indeed, the binding of Ca^{2+} ions by proteins is pivotal in a variety of important processes, and it is, therefore, worthwhile to compile the methods used to study Ca^{2+}-binding proteins in a volume such as this.

References

1. Bronner, F. and Petpulik, M. (1992) Extra- and intracellular calcium and phosphate regulation in: *Basic Research to Clinical Medicine*, CRC Press, Boca Raton, Florida.
2. Holt, C. (1992) Structure and stability of bovine casein micelles. *Adv. Protein Chem.* **43,** 63–151.
3. Nelsestuen, G. L. and Gaul Ostrowski, B. (1999) Membrane association with multiple calcium ions: vitamin-K-dependent proteins, annexins and pentraxins. *Curr. Opin. Struct. Biol.* **9,** 433–437.
4. Sunnerhagen, M., Forsén, S., Hofren, A. M., Drakenberg, T., Teleman, O., and Stenflo, J. (1995) Structure of the Ca^{2+}-free Gla domain sheds light on membrane binding of blood coagulation proteins. *Nat. Struct. Biol.* **2,** 504–509. Erratum (1996) *Nat. Struct. Biol.* **3,** 103.
5. Selander-Sunnerhagen, M., Ullner, M., Persson, E., Teleman, O., Stenflo, J., and Drakenberg, T. (1992) How an epidermal growth factor (EGF)-like domain binds calcium. High resolution NMR structure of the calcium form of the NH2-terminal EGF-like domain in coagulation factor X. *J. Biol. Chem.* **267,** 19,642–19,649.
6. Maurer, P., Hohenester, E., and Engel, J. (1996) Extracellular calcium-binding proteins. *Curr. Opin. Cell Biol.* **8,** 609–617.
7. Downing, A. K., Knott, V., Werner, J. M., Cardy, C. M., Campbell, I. D., and Handford, P. A. (1996) Solution structure of a pair of calcium-binding epidermal growth factor-like domains: implications for the Marfan syndrome and other genetic disorders. *Cell* **85,** 597–605.
8. Muranyi, A., Finn, B. E., Gippert, G. P., Forsen, S., Stenflo, J., and Drakenberg, T. (1998) Solution structure of the N-terminal domain from human factor VII. *Biochemistry* **37,** 10,605–10,615.
9. McKenzie, H. A. and White, F. H., Jr. (1991) Lysozyme and α-lactalbumin: structure, function, and interrelationships. *Adv. Protein Chem.* **41,** 173–315.
10. Aramini, J. M., Drakenberg, T., Hiraoki, T., Ke, Y., Nitta, K., and Vogel, H. J. (1992) Calcium-43 NMR studies of calcium-binding lysozymes and α-lactalbumins. *Biochemistry* **31,** 6761–6768.
11. Aramini, J. M., Hiraoki, T., Nitta, K., and Vogel, H. J. (1995) Admium-113 NMR studies of metal ion binding to bovine and human a-lactalbumin and equine lysozyme. *J. Biochem.* **117,** 623–238.
12. Pan, C. Q. and Lazarus, R. A. (1999) Ca^{2+}-dependent activity of human DNase I and its hyperactive variants. *Protein Sci.* **8,** 1780–1788.
13. Smith, C. A., Toogood, H. S., Baker, H. M., Daniel, R. M., and Baker, E. N. (1999) Calcium-mediated thermostability in the subtilisin superfamily: the crys-

tal structure of Bacillus Ak. 1 protease at 1.8 Å resolution. *J. Mol. Biol.* **294,** 1027–1040.
14. Hosfield, C. M., Elce, J. S., Davies, P. L., and Jia, Z. (1999) Crystal structure of calpain reveals the structural basis for Ca^{2+}-dependent protease activity and a novel mode of enzyme activation. *EMBO J.* **24,** 6880–6889.
15. Strobl, S., Fernandez-Catalan, C., Braun, M., Huber, R., Masumoto, H., Nakagawa, K., Irie, A., Sorimachi, H., Bourenkow, G., Bartunik, H., Suzuki, K., and Bode, W. (2000) The crystal structure of calcium-free human m-calpain suggests an electrostatic switch mechanism for activation by calcium. *Proc. Natl. Acad. Sci. USA* **97,** 588–592.
16. Yin, H. L. and Stossel, T. P. (1980) Purification and structural properties of gelsolin, a Ca^{2+}-activated regulatory protein of macrophages. *J. Biol. Chem.* **255,** 9490–9493.
17. Nunnally, M. H., Powell, L. D., and Craig, S. W. (1981) Reconstitution and regulation of actin gel-sol transformation with purified filamin and villin. *J. Biol. Chem.* **256,** 2083–2086.
18. Maekawa, S. and Sakai, H. (1990) Inhibition of actin regulatory activity of the 74-kDa protein from bovine adrenal medulla (adseverin) by some phospholipids. *J. Biol. Chem.* **265,** 10,940–10,942.
19. Yin, H. L. and Stossel, T. P. (1979) Control of cytoplasmic actin gel-sol transformation by gelsolin, a calcium-dependent regulatory protein. *Nature* **281,** 583–586.
20. Burtnick, L. D., Koepf, E. K., Grimes, J., Jones, E. Y., Stuart, D. I., McLaughlin, P. J., and Robinson, R. C. (1997) The crystal structure of plasma gelsolin: implications for actin severing, capping, and nucleation. *Cell* **90,** 661–670.
21. Robinson, R. C., Mejillano, M., Le, V. P., Burtnick, L. D., Yin, H. L., and Choe, S. (1999) Domain movement in gelsolin: a calcium-activated switch. *Science* **286,** 1939–1942.
22. Vogel, H. J. (1994) Calmodulin: a versatile calcium mediator protein. *Biochem. Cell Biol.* **72,** 357–376.
23. Means, A. R., VanBerkum, M. F. A., Bagchi, I., Lu, K. P., and Rasmussen, C. D. (1991) Regulatory functions of calmodulin. *Pharmac. Ther.* **50,** 255–270.
24. Williams, R. J. P. (1992) Calcium fluxes in cells: new view on their significance. *Cell Calcium* **20,** 87–93.
25. Fraústo da Silva, J. J. R. and Williams, R. J. P. (1991) *The Biological Chemistry of the Elements*. Oxford University Press, Toronto.
26. Lippard, S. J. and Berg, J. M (1994) *Principles of Bioinorganic Chemistry*. University Science Books, Mill Valley, CA.
27. Dubois, T., Oudinet, J. P., Mira, J. P., and Russo-Marie, F. (1996) Annexins and protein kinases C. *Biochim. Biophys. Acta* **1313,** 290–294.
28. Nalepski, E. A. and Falke, J. J. (1996) The C2 domain calcium-binding motif: structural and functional diversity. *Protein Sci.* **5,** 2375–2390.
29. Rizo, J. and Südhof, T. C. (1998) C2-domains, structure and function of a universal Ca^{2+}-binding domain. *J. Biol. Chem.* **273,** 15,879–15,0882.
30. Strynadka, N. C. J. and James, M. N. G. (1989) Structures of the helix-loop-helix calcium binding proteins. *Annu. Rev. Biochem.* **58,** 951–998.

31. Ikura, M. (1996) Calcium binding and conformational response in EF-hand proteins. *Trends Biochem. Sci.* **21,** 14–17.
32. Kawasaki, H., Nakayama, S., and Kretsinger, R. H. (1998) Classification and evolution of EF-hand proteins. *Biometals* **11,** 277–295.
33. Kretsinger, R. H. and Nockolds, C. E. (1973) Carp muscle calcium-binding protein II. Structure determination and general description. *J. Biol. Chem.* **248,** 3313–3326.
34. Vijay-Kumar, S., and Cook, W. J. (1992) Structure of a sarcoplasmic calcium-binding protein from *Nereis diversicolor* refined at 2.0 Å resolution. *J. Mol. Biol.* **224,** 413–426.
35. Tanaka, T., Ames, J. B., Harvey, T. S., Stryer, L., and Ikura, M. (1995) Sequestration of the membrane-targeting myristoyl group of recoverin in the calcium-free state. *Nature* **376,** 444–447.
36. Smith, S. P. and Shaw, G. S. (1998) A change in hand mechanism for S100 signalling. *Biochem. Cell Biol.* **76,** 324–333.
37. Heizmann, C. W. and Cox, J. A. (1998) New perspectives on S100 proteins: a multifunctional Ca^{2+}-, Zn^{2+}-, and Cu^{2+}-binding protein family. *Biometals* **11,** 383–397.
38. Donato, R. (1999) Functional roles of S100 proteins, calcium binding proteins of the EF-hand type. *Biochim. Biophys. Acta* **1450,** 191–231.
39. Haeseleer, F., Sokal, I., Verlinde, C. L. M., Erdjument-Bromage, H., Tempst, P., Pronin, A. N., et al. (2000) Five members of a novel Ca^{2+}-binding protein (CABP) subfamily with similarity to calmodulin. *J. Biol. Chem.* **275,** 1247–1260.
40. Méhul, B., Bernard, D., Simonetti, L., Bernard, M. A., and Schmidt, R. (2000) Identification and cloning of a new calmodulin-like protein from human epidermis. *J. Biol. Chem.* **275,** 12,841–12,847.
41. Farah, C. S. and Reinach, F. C. (1995) The troponin complex and regulation of muscle contraction. *FASEB J.* **9,** 755–767.
42. Gagné, S. M., Li, M. X., McKay, R. T., and Sykes, B. D. (1998) The NMR angle on troponin C. *Biochem. Cell Biol.* **76,** 302–312.
43. Filatov, V. L., Katrukha, A. G., Bulargina, T. V., and Gusev, N. B. (1999) Troponin: structure, properties, and mechanism of functioning. *Biochemistry* (Moscow) **64,** 1155–1174 (translated from Russian).
44. Zhang, M. and Yuan, T. (1998) Molecular mechanisms of calmodulin's functional versatility. *Biochem. Cell Biol.* **76,** 313–323.
45. Crivici, A. and Ikura, M. (1995) Molecular and structural basis of target recognition by calmodulin. *Annu. Rev. Biophys. Biomol. Struct.* **24,** 85–116.
46. Babu, Y. S., Bugg, C. E., and Cook, W. J. (1988) Structure of calmodulin refined at 2.2 Å resolution. *J. Mol. Biol.* **204,** 191–204.
47. Chattopadhyaya, R., Meador, W. E., and Means, A. R. (1992) Calmodulin structure refined at 1.7 Å resolution. *J. Mol. Biol.* **228,** 1177–1192.
48. Heidorn, D. B. and Trewhella, J. (1988) Comparison of the crystal and solution structures of calmodulin and troponin C. *Biochemistry* **27,** 909–915.
49. Ikura, M., Spera, S., Barbato, G., Kay, L. E., Krinks, M., and Bax, A. (1991) Secondary structure and side-chain ^1H and ^{13}C resonance assignments of calmodulin in solution by heteronuclear multidimensional NMR spectroscopy. *Biochemistry* **30,** 9216–9228.

50. Barbato, G., Ikura, M., Kay, L. E., Pastor, R. W., and Bax, A. (1992) Backbone dynamics of calmodulin studied by ^{15}N inverse detected two-dimensional NMR spectroscopy: the central helix is flexible. *Biochemistry* **31,** 5269–5278.
51. Strynadka, N. C. J. and James, M. N. G. (1989) Structures of the helix-loop-helix calcium binding proteins. *Annu. Rev. Biochem.* **58,** 951–998.
52. Marsden, B. J., Shaw, G. S., and Sykes, B. D. (1990) Calcium binding proteins. Elucidating the contributions to calcium affinity from an analysis of species variants and peptide fragments. *Biochem. Cell Biol.* **68,** 587–601.
53. Reid, R. E., Gariépy, J., Saund, A. K., and Hodges, R. S. (1981) Calcium-induced protein folding. Structure-affinity relationships in synthetic analogs of the helix-loop-helix calcium binding unit. *J. Biol. Chem.* **256,** 2742–2751.
54. Gariépy, J., Sykes, B. D., Reid, R. E., and Hodges, R. S. (1982) Proton nuclear magnetic resonance investigation of synthetic calcium binding peptides. *Biochemistry* **21,** 1506–1512.
55. Andersson, A., Forsén, S., Thulin, E., and Vogel, H. J. (1983a) Cadmium-113 nuclear magnetic resonance studies of proteolytic fragments of calmodulin: assignment of strong and weak cation binding sites. *Biochemistry* **22,** 2309–2313.
56. Brokx, R., D. and Vogel, H. J. (2000) Peptide and metal ion dependent association of isolated helix-loop-helix calcium binding domains: studies of thrombic fragments of calmodulin. *Protein Sci.* **9,** 964–975.
57. Kuboniwa, H., Tjandra, N., Grzesiek, S., Ren, H., Klee, C. B., and Bax, A. (1995) Solution structure of calcium-free calmodulin. *Nat. Struct. Biol.* **2,** 768–776.
58. Zhang, M., Tanaka, T., and Ikura, M. (1995) Calcium-induced conformational transition revealed by the solution structure of apo calmodulin. *Nat. Struct. Biol.* **2,** 758–762.
59. Rhoads, A. R. and Friedberg, F. (1997) Sequence motifs for calmodulin recognition. *FASEB J.* **11,** 331–340.
60. Zhang, M. and Vogel, H. J. (1997) Interaction of a partial calmodulin-binding domain of caldesmon with calmodulin. *Prot. Pept. Lett.* **4,** 291–297.
61. Yuan T. and Vogel H. J. (1998) Calcium-calmodulin-induced dimerization of the carboxyl-terminal domain from petunia glutamate decarboxylase. A novel calmodulin-peptide interaction motif. *J. Biol. Chem.* **273,** 30,328–30,335.
62. Ikura, M., Clore, M., Gronenborn, A. M., Zhu, G., Klee, C. B., and Bax, A. (1992) Solution structure of a calmodulin-target peptide complex by multidimensional NMR. *Science* **256,** 632–638.
63. Meador, W. E., Means, A. R., and Quiocho, F. A. (1992) Target enzyme recognition by calmodulin: 2.4 Å structure of a calmodulin-peptide complex. *Science* **257,** 1251–1255.
64. Meador, W. E., Means, A. R., and Quiocho, F. A. (1993) Modulation of calmodulin plasticity in molecular recognition on the basis of X-ray structures. *Science* **262,** 1718–1721.
65. Elshorst, B., Hennig, M., Försterling, H., Diener, A., Maurer, M., Schulte, P., et al. (1999) NMR solution structure of a complex of calmodulin with a binding peptide of the Ca^{2+} pump. *Biochemistry* **38,** 12,320–12,332.

66. Osawa, M., Tokomitsu, H., Swindells, M. B., Kurihara, H., Orita, M., Shibanuma, T., et al. (1999) A novel calmodulin target recognition revealed by its NMR structure in complex with a peptide derived from Ca^{2+}-calmodulin dependent protein kinase kinase. *Nat. Struct. Biol.* **6,** 819–824.
67. Vogel, H. J. and Zhang, M. (1995) Protein engineering and NMR studies of calmodulin. *Mol. Cell Biochem.* **149/150,** 3–15.
68. Yuan, T., Ouyang, H., and Vogel, H. J. (1999) Surface exposure of the methionine side chains of calmodulin in solution. *J. Biol. Chem.* **274,** 8411–8420.
69. Goldberg, J., Nairn, A. C., and Kuriyan, J. (1996) Structural basis for the autoinhibition of calcium/calmodulin-dependent protein kinase I. *Cell* **84,** 875–887.
70. Koradi, R., Billeter, M., and Wuthrich, K. (1996) MOLMOL: a program for display and analysis of macromolecular structures. *J. Mol. Graphics* **14,** 29–32.

2

Calcium

Robert J. P. Williams

1. Introduction

When one considers the chemistry of any element in the context of biological organisms, it is exceedingly important to observe it in relationship to the chemistries of all the other elements in the environment that are used, or even not used, by cells. One major concern is the limitation of the availability of the element concerned, in part because of the presence of other elements, which may have changed during the evolution of Earth and its organisms. This is a geochemical problem, but even when the element has an easily available form in a compound, the accessibility to biology may be restricted through this very combination. One or two simple points then need stressing. Elements such as H, C, O, and, to lesser extent, N, S, and P, the major elements of bioorganic chemistry, are all abundant in the universe and are geochemically available. However, unfortunately, several of these elements are locked up in compounds so their accessibility for transformation into manipulatable atomic elements in cells is very restricted — consider H in H_2O, C in CO_2, N in N_2, and S in SO_4^{2-}, where in each case the respective elements, H, C, N, and S are difficult for organisms to obtain. Only O and P as O_2 today and HPO_4^{2-} are genuinely available and in a suitable form for immediate use in a cell. By way of comparison, metal ions such as calcium and a few nonmetals such as chlorine (as chloride) are quite abundant and are relatively freely available as ions in the forms in which organisms use them. Note that calcium concentrations are restricted by the presence of carbonate. However, for the simple purpose of the essential cellular organic synthesis of H, C, N, O, S, and P compounds, these two elements, calcium (Ca^{2+}) and chlorine (Cl^-) together with sodium (Na^+) are of little value and, in fact, are deleterious to the general stability of cell life in the condition in which they are available in the sea. That is, 10 mM Ca^{2+} and 500 mM

Table 1
The Essential Elements of Life[a]

			H								
–	–	–	C	N	O						
Na	Mg	–	(Si)	P	S	Cl					
K	Ca	–	–	(V)	–	Mn	Fe	Co	Ni	Cu	Zn
				(As)	Se	(Br)					
(Sr)					Mo	(I)					
(Ba)					(W)						

[a]Elements in brackets are essential in some organizms.

Table 2
Physical Properties of Calcium and Other Divalent Ions

Ion	Ca^{2+}	Mg^{2+}	Sr^{2+}	Ba^{2+}	Mn^{2+}	Fe^{2+}	Zn^{2+}	Cd^{2+}
Radius Å	1.00	0.65	1.13	1.35	0.75	0.70	0.65	0.90
Electron Affinity (eV)	18.00	22.7	16.7	15.2	23.1	24.1	27.4	25.9

Na^+ and Cl^-. Ca^{2+} at this level tends to precipitate many vital anions, whereas Na^+ and Cl^- at the levels in the sea would cause osmotic problems if allowed to enter cells freely. Even Mg^{2+} 30 mM and SO_4^{2-} 20 mM in sea water are too concentrated to be allowed free access to cells. On the other hand, in fresh water, almost all 20 essential elements of life (*see* **Table 1**) have to be taken up into cells because their concentrations are so low. In particular, note that in soil water the retention of calcium by soil silicates can be strong. The problems of handling calcium in organisms therefore parallel the handling of many other elements because each element of some 20 is required in a certain amount and both deficiency and excess cause problems for the organism. We turn now directly to problems concerning calcium.

The sea was the original source of life, so we first describe sea water conditions here. If the sea was initially somewhat more acidic than it is today (pH = 8.0), then calcium could have been even more readily available and greater protection against it was necessary for cells. Before going further into the biological chemical problems, we need to look at calcium and its basic chemistry.

2. The Character of the Calcium Ion

This chapter is a combination of earlier literature on calcium biochemistry as given in somewhat more detail in two books *(1,2)* and several recent review chapters *(3–5)*. This chapter attempts to give an overall view of calcium bio-

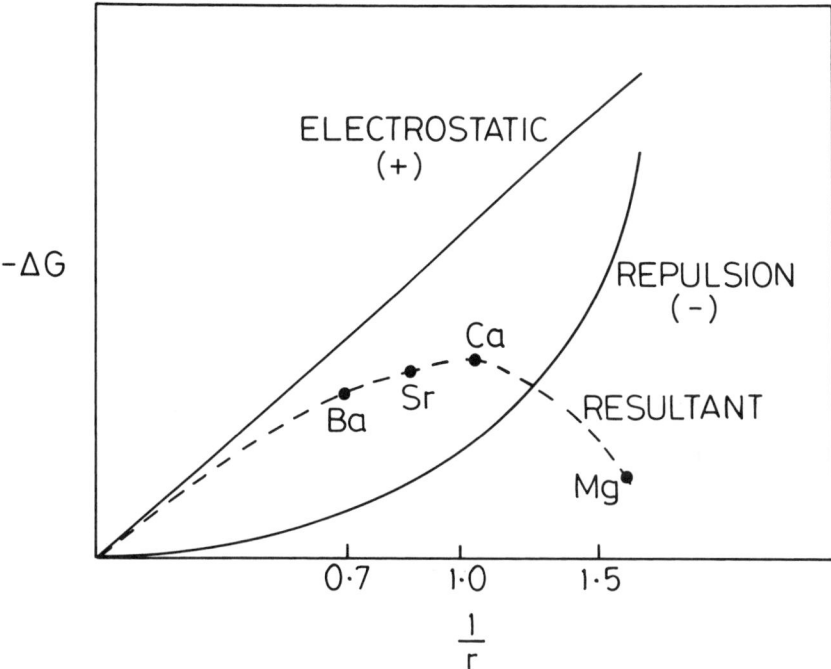

Fig. 1. The changes in attractive electrostatic energy ($1/r$) and repulsive energy ($1/r^n$) as a large number of anions are brought closer to a central cation. The resultant energy has a maximum at a certain cation size for a given anion size.

logical activity as seen by a chemist and more detailed and sophisticated views of particular features must be found in later chapters in this volume.

Calcium exists in nature in one form only — the Ca^{2+} ion. Its character relative to other ions of its own charge type is shown in **Table 2** and **Fig. 1**. It is different from all other ions, except Sr^{2+}, by a considerable margin. We must note, however, that it is the same size as Na^+. The other readily available ions, which are somewhat similar in size and/or charge, are Mg^{2+} and Mn^{2+}. There are interesting parallels and differences in the resulting chemistries of these elements.

3. The Chemistry of the Calcium Ion: Principles and Precipitation

We shall consider that all calcium–ion interactions are electrostatic with no covalent contribution and that, although this is also true for Mg^{2+}, Na^+, and Sr^{2+}, for example, it is less and less the case in the series of divalent ions which show increasingly covalent bonding

$$Ca^{2+} (Mg^{2+}) < Mn^{2+} < Fe^{2+} < Co^{2+} < Ni^{2+} < Cu^{2+} < Zn^{2+}.$$

The upshot of this ionic character is that as far as biochemical relevance is concerned Ca^{2+}, Mg^{2+}, Sr^{2+}, K^+, and Na^+ only bind to ligating oxygen (O) donors such as H_2O, RCO_2^-, R_2CO, R_2O, and RCH_2OH and not to nitrogen (N) or sulphur (S) donors. Two major factors in cells then affect the binding ability of the calcium ion in solids or solution in competition with other ions: 1) the ability of N- and S-donors to remove competing ions from Mn^{2+} along the above series to Zn^{2+}; and 2) the ability of Ca^{2+} to bind preferentially relative to Mg^{2+} to particular O-donor ligands because of its size and ligating capability, which is less restricted sterically than is Mg^{2+} binding.

Calcium ion binding must then be seen in the context of its size, and the possible oxygen atom donor arrangements around it. Turning first to solids where oxygen anions or atoms are close-packed as in say, oxide lattices, there are generated octahedral holes. The holes of radius 0.6 Å fit a magnesium ion excellently, but calcium is too large and forms an 8-coordinate oxide lattice. This is the same distinction as applies between NaCl (6-coordination) and CsCl (8-coordination) lattices. Hence, with the small anions, such as OH^- and F^-, which can pack tightly around small cations, we find that $Mg(OH)_2$ and MgF_2 are more insoluble than $Ca(OH)_2$ and CaF_2. However, as anions become larger, their packing gives rise to larger and larger holes and the balance in the equation

$$M(H_2O)_n^{2+} + X^2 \rightleftarrows MX + nH_2O \tag{1}$$

changes. Because Mg^{2+} interacts extremely well with six (small) water molecules in $[Mg(H_2O)_6]^{2-}$, larger anions do not replace the water readily to give precipitates. The larger Ca^{2+} ion binds more strongly to those anions large relative to water and, hence, these anions remove Ca^{2+} readily, but not Mg^{2+} to give precipitates (*see* **Table 3**). Typically large anions, CO_3^{2-}, and PO_4^{2-}, precipitate with calcium at lower metal ion concentrations than with magnesium. As a consequence, Ca^{2+} has a lower concentration in sea water and we find deposits, both geochemical and biochemical, of $CaCO_3$, $CaSO_4$, and $Ca_3(PO_4)_2$, but not of the corresponding magnesium salts except mixed in the calcium salts. Sulphate is so large in fact that the order of precipitation is

$$Ba^{2+} > Sr^{2+} > Ca^{2+} \gg Mg^{2+}.$$

The common form of Ba^{2+} and Sr^{2+} in geological or biological hard structures is as $BaSO_4$ and $SrSO_4$, respectively (*see* **Fig. 1**).

The logic extends to organic anions, which by their nature, are generally large, e.g., organic carboxylates and phosphates. For example, calcium has an insoluble oxalate, often found in plant tissue, but magnesium does not. Moreover, calcium tends to precipitate many polyanions, such as DNA, RNA, and some acidic proteins. There is difficulty then in keeping calcium and accompanying anions in solution inside cells and as a protection calcium is rejected. We

Table 3
Solubility-Product Constants (SP)[a]

Substance	Formula	Solubility product
Aluminum hydroxide	Al(OH)$_3$	2×10^{-32}
Barium carbonate	BaCO$_3$	5.1×10^{-9}
Barium oxalate	BaC$_2$O$_4$	2.3×10^{-8}
Barium sulfate	BaSO$_4$	1.3×10^{-10}
Cadmium hydroxide	Cd(OH)$_2$	5.9×10^{-15}
Cadmium oxalate	CdC$_2$O$_4$	9×10^{-8}
Cadmium sulphide	CdS	2×10^{-28}
Calcium carbonate	CaCO$_3$	4.8×10^{-9}
Calcium fluoride	CaF$_2$	4.9×10^{-11}
Calcium oxalate	CaC$_2$O$_4$	2.3×10^{-9}
Calcium sulfate	CaSO$_4$	2.6×10^{-5}
Magnesium ammonium phosphate	MgNH$_4$PO$_4$	3×10^{-15}
Magnesium carbonate	MgCO$_3$	1×10^{-5}
Magnesium hydroxide	Mg(OH)$_2$	1.8×10^{-11}
Magnesium oxalate	MgC$_2$O$_4$	8.6×10^{-5}
Manganese(II) hydroxide	Mn(OH)$_2$	1.9×10^{-15}
Manganese(II) sulphide	MnS	3×10^{-15}
Strontium oxalate	SrC$_2$O$_4$	5.6×10^{-8}
Strontium sulfate	SrSO$_4$	3.2×10^{-7}
Zinc hydroxide	Zn(OH)$_2$	1.2×10^{-17}
Zinc oxalate	ZnC$_2$O$_4$	7.5×10^{-9}
Zinc sulphide	ZnS	4.5×10^{-24}

[a]Data from *Stability Constants* (1964), *Spec. Pub. No. 7*. The Chemical Society, London. Note the preferred use of Roman labels of oxidation states.

can, therefore, ascribe the selective use of calcium in biology both in minerals and in the crosslinking of many extracellular matrices to the size of this doubly charged cation.

It is also of importance to compare Ca^{2+} precipitation with that of one important trivalent ion, Al^{3+}. Al^{3+}, a small cation, like Mg^{2+}, but more so, has an insoluble hydroxide and a somewhat insoluble phosphate. It is often found in Ca^{2+} phosphate precipitates, e.g., bone and with silica in what may be aluminosilicates outside cells. Silicic acid does not precipitate with Ca^{2+} at pH = 7.0. Because of its propensity to precipitate organic anions as well as inorganic anions Al^{3+}, like *all* trivalent ions and Ca^{2+}, is prevented from entering all cells. To the best of the author's knowledge this includes not only Al^{3+} and Sc^{3+}, but also Fe^{3+}. Although none of these ions as free ions in solution exceeds 10^{-11} M at pH = 7.0, they could compete for certain external anionic sites, which can also bind Ca^{2+}, *see* **Subheading 11**.

4. Calcium Salts Precipitated in Organisms

Table 3 shows the solubility products of calcium salts. It also shows the effective solubility products of some salts of magnesium and aluminium at the biological pH = 7.5 of most intracellular fluids indicating why ions such as Ca^{2+} must be excluded. The sea is at a high pH of 8.4 and it is close to a saturated $CaCO_3$ solution so that many unicellular and multicellular rather simple organisms in the sea make $CaCO_3$ external shells quite easily. Notice that the precipitation of phosphates in the sea is less probable than carbonates because of the low level of phosphate. Now some fresh water organisms can also make $CaCO_3$ shells, which means that the precipitation has to be carried out in an environment of Ca^{2+} + HCO_3^-, which is made by the living system. Vertebrates have an extracellular pH = 7.0 and precipitate calcium phosphate $Ca(HPO_4)_2$ or $Ca_3(PO_4)_2$ before it is transformed into the more insoluble $Ca_2(OH)(PO_4)$, an approximate formula for hydroxyapatite. Note that in circulating liquids of these organisms, HPO_4^{2-} and free Ca^{2+} concentrations are not far from 1.0 mM so that phosphate, and not carbonate, is precipitated. As stated, the free [Ca^{2+}] in the sea is much higher approaching 10 mM, but phosphate is much lower.

It is easy to see how a biological mineral, such as bone, which is a composite of polymer and crystals, can redissolve by making the extracellular fluid of pH << 7.0 locally. In the body of higher animals there are special cells, osteoblasts, for this dissolution. The cells bind to the bone trapping a small aqueous volume between the bone and themselves (*see* **Fig. 2**). They then release into this volume acid to solubilize the phosphate plus enzymes for destroying the bone polymers.

There are other factors that affect solubility, because a compound such as apatite readily incorporates other ions than Ca^{2+}, HPO_4^{2-}, (PO_4^{3-}), and OH^-. Most notable are F^-, Mg^{2+}, CO_3^{2-}, and Al^{3+}. The effect of fluoride is to harden apatite, whereas if anything, Mg^{2+} and CO_3^{2-} weaken it. The uptake of Al^{3+} into bone needs special comment (*see later*). Of course, both $CaCO_3$ and $Ca_2(OH)PO_4$ are more soluble in acid so that acid-producing bacteria decay teeth.

One insoluble calcium salt not affected by acidity until below pH = 5.0 is calcium oxalate. Plant extracellular fluids are at about pH = 5.5 and this pH prevents carbonate or phosphate precipitation. Several plant species then generate calcium oxalates, e.g., rhubarb leaves, which are poisonous to humans as they redissolve. The formation of oxalates appears to be both protective, as is shell and bone, but also may be a way of eliminating excess calcium. A particu-

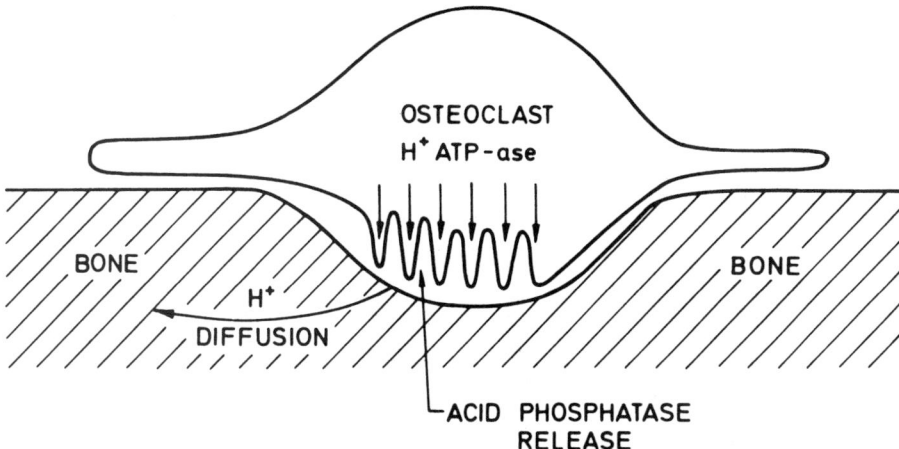

Fig. 2. The interaction of an osteoclast cell with a bone surface. The cavity is extended as acid and enzymes are ejected into the small local space shown trapped between cell and bone surface. NB: protons diffuse readily in bone.

larly fascinating example of oxalate precipitation is found in the hairs of the stinging nettle.

5. Solubility and Precipitation Mechanisms

Solubility depends upon the size of a crystal and it is the case that smaller crystals are more soluble. Two factors affect solubility relative to size:

1. The internal free energy of the salt excluding any surface effects, $-\Delta G_i$ per mole, is approximately independent of crystal size.
2. The surface free energy, $-\Delta G_s$, which is a larger contribution per mole the smaller the crystal, and is less than $-\Delta G(\text{solution})$.

Now $-\Delta G_s < -\Delta G_i$ (NB for $-\Delta G$ the more negative means the more stable) so that the weighted sum for a given crystal size per mole is less negative for small crystals and the smaller the crystal the greater the solubility. It takes a degree of supersaturation to cause precipitation initially. The picture changes in the presence of an organic polymer if the polymer binds to the crystal surface when there is a new term stabilizing the surface $-\Delta G_p$. When $-(\Delta G_s + \Delta G_p)$ approaches $-\Delta G_i$ crystallization can occur without any supersaturation. Furthermore, if $-(\Delta G_s + \Delta G_p) > -\Delta G_i$, then the insolubility of the small crystallites exceeds that of the large crystals. An obvious way of stabilizing a small crystal is then to build a small cavity to stabilize all surfaces. The polymers of the cavity then

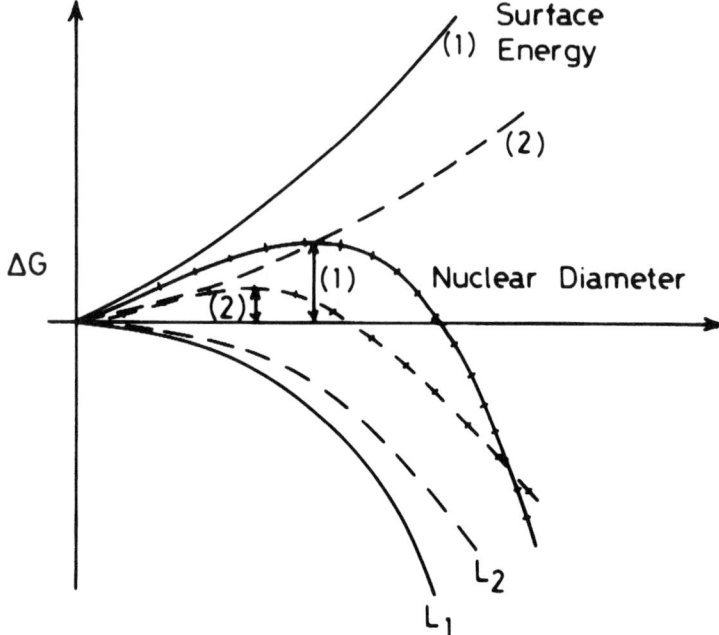

Fig. 3. The free energy per mole relative to the hydrated ions on forming the interior of a crystal, L_1 and L_2, or a pure surface layer (1) and (2). The second is unstable and crystals can only grow when $(L_n - (n))$ exceeds the points indicated by the double arrows $\updownarrow (n)$. The figure shows that despite a lower internal energy (lattice energy) and ultimately a lower total stability and hence a higher solubility the crystals L_2 will crystallize more rapidly because the nucleation barrier $\updownarrow (n)$ for the weaker and more soluble crystals is lower.

protect the crystal from dissolution both in a thermodynamic and a kinetic sense (*see* **Fig. 3**). This is undoubtedly the case in the making of shells and bones.

Now opposing crystallization are small molecule inhibitors. Consider a newly formed nucleus of a crystal that has exposed surfaces. $-\Delta G_i - \Delta G_s - \Delta G_p$ is not favorable to growth until the crystal reaches a given size. Before that size is reached, dissolution can occur. A small molecule (inhibitor) that blocks growth points on the surface will favor rate of dissolution relative to growth rate and hence prevents crystal growth. Some possible inhibitors are listed in **Table 4**. Cells undoubtedly use both in the control of precipitate formation.

6. Complex Ion Formation

The same principles of binding strengths relative to ion sizes apply to equilibrium in solution. The simplest equilibrium reaction is

Table 4
Inhibitors of Crystal Growth

Mineral	Inhibitor (examples)
Calcium carbonate	F⁻, HPO, oxalate
Calcium phosphate	F⁻, organic phosphonates, malonate

Table 5
Some Stability Constants, log$K$, of Magnesium, Calcium, and Manganese Complexes at Zero Ionic Strength and T = 25°C

Ligand	Log$K$ Mg^{2+}	Ca^{2+}	Mn^{2+}
Acetate	0.8	0.8	approx 1.0
Glycine[a]	approx 3.0	1.5	3.5
Iminodiacetate[a]	3.7	3.4	
EDTA[a]	9.1	11.0	14.1
EGTA[a]	5.5	11.2	12.5
4,5 Dihydroxybenzene 1,3 disulphonate[a]	5.9	6.0	8.6

[a]Small donor centers, N or O, assist Mg^{2+} binding relative to Ca^{2+}, e.g., unsaturated amines and phenols. Data for pH = 7.0 are >10^3-fold lower.

$$M^{2+} + X^- \rightleftarrows MX^+ \quad (2)$$

We write in place of a solubility product an associative stability constant K.

$$K = |MX| / |M||X| \quad (3)$$

The stability constant is dependent upon the type and charge of the donor groups of X, but also on the steric constraints that impose themselves as several groups crowd around the Ca^{2+} ion, **Fig. 1**. Steric hindrance can be in the coordination sphere or arise as interference generated internally in a large unit such as ethylene glycol-bis N,N,N',N'-tetraacetic acid (EGTA), **Fig. 4**, as it folds. Notice how calcium ions cause the organic moiety to fold in a particular way so that obviously the outside of the complex has more selective recognition features than the bare ion. Some examples of steric factors influencing stability constants are given in **Table 5**. The coordination in all these complexes remains as dominantly through O-atom donors. Notice that Mg^{2+} does not bind strongly to these particular centers, but preferential reagents for Mg^{2+} (and Al^{3+}) can be developed using small RO⁻ centers such as phenolates. RO⁻ here is a small anion in the sense that only -O⁻ presents itself in the coordination sphere.

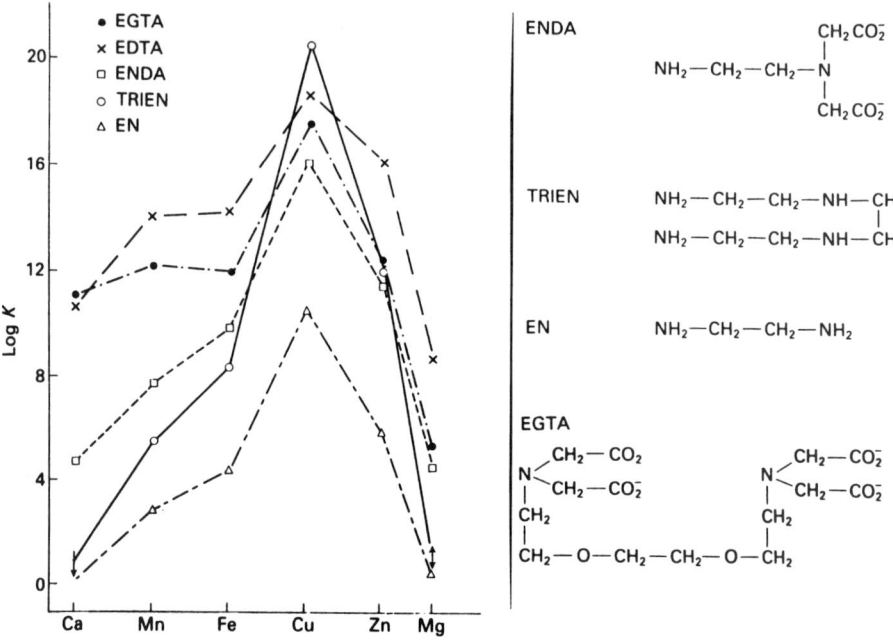

Fig. 4. The stability constants of some complexes of the divalent cations. EDTA has a similar formula to EGTA but with the central -O.CH$_2$.CH$_2$.O- unit missing. Values for Mg^{2+} are EDTA 10.5 and EGTA 6.5. Note how the larger ligands which are anions stabilize Ca^{2+} and Mn^{2+}.

Now complex ion formation may not be in thermodynamic equilibrium between free Ca^{2+} ions and bound species. We can illustrate this problem best in the case of Mg^{2+} binding to chlorophyll in proteins. The Mg^{2+} ion is found to be held by 5 N-donor atoms, four from the chlorin and one from the histidine of the protein, **Fig. 5**. This is a very unusual site to be occupied by Mg^{2+} especially when it is remembered that in biological cells there are competing ions such as Fe^{2+}, Co^{2+}, and Ni^{2+} all of which bind ring chelates related to chlorin extremely strongly. We know that in this case Mg^{2+} is forced in chlorin by an insertion reaction and that the hydrophobic nature of the chlorophyll forces this molecule into a protein leaving the Mg^{2+} adjacent to a histidine N-donor. We do not know of any parallel example in calcium chemistry as yet, but the assumption that equilibrium holds may not be universally true.

In **Eq. 3**, we have written the free concentrations of both Ca^{2+} and X$^-$ as determining factors in forming a complex. Hence, we must discover the levels of free Ca^{2+} and of other cations and of the ligands, X$^-$, in order to see how selectivity of association is managed. In cells, the free-calcium concentrations

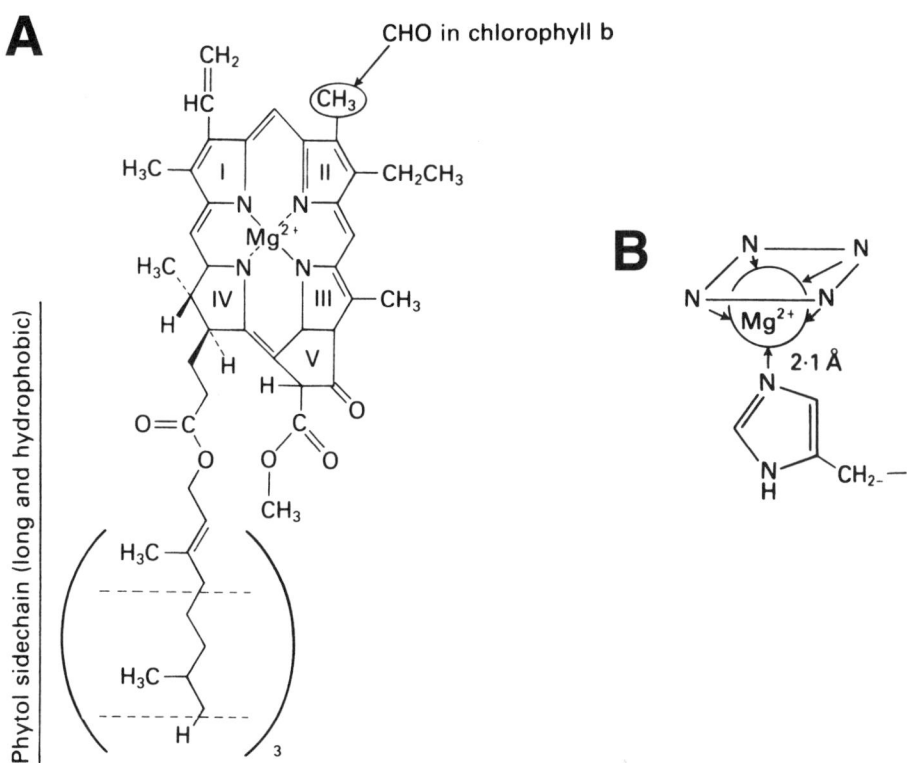

Fig. 5. The formula of chlorophyll and an illustration of how Mg^{2+} nearly fits the central cavity. N- and 5-coordination is forced upon the Mg^{2+} ion.

are manipulated by pumps in membranes and the free concentrations of binding ligands, particularly proteins, are also controlled by pumping across membranes in addition to the controls over their syntheses. This pumping reduces the Ca^{2+} ion concentration to 10^{-8} M in the cytoplasm of resting cells. It is the reaction of such low Ca^{2+} concentrations that have to be protected from competition from the ions in the Mn^{2+} to Zn^{2+} series. Ca^{2+} concentration in vesicles or in extracellular fluids is usually close to 10^{-3} M. Ligands to which Ca^{2+}, rather than other cations, can bind depends on the competition from other cations and ligands *in the same compartment*. Outside the cell, the design of binding agents has to allow Ca^{2+}, but not Al^{3+} binding and complex ion formation. The competition between cations for a given ligand is then dependent on the condition of the solution, which includes all other anions, pH, and cations. The constant for binding is then called a conditional binding constant.

It is now worth stressing that the Mg^{2+} concentration in the cell cytoplasm is 10^{-3} M so that without giving precipitates it can bind to many organic anions, especially phosphates, e.g., adenosine triphosphate (ATP). It is not the difference in stability constant that prevents Ca^{2+} binding, because its absolute constants are very closely similar to those of Mg^{2+}, but the difference in the permitted concentration, 10^5, between these two divalent ions in the cytoplasmic solution.

7. Ca^{2+} and Transition Metal Binding

All simple ligands bind metal ions in the order

$$(Ca^{2+}<) Mn^{2+} < Fe^{2+} < Co^{2+} < Ni^{2+} < Cu^{2+} < Zn^{2+},$$

and all these metal ions bind all types of donor more strongly than Ca^{2+}. Without constraints on the free metal ion concentrations in cells and on the steric demands of complicated ligands Ca^{2+} would not be able to bind to any organic molecule in cells. In the first instance the concentrations of competing metal ions in the cytoplasm is reduced by binding to stronger donors so that the free ion levels are approximately

	Mn^{2+}	Fe^{2+}	Co^{2+}	Ni^{2+}	Cu^{2+}	Zn^{2+}	Cd^{2+}	Ca^{2+}
$\log[M^{2+}_{free}]$	−7	−8	−9	−11	−15	−12	−15	−8

At these concentration levels, none of these ions can bind to the calcium-binding sites because the structures generated by the protein folds that hold calcium do not allow a collapse of ligand donor groups to give a smaller hole size than a radius of 1.0 Å and so chelation of the resultant O-donors is relatively weak. The radius of Ca^{2+} is 1.0 Å, whereas the other ions have the following average radii (Å)

Mn^{2+}	Fe^{2+}	Co^{2+}	Ni^{2+}	Cu^{2+}	Zn^{2+}	Cd^{2+}
0.75	0.70	0.68	0.65	0.60	0.65	0.90

We see the effect of size very clearly when we compare ethylenediaminetetracetic acid (EDTA) with EGTA binding, **Fig. 4**. Note that Cd^{2+} is removed overwhelmingly by its preferential binding to strong RS^--containing proteins such as metallothionein because otherwise it would be a serious competitor. Using selective chelation we can see how a cell can devise stratagems for even leaving a good calcium conditional-binding site, $K = 10^7$, open and unoccupied in the presence of other metal ions at the concentrations of free ions found in cells.

Now, whereas we have described the simplest equilibria

$$Ca^{2+} + X^- \rightleftarrows CaX^+,$$

there are many examples of more complex reactions of the kind

$$nCa^{2+} + X^{2-} \rightleftarrows Ca_nX^{(2n-2)+} \tag{4}$$

Calcium 33

and
$$Ca^{2+} + X^- + Y^- \rightleftarrows CaXY, \quad (5)$$

or even
$$nCa^{2+} + mX^- + pY^- \rightleftarrows Ca_nX_mY_p \quad (6)$$

These cases can result in cooperative or anticooperative combinations where the ultimate extent of cooperation is precipitation. You will find throughout this volume many examples of cooperative binding of several calcium ions to one protein, e.g., in calmodulin, and of extensive crosslinking by Ca^{2+} to give CaXY chains as in fibrillin. The exact distribution of species in these cases is often extremely difficult to evaluate.

8. Condensation Equilibria

Now there is an intermediate condition between simple-complex ion formation, **Eq. 1**, and precipitation. Consider a large polymer such as DNA with a distribution of negative charge all along its backbone. There is association with cations, but it is observed that although the DNA remains in solution, it can collapse and condense into a very small volume at a critical concentration of the cation. It is not necessary for the cation to be bound directly to the polymer, but it is held by the high-surface potential generated by the fold. We believe that this is the mode of retention of Ca^{2+} by sequestrin (also, *see* Chapter 18, in this volume. Here, the protein surface has a high concentration of carboxylate groups. Segments of it collapse, or rather now condense, around some 40 Ca^{2+} ions so that effectively they all have the same binding constant. Because the Ca^{2+} is in equilibrium with free ions, it is readily released on exposure to solutions of low free Ca^{2+}.

9. Calcium, Membranes, and Walls

The membranes of all cells are negatively charged and as such attract Ca^{2+} ions. Now there is no requirement for either an even distribution of Ca^{2+} across membranes or along membranes so that Ca^{2+} together with other cations may form a pattern of cations lying under the membrane. This binding can also cause a membrane to collapse. In fact, membranes are supported by an underlying structure of proteins that prevent condensation as described. They are likely to be localized differences both along a membrane and across it of its negatively charged headgroups. Together, with localized membrane curvature, these differences will lead to corresponding variations in associated calcium-ion concentration. A recent atomic force microscopy study shows the way Ca^{2+} induces coalescence of lipids into domains (*see* **ref. 9**). However, because the calcium ions have pumps and channels forcing or allowing Ca^{2+} ion flow, the

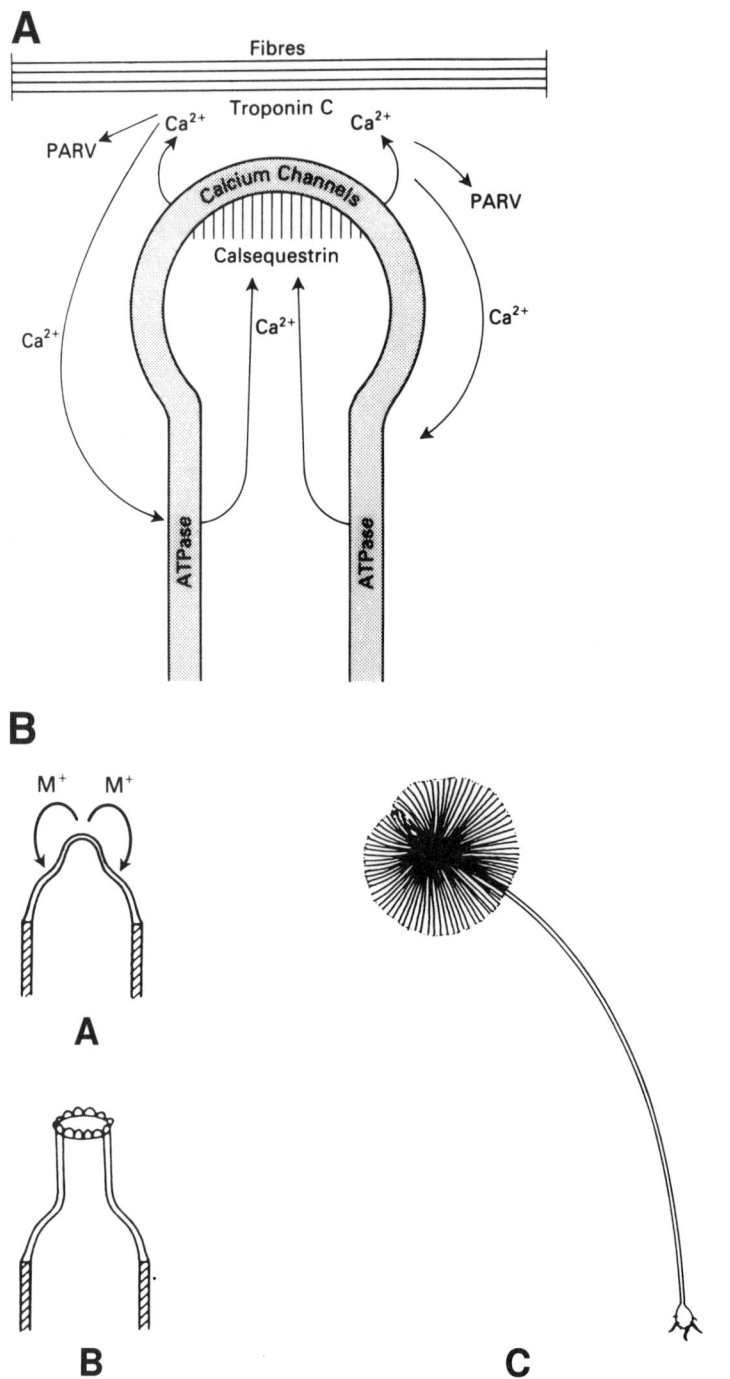

Fig. 6. The calcium currents around the terminal of a T-tubule of the sarcoplasmic reticulum.

**Table 6
Variety of Ca^{2+}-Binding Proteins**

Class	Examples	Strength (log$K$)
Classical EF-Hand	Calmodulin, Parvalbumin Troponin-C	7
Nonclassical EF-Hand	S-100, Calbindins	7
Others	Annexins (+ Lipids)	3–4(7)
	C$_2$ Domains (+ Lipids)	3–4(7)
	Phospholipases (+ Lipids)	3–4(7)

**Table 7
Extracellular Binding Proteins**

Protein Class	Binding Constant
Digestion	Proteases, saccharases, nucleases (Ca^{2+} often activates)
Bone proteins	Osteocalcin, phosphophoryn
Storage	Calsequestrin

association of membranes with Ca^{2+} must then produce Ca^{2+} currents locally (*see* **Fig. 6A,B**).

10. The Fixed Binding Constants for Calcium Ions in the Cytoplasm

So far we have described proteins available for calcium binding $K = 10^7\ M^{-1}$, but most of them remain unbound in the resting state of the cell when [Ca^{2+}] is less than $10^{-7}\ M$. On excitation, the Ca^{2+} rises to $10^{-6}\ M$. An amazing feature of the binding by intracytoplasmic proteins on excitation is therefore that, despite their variety, **Table 6**, the binding constants in the precise place where they are used, are virtually identical and close to $10^7\ M^{-1}$. This is a necessary feature because they must all respond to the *same input* concentration of calcium from stores. To manage this result for many different sites, for example, calmodulins and S-100, and even many combinations of sites, for example, annexins or phospholipases plus phospholipids, is a remarkable achievement of evolution. It implies that unless proteins have different time constants of action then there will be intense cooperation between many Ca^{2+} triggered events. A parallel achievement is found in the extracellular fluids where the binding constants for calcium of many different sites is close to $10^3\ M^{-1}$ compared with $10^7\ M^{-1}$ for the internal cytoplasmic sites, **Table 7**. Here, external sites include those of the extracellular fluids and those of the endoplasmic reticulum. In each case, too, the amount of protein present in the case of mammals is controlled so that equilibrium is balanced close to the solubility product of bone, which gives a precise standing concentration of calcium around 10^{-3} M.

Concentration gradients of 10^4 across both the exterior cytoplasmic membrane and the interior endoplasmic reticulum membrane are maintained by ATPase pumps rejecting calcium from the cytoplasm. The pumps stop once the internal calcium is below the binding constant of the internal proteins, e.g., calmodulins. This is an example of feedback control, which is very common in cells. In animals, bone buffers the external fluids while excess calcium is rejected by the kidney and epithelial cells manage the intake. An animal then has a tight homeostasis relative to that which a plant can achieve outside cells. If the Ca^{2+} concentration rises to 10^{-5} M internally, the cell is killed and this may be part of the process of apoptosis.

In the discussion of Ca^{2+} binding, the general effect of physical forces as opposed to chemical binding must not be forgotten. The membranes of biological cells usually carry a potential because of the overall directional pumping of ions. The negatively charged side of the membrane concentrates all cations near it while the positively charged side repels them. When Ca^{2+} ions move through membranes, physical fields are of considerable consequence.

The implication of this vast accumulation of data concerning calcium binding to a diversity of proteins inside cells, outside cells, and in vesicles and organelles is that much, if not all, the calcium is bound in the resting state of a cell at equilibrium. The need to protect the inside cellular DNA from calcium and yet use the calcium gradient in signaling (allowing Ca^{2+} to enter the cytoplasm) has demanded a free Ca^{2+} of 10^{-8} M so as to allow the evolution of series of Ca^{2+} binding *systems* all with binding constants close to 10^7. At the same time, the need to protect external fluids from general damage, while allowing specified precipitation, demanded a maximum concentration of some 10^{-3} M Ca^{2+}. So that external signaling should be operationally effective, for example, the onset of digestion by release of hydrolases to the external calcium, the concentration in external fluids and in vesicles could not fall below 10^{-4} M Ca^{2+}. To secure both buffering and further value of Ca^{2+} ion properties in these external fluids, demands a second vast series of Ca^{2+}-binding proteins to which Ca^{2+} bound with log$K$ approx 10^4. These features began to appear as cellular life appeared and then developed extreme degrees of sophistication (*see* **Fig. 7** and other chapters in this volume).

11. Calcium and Aluminium (Acid Rain)

We have stated most calcium salts are more soluble in acid solution and this applies also to the salts of aluminium. Of particular importance is the release of Al^{3+} from clay minerals of soil. The minerals are aluminosilicates with a vari-

Calcium

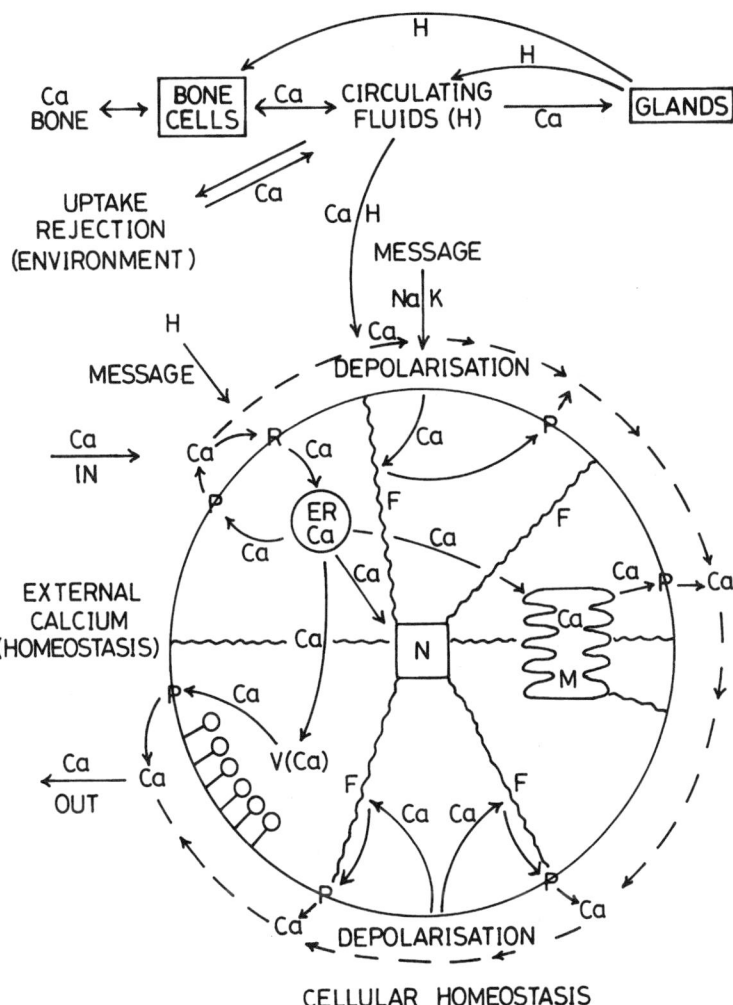

Fig. 7. **(A)** The vast complexity of calcium current flow in advanced cells (N-nuclease; ER, endoplasmic reticulum; M, mitochondria; H, hormone; F, filament), compare the simpler flow in **(B)** the tip of acetobalaria.

ety of other cations. When soil releases aluminium to waters, this metal ion damages plant life. We need to see how.

The external surfaces of most plant roots and indeed the exposed surfaces of most cells from bacteria upward are made of anionic polymers. The anions are neutralized to some extent by calcium which then crosslinks the extracellular matrix. The polymers are anionic in fair part because of the presence of carboxylate groups with a pK_a < 5 so that protons do not compete with Ca^{2+} until

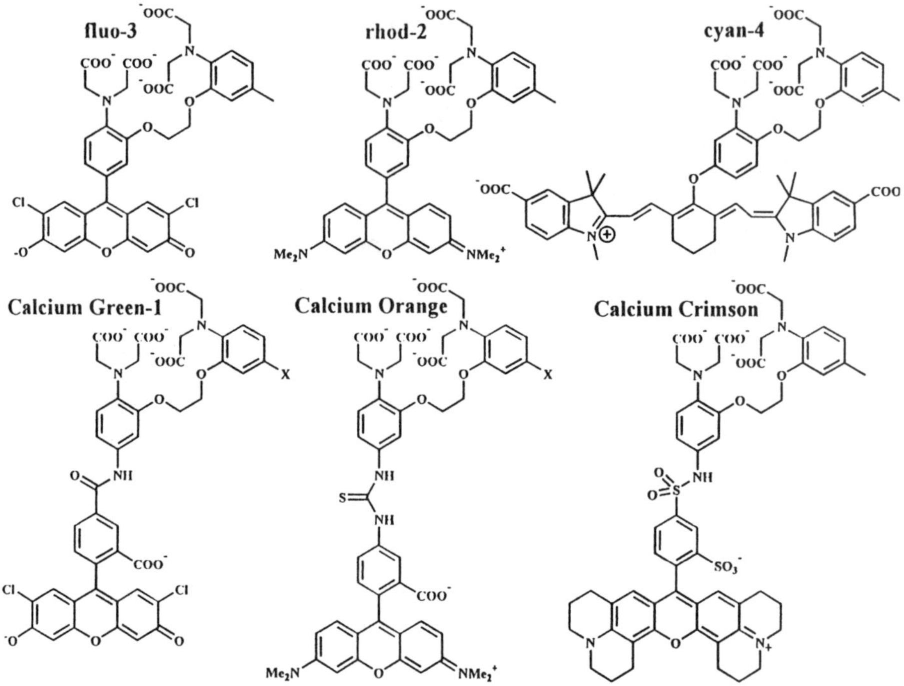

Fig. 8. The formulas of some calcium dyes and indicators.

extreme and very unusual acid conditions. Ca^{2+} binding is then pH independent from pH = 7.5 down to 5.0. Aluminum in silicates of the soil is extremely pH dependent and even the favored groups of organic matrices at pH = 7.5 for aluminum binding, phenolates, have a high pK_a (>10.0) so that they too hold Al^{3+} less and less as pH is lowered. Decrease in pH, acidity, then causes free Al^{3+} to increase and Ca^{2+} to be displaced by Al^{3+}, which, once released from soil or organic phenolates, will bind better to carboxylates.

There is in humans a disease called dialysis dementia, which is caused by Al^{3+} getting to the brain and destroying cells. It could well be that because the brain fluids are very low in calcium and there is no bone in contact with the fluid, once Al^{3+} enters these fluids it destabilizes cells by effectively destroying Ca^{2+} crosslinks. As stated earlier, Al^{3+} is also accumulated in bones.

12. Calcium Indicators

Indicators of free-calcium concentration are based upon selective calcium binding to fluorescent dyes or through the use of selective calcium electrodes. The commonly used dyes, **Fig. 8**, mimic the selective binding by EGTA. Their effective binding constants are similar to that of EGTA at pH = 7.5 but notice that the use of an aniline-N and not an ammonia N in the construction lowers the pK_a from around 10 to about 6.5. Thus, at pH = 7.5 the binding by these dyes is pH independent. (Of course they have an absolute binding constant approx 10^2 to 10^3 lower than EGTA.)

The dyes do bind Mg^{2+} much as does EGTA, but with a binding constant of around 10^3. Hence, Mg^{2+} binding only interferes slightly with Ca^{2+} estimations. Although free Ca^{2+} variations in cells are quite dramatic, free Mg^{2+} concentrations hardly vary so that calibration of the dye-stuffs response to calcium changes should be relatively easy, though there is much controversy. Notice that interference from Mg^{2+} is avoided again by the steric constraints of a large chelation system.

With respect to interference, a better reagent would be a pure O-donor ligand. In fact, biological proteins of the required kind are available as indicators. The best known and most used is aequorin, which is, in effect, a fluorescent calmodulin. It does not suffer from Mg^{2+} or other metal ion competition in its binding with calcium.

Nothing said so far has related to the kinetics of Ca^{2+} ion reactions.

13. Diffusion Rates

The rate of diffusion of a small unit such as the calcium ion is dependent upon a constant, k_{diff}, multiplied by the ion concentration. The limiting diffusion rate in water for an ion is related to the dissociation of water molecules from around the ion. The calcium ion has an almost optimal diffusion rate of 1.0×10^9 m^2/s^{-1} and an almost optimal water exchange rate of 10^8 to $10^9/s^{-1}$. This means that at high concentration there is no difficulty for the equilibration of calcium ions in water solutions of considerable volume, e.g., a millimolar solution in a beaker of water with gentle stirring. The problem in biological solutions is very different. In cells, the $[Ca^{2+}]$ is $< 10^{-7}$ M. Distribution over µm distances is now restricted. It is further restricted by the many negative surfaces of membranes and proteins. In fact, space is very crowded in cells (see **Fig. 9**). To overcome this problem, cells have evolved carrier molecules: proteins, which being negatively charged and with a high affinity for calcium can reequilibrate calcium concentrations. They are at a concentration of 10^{-3} M and redistribute Ca^{2+} faster than Ca^{2+} at 10^{-8} M can diffuse by itself. We believe this is the function of the protein, calbindin, which is described in Chapter 10, in this volume. The Ca^{2+} effectively takes a ride on the back of the protein.

14. Time Constants and Ca^{2+} Action

Certain cellular operations must be triggered by Ca^{2+} release and then relaxed while much other cell activity is not perturbed. There are two ways of managing

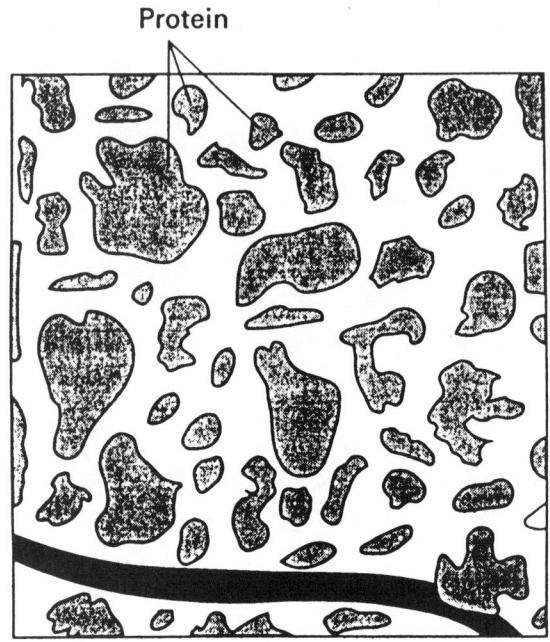

Fig. 9. The packing of molecules in cells indicating how crowded is biological space.

isolation of an activity. The first is to have some binding reactions and their relaxations much faster than others, whereas the second concerns the localized release of calcium ions within the total cell volume. Binding rates must be controlled at the on-rate step, which implies that some proteins accept Ca^{2+} more slowly than others. Two mechanisms are open. 1) The calcium site is blocked by an ion that leaves the site slowly. The obvious choice is Mg^{2+}, which, in fact, blocks access to the Ca^{2+} sites of parvalbumin. In a fast muscle, action is as follows:

$$\text{Troponin C} + Ca^{2+} \underset{}{\overset{\text{fast}}{\rightleftarrows}} \text{Action}$$
$$\downarrow\uparrow \text{slow}$$
$$Mg^{2+} \text{ Parvalbumin} + Ca^{2+} \rightleftarrows Ca^{2+} \text{ Parvalbumin} + Mg^{2+}$$

The effective binding constants are the same for the two proteins, troponin C and parvalbumin, but binding is slower to the second. A final slower step is the outward pumping of Ca^{2+} by the membrane ATPase.

The second possibility 2) relies on the fact that Ca^{2+} release into a cell at a certain point may be buffered locally so quickly that Ca^{2+} fails to reach most of

Table 8
Calcium Light Isotope Enrichment

Tissue	$^{40}Ca/Ca^{44}$ Enrichment[a]
$CaCO_3$ of foraminiferae formed directly from sea water	1.00+
$CaCO_3$ of coccoliths formed by internal precipitation of calcium in vesicles	2.15+
$Ca_2(OH)PO_4$ of deer bones. Note passage of calcium through many steps in plants and animals	2.50++

[a]Enrichment of lighter isotope in thousandths. Data from **refs.** *7* and *8*.

the cell. For example, at a nerve synapse, the calcium pulse does not influence the center of the cell.

15. Calcium Exchange

If our understanding of calcium binding and kinetics are correct, then calcium should exchange rapidly with its surroundings until it is trapped within a permanent structure. In such a case, the calcium binding in living organisms would equilibrate quickly with external calcium, but at a level manipulated by energy input at pumps. However, if calcium becomes trapped during the processes of metabolism, say as shell or bone, then the calcium no longer equilibrates. The study of the fractionation of isotopes of light elements such as H, C, and N indicate that the fractionation in such a thoroughly equilibrated process is likely to be small. However, in a series of rate-limited steps leading to a trapped atom, then much larger kinetic isotope fractionation can occur. It is possible today to apply this analysis to heavier elements. For example, the analysis of sulphur isotope ratios has been used to distinguish sulphide $\rightleftarrows$ sulfate exchange reactions of mineralogical as opposed to biological origin. The same approach can be used to analyze calcium fractionation.

16. Ca^{2+} Isotope Distribution and Evolution

Recently it has become possible to follow the kinetics of calcium flow using isotope fractionation. As is usual, lighter isotopes pass over barriers more easily than heavier ones. In the case of calcium, this means that the $^{40}Ca/^{44}Ca$ ratio increases on calcium passage through rate-limited chemical transfer steps. **Table 8** shows that the light isotope enrichment is greatest in those tissues that are made by passing calcium through the greatest number of steps. In evolution, the number of steps through which calcium may pass has increased with the complexity of organisms. The determination of a isotope ratio in a sediment indicates the type of organism that existed at a given time. Perhaps as expected the ratio is greatest in the order formation of external precipitates (coral) < formation of crystalline shells (coccoliths) < formation of bone (deer).

17. Calcium in Evolution

We introduce this topic here because it is necessary background to the use of Ca^{2+} isotope distribution. A major development in evolution, probably prior to the development of all, but very low levels of dioxygen, was the involvement of calcium described above, *see* **Fig. 7**. At first, calcium ions only strengthened exterior membranes or formed crudely constructed external precipitates as seen today in algal coral reefs. The next development in single-cell eukaryotes is seen in the *foraminiferas* (dating back to more than 10^9 yr ago) which have well-ordered crystalline *externally formed shells*, and in coccoliths, which now form shells from *units made in internal vesicles*, in fossil deposits. It is also notable that features common to many unicellular protozoa, depend on calcium. Examples are the cilia beating of *paramecium*, the stalk contraction of *vorticellia*, and cytoplasmic streaming in *physarium*. Again yeasts, which are fungi and separated from both bacteria and the antecedents of plants and animals around 1–2 billion years ago, have well-developed calcium control systems. In all these early examples, calcium gradients are associated with slow development, or even constant activity, so that it would appear that the stress of today's scientific investigation of fast triggering by calcium *in higher animals*, is misplaced, if applied to the whole of calcium functions. Calcium concentration changes are an essential feature of differential growth and only later do they become used in fast triggering useful for mobile animals. Growth with development is, of course, a continuous feature of all eukaryotes, while prokaryotes just divide. The final extension to the use of fast changes comes in animals, but we should see first the developing actions of calcium in plants, especially perhaps in primitive plants.

Undoubtedly, the first multicellular organisms were plants. Hence, it is to plant life we turn to appreciate how the functions of calcium developed after those in eukaryote single cells. Immediately we see that large calcium gradients were maintained across the cytoplasmic membrane, but the extracellular fluids were not particularly well controlled. The endoplasmic reticulum and the vacuole of plants quite generally were high in calcium also. The novel functional value of calcium lay now in the relationship through its concentration in the cytoplasm to the level of the newly acquired set of organic messengers, which went between cells — the plant hormones and control molecules such as phytochromes. It is these molecules which to this day control the path of differentiation during growth. Thus, calcium became a major player in linking the

Calcium

outside environmental changes with the internal metabolism and differentiation into organs during the plant life cycle through a sequence of events in which the hormone 1) could or 2) could not penetrate the outer membrane. Only in case 2) was a calcium message triggered.

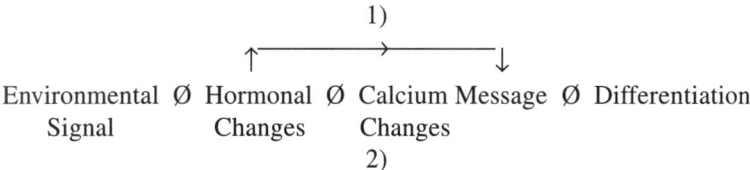

Environmental Ø Hormonal Ø Calcium Message Ø Differentiation
Signal   Changes   Changes

2)

A very interesting example seen today is the adjustment of the steady-state cellular calcium concentration with the exposure to light and even light of different wavelengths. It is thought that the changes in calcium alter the structural organization of filamentous constructs such as actino-myosin and tubulin so allowing chloroplast reorganization and streaming and also changing development. Additionally, the new calcium levels act to adjust phosphorylation, via kinases, and then to alter gene expression as in 2) above. A very similar response in plant roots is that to Earth's gravitational field. Here, the calcium current circulating at the root tip (*see* **Fig. 8**) is apparently affected by gravity so that the internal calcium levels change and then affect the downward growth of the root.

Calcium is again not to be seen as a trigger for immediate fast response, but as the intermediary in the slow responses of development during growth and opposite a changing environment. (Of course, some fast responses such as the closing of a fly-trap plant are related to calcium also). It may be that calcium circulates constantly through the cytoplasm of all cells, maintaining homeostasis locally in different ways according to the environment, and the placings of calcium pumps and channels. Circulation changes of flow adjust the calcium input to the cytoplasm and the new stationary level of this ion then is relayed via phosphorylation to bring about cellular development. It is probable too that the unavoidable rise in basal level calcium as repeated waves of free calcium sweep cells (*see* **Fig. 10**) is, in part, responsible for differentiation. This can be seen in the development of an egg after fertilization. It is becoming apparent that calcium is involved in a vast number of steps in cells connecting the external conditions to cell development. We then must use network diagrams to illustrate the function of calcium in assisting cell homeostasis as well as triggering (*see* **Fig. 11**).

The final step in calcium circuitry in animals adds to the sophistication of the circuits of plants. Here, the external calcium experienced by cells is held constant and so is the temperature in the body's external solutions. This has allowed tight control over calcium injection into cells and has allowed rapid

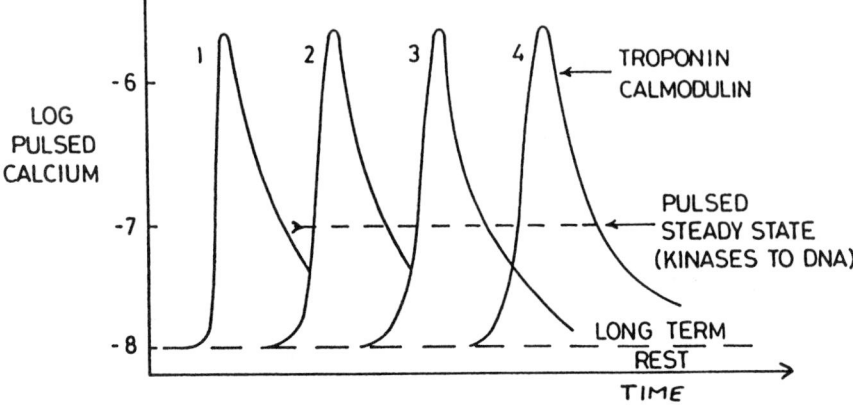

Fig. 10. The effect of a series of calcium waves is to raise the base level of free calcium. Hence, one wave triggers very fast responses, but series of waves in quick succession trigger slow responses of great sensitivity to calcium.

responses to be separate from slower activities. The best control of these external fluids is by bone as illustrated in the final paragraphs.

18. Calcium Circuits

All ions can carry currents much in the same way as electrons do but at lower speed. Hence, ions can be used to form circuits just as electrons can. A circuit needs, of course, a power supply and a conducting medium. For calcium in biological cells, the power supply is the ATPase pump, which gives the gradients across membranes. The conducting medium is water in bulk or in membrane channels. A circuit can be connected to many devices such as switches on channels, such as calmodulin for example, or condensers, such as sequestrin in the endoplasmic and sarcoplasmic reticula. Now to link the many parts of a circuit in a harmonious activity, it is necessary to introduce feedback flow of the current carrier, here calcium. This has been discussed in many publications. Given the large number of devices in a cell all connected to calcium flow it is probable that all eukaryote cells have a constant, as well as a stimulated, calcium current monitoring and keeping activity coherent. A possible view of evolution is that the change of prokaryote to eukaryote involved the development of just such a calcium circuit. Thus, although suitable organic polymers evolved in cells to make structural and functional machinery, their control circuitry appeared through the use of internal connections in the cytoplasm based initially in prokaryotes on proton, electron, and phosphate, as well as on some substrate metabolism and iron internal flows. Later in eukaryotes all these internal circuits became connected to Ca^{2+} flow across membranes that allowed

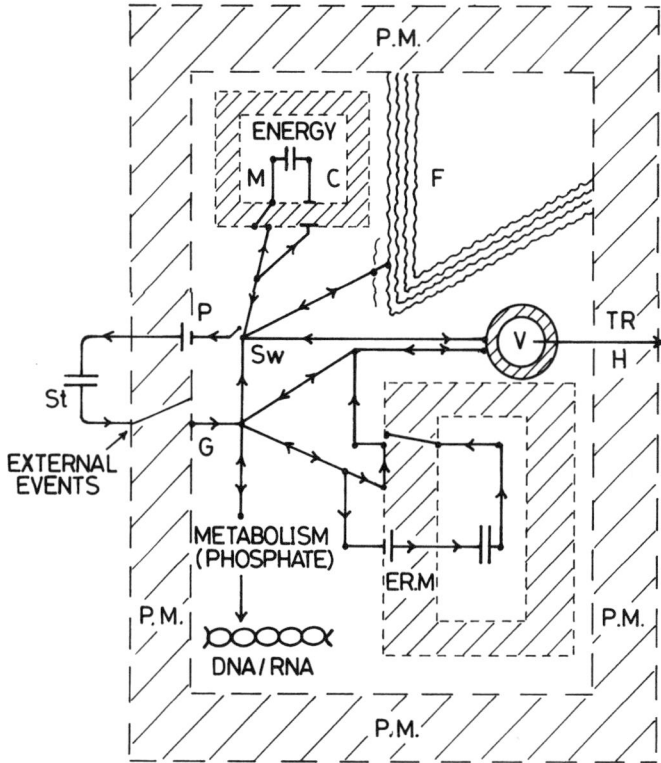

Fig. 11. A circuit diagram for calcium flow connecting a large number of internal activities to the external environment. M, mitochondria; C, chloroplast; V, vesicle or vacuole perhaps including oxalate production, as well as exocytosis; ER.M. endoplasmic reticulum membrane; P.M., plasma membrane; P, pump; G, gate; Sw, switch; F, filament; St, external store; TR, transmitter; H, hormone.

an interaction between the external environment and the cytoplasm. The further development of multicellular organisms depended upon the evolution of organic (current) carriers in circuits; the brain appeared once Na^+/K^+ circuits were developed, and evolutionary use of ion flow in nerves similar to, but simpler than, that of Ca^{2+} and based upon the production of Na^+/K^+ ion gradients for earlier and quite other reasons, namely osmotic and electrolyte balance (*see* **Table 9**). Finally, external circuits arose once man discovered how to use the electron.

19. Calcium Pumps and Buffers

To create the calcium circuits, the power system for the ion, the calcium pump, has to be selective against other ions. We know that the membrane part

Table 9
Evolution of Simple Ionic Equilibrium Signals

Primitive Organisms Prokaryotes	$Mg^{2+}/ATP^{4-}/HP^{2-}$ controls phosphorylations Fe^{2+} controls redox equilibrium $Na^+/K^+/Cl^-$ control osmotic pressure
Single cell Eukaryotes	Ca^{2+} controls activated states and relationship to environment Mn^{2+} controls development of plant-related organisms
Multicellular Organisms	Zn^{2+} controls hormonal responses relating to growth and development and connective tissue $Cu^+(Cu^{2+})$ controls connective tissue responses. Extended use of Ca^{2+} in excited states. Generation of Na^+ (K^+) signaling and the evolution of the nervous system.

of the pump contains a receptor site for calcium with one carboxylate group and several carbonyls. To be useful in function, this receptor has to switch its binding strength from an uptake $\log K = 7$ (inside) to a release $\log K = 3$ (outside) as the calcium is rejected from the cell. In other words, because of the input of energy from ATP (or a coupling of a reverse gradient of Na^+ or H^+) the conventional strong binding of the cytoplasm has to be weakened to that of the binding in extracellular fluids. Obviously, an energized conformational switch is necessary. Although a crystal structure of the pump will soon be available (*see* **ref. *9***), I will describe a working hypothesis here, as shown in **Fig. 12**. This pump has a feedback switch-off caused by calcium binding to an ancillary protein of the pump, calmodulin. Thus, the pump operates at a basal level of flow opposite $10^{-7}\ M\ Ca^{2+}$, as do all the calcium-based components of the Ca^{2+} circuit. Note how on rise of Ca^{2+} through activated channels, all parts of the circuit are based on interaction of binding constant, 10^6 to 10^7. This allows a rapid change of state provided relaxation is fast. However, multiple rapid pulsing or the absence of proteins for *fast* relaxation lead to a different bias in the whole circuit. This, in turn, can communicate to other circuits, e.g., that of phosphate making one communication network; that of Ca^{2+}, link to another; that of phosphate, and then to a third; for example, the proton, and so on.

20. Calcium Exchangers

Cells have always required ways of lowering calcium concentration in the cytoplasm because calcium at high concentrations $>10^{-5}\ M$ tends to coagulate many biological polymers. The primitive mechanism for reducing $[Ca^{2+}]$ appears to be using the Ca^{2+}/H^+ exchanger because this is found in all bacteria even where the Ca^{2+} ATPase is not observed. The importance of the ATPases

Calcium

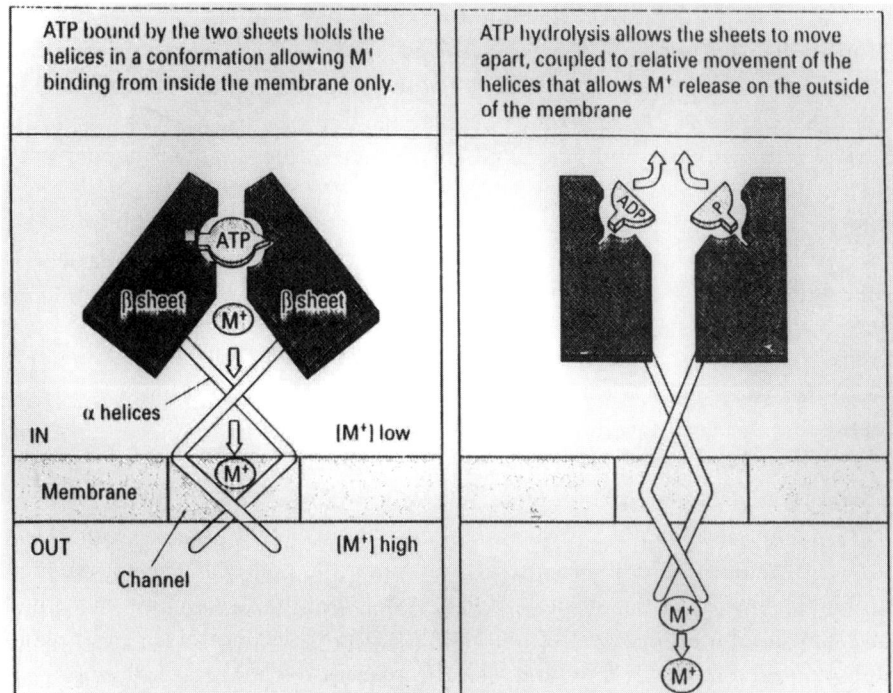

Fig. 12. An idealized version of an ion pump. The ATPase acts much like any kinase in a hinge-bending mode. The pumping is caused by cyclical transformations of the helices of the membrane causing gating of Ca^{2+} flow. The activity is different from the proton ATP-synthase and is probably more primitive being related to the pyrophosphate synthase.

develops as speedier response and relaxation are required in eukaryotes. The earliest organisms may not have needed a very low calcium concentration, that is below 10^{-5} M, as indicated by the level of tolerance of bacteria to high external calcium and the viable character of organelles to quite a high Ca^{2+} intake. Notice also that the process of sporulation common in lower organisms frequently requires uptake of considerable amounts of calcium without killing the organism when making a spore.

21. Calcium Networking between Compartments and Organelles

The flow of calcium into the cytoplasm from external sources is coupled to the flow from the endoplasmic reticulum. This flow raises the level of free calcium, which during relaxation, is pumped out of the cell and back into the reticulum or it is pumped into mitochondria or chloroplasts. The effect of calcium in these organelles is to increase 1) the metabolism of substrates in mitochondria via activation of dehydrogenases, or 2) to generate increase photochemical activity in chloroplasts. Both these changes lead to increase in proton gradient activity and then increase in ATP production. Owing to the restricted diffusion of both Ca^{2+} and H^+, this upgrading of ATP occurs only locally, but of course, it aids the rapid recovery of the cell because ATP is required to pump calcium into vesicles (ER) and out of the cell. Other branches of the ER may be responsible for the delivery of proteins or lipids to mitochondria or chloroplasts. It is observed again that there are local Ca^{2+} stimulated zones for these uptakes.

22. Summary

This chapter describes the chemical and biological value of the calcium ion. In calcium *chemistry*, our main interest is in equilibria within static, nonflowing systems. Hence, we examined the way calcium formed precipitates and complex ions in solution. We observed thereafter its uses by humankind in a vast number of materials such as minerals, e.g., marble, concrete, mortars, which parallel the biological use in shells and bones. In complex formation, we noted that many combinations were of anion interaction with calcium for example in the uses of detergents and medicines. The rates of exchange of calcium from bound states were noted but they had little application. Calcium ions do not act as catalysts of organic reactions.

In biological systems, interest is in the above chemistry, but extends to the fact that Ca^{2+} ions can carry information by flowing in one solution or from one solution to another through membranes. Hence, we became interested in the details of rates of calcium exchange. The fast exchange of this divalent ion from most organic binding sites has allowed it to develop as the dominant second messenger. Now the flow can be examined in vitro as calcium binds particular isolated proteins, which it activates as seen in physical mechanical changes or chemical changes and this piece-by-piece study of cells is common. Here, however, we have chosen to stress the whole circuit of Ca^{2+} action indicating that the cell is organized both at a basal and an activated state kinetic level by the steady state flow of the ion (*see* **Fig. 11**). Different time constants of exchange utilizing very similar binding constants lead to: 1) fast responses as in the muscle of an animal; or 2) slower change as in differentiation of an egg or seed. Many other changes of state may relate to Ca^{2+} steady-state levels of flow in the circuitry and here we point to two: 1) dormancy in reptiles and animals; and 2) sporulation in both bacteria and lower plants.

In the other chapters of this volume many components of the overall circuitry will be described. The reader should try to marry these into the overall activity of the cell for on top of molecular biology there is the cooperative system molecular biology of cells. To give an analogy, whereas much can be understood from the analysis of the properties of single-isolated water molecules, even examining their interaction in ice, this study alone cannot lead to an appreciation of the melting or boiling points of bulk water.

References

1. Frausto da Silva, J. R. R. and Williams, R. J. P. (revised 2001) Calcium-controls and triggers, In: *The Biological Chemistry of the Elements*. Oxford University Press, Oxford, pp. 268–298.
2. Williams, R. J. P. and Frausto da Silva, J. R. R. (1996) Organisation in advanced organisms, In: *The Natural Selection of The Chemical Elements*. Oxford University Press, Oxford, pp. 3–27.
3. Williams, R. J. P. (1998) Calcium in health and disease: meeting report, In: *Cell Calcium* **24**, 233.
4. Williams, R. J. P. (1999) Calcium-outside/inside homeostasis and signalling, In: *Biochim. Biophys. Acta* **1448**, 153.
5. Williams, R. J. P. (1999) Calcium: the developing role of its chemistry in evolution, In: *Calcium as a Cellular Regulator* (Carafoli, E. and Klee, C., eds.), Oxford University Press, Oxford, pp. 1–27.
6. Shao, Z. and Yang, J. (1995) Progress in high resolution atomic force microscopy in biology, In: *Quart. Rev. Biophys.* **28**, 195–251.
7. Zhu, P. and MacDougall, J. D. (1998) Calcium isotopes in the marine environment and oceanic calcium cycle, In: *Geochim. Cosmochim. Acta* **62**, 1691–1698.
8. Skulan, J., DePaolo, D. J., and Owens, T. L. (1997) Biological control of calcium isotopic abundances in the global calcium cycle, In: *Geochim. Cosmochim. Acta* **61**, 2505–2510.
9. Presented at the Calcium-Binding proteins in Health and Disease (1999) Kisaraza, Japan.

3

Crystal Structure of Calpain and Insights into Ca^{2+}-Dependent Activation

Zongchao Jia, Christopher M. Hosfield, Peter L. Davies, and John S. Elce

1. Introduction

Calpains are the only known enzymes that combine protease activity with a dependence on Ca^{2+}-binding to EF-hands in one molecule. The µ- and m-calpains are cytosolic cysteine proteases that are ubiquitously expressed and differ in their sensitivity to Ca^{2+}. They consist of an isoform-specific catalytic approx 80-kDa subunit and a common regulatory approx 28-kDa subunit. Although the exact physiological roles of calpains remain to be elucidated, their functional characteristics and wide distribution suggest that they have important cellular roles, which have been reviewed elsewhere *(1–4)*.

Calpain has also been identified as a target for therapeutic drug design because of its implication in several pathological conditions, including cerebral and cardiac ischemia, cataract formation, and Alzheimer's disease *(5)*. These conditions, typified by altered Ca^{2+}-homeostasis resulting from some sort of cellular stress, appear to result in disrupted regulation and excessive activation of calpain.

These considerations made calpain an attractive target for crystallographers, and many efforts have been made to crystallize calpains from natural sources. Satisfactory crystals were not obtained from these sources, probably because of heterogeneity arising either from partial autolysis, or possibly from phosphorylation. In practice, crystallization succeeded only with recombinant calpain modified with a histidine-tag to simplify purification *(6,7)*.

Within the cell, proteases are synthesized as zymogens, or inactive precursors, to prevent inappropriate or premature degradation of substrates *(5,8)*. Zymogen conversion to the active enzyme is triggered by a variety of factors such as a change in pH. In trypsin-like enzymes, activation involves very limited cleavages, which permit critical adjustments in the active site. The activation of profactor D, which is otherwise a normal serine protease, follows a unique pathway, involving reorientation of residues to form a self-inhibited mature enzyme *(9)*. In all known cysteine proteases, activation involves autolytic removal of a propeptide to generate the mature protease, and the inactivity of the zymogen is due to the occupation of the active site by the propeptide, providing a steric block for other substrates *(8)*. The active sites of almost all zymogens, and their related active proteases are otherwise virtually indistinguishable, although proplasmepsin II is an exception *(10)*. Because calpain appears to have some zymogen-like characteristics, it is usually not referred to as a zymogen as it clearly does not function in the typical fashion.

In order to relate the crystal structure to biological activities of the enzyme, it is important to discuss several aspects of calpain regulation. In addition to the absolute requirement for Ca^{2+} to produce a functional protease, calpain activity is carefully controlled by several other mechanisms. It is known that the concentrations of Ca^{2+} required for activity in vitro (approx 10–50 μM in μ-calpain and approx 300–500 μM in m-calpain) are considerably higher than physiological Ca^{2+} concentrations, which are generally less than 1 μM *(11)*. This suggests a requirement for endogenous activators, which have indeed recently been reported *(12)*. In contrast, the calpain inhibitor calpastatin found in all cells acts specifically to block calpain activity. Because many preferred calpain substrates are associated with the cytoskeletal matrix, Ca^{2+}-dependent translocation to the cell membrane is often suggested as a mechanism to allow escape from inhibition by calpastatin, whereas utilizing cellular localization as one factor affecting substrate specificity *(2,13)*. A further regulatory factor is autolytic cleavage, by which a short N-terminal peptide is cleaved from the large subunits of μ- and m-calpain very rapidly after addition of Ca^{2+}. The enzyme is active in the presence of Ca^{2+} before cleavage of this N-terminal anchor, although this cleavage is very rapid *(14)*. Conversely, cleavage of the peptide is not alone sufficient to activate the protease *(8)* because calpain still requires Ca^{2+} after autolysis, although at a lower level. It is well established that cleavage after residue nine (and later possibly after residue 19) in the catalytic subunit of m-calpain greatly increases the in vitro sensitivity of the enzyme towards Ca^{2+} *(11)*. In the regulatory subunit autolytic cleavage of the N-terminal 86 residues which constitute D-V is yet another process accompanying activation, although it does not by itself affect the Ca^{2+} requirement, and its significance is not well understood *(15)*. Finally, an interesting regulatory mechanism involving dissociation of the regulatory subunit has been proposed *(16–18)*. In this model, the calpain subunits dissociate on exposure to Ca^{2+}, thereby liberating a catalytic subunit with increased Ca^{2+}-sensitivity and a capacity for membrane-association. There remains some uncertainty about this model, because it was not supported by various studies involving immunoprecipitation and affinity chromatography of natural and mutant calpains in the presence of Ca^{2+} *(19,20)*. However the question should be considered to remain open, because of the great technical difficulties arising from the characteristic autolysis and aggregation of calpain in the presence of Ca^{2+}. Clearly, the mechanisms governing activation and inhibition of calpain in vivo are complex, and further experiments will be required for a full understanding of calpain regulation.

2. Results and Discussion
2.1. Overall Structure of m-Calpain

The molecule is an elongated, multidomain assembly, with dimensions of approx $100 \times 60 \times 50$ Å, which is illustrated as a ribbon diagram in **Fig. 1A**. The catalytic subunit consists of four distinct structural domains (D-I to D-IV) as well as an α-helical N-terminal anchor of 19 residues, which does not constitute a true domain. This structure, which is the first reported for any calpain *(21)*, requires a new domain classification (**Fig. 1B**) that differs somewhat from that defined solely from the primary sequence *(22)*. As in other papain-like cysteine proteases, the catalytic triad residues (*Cys*105 in D-I, *His*262 and *Asn*286 in D-II) are located in a cleft at the interface of D-I and D-II *(23,24)*. D-I (residues 20–210) has a fold unrelated to any previously observed, consisting of a central helix flanked on three faces by a cluster of α-helices, and two anti-parallel β-sheets. Apart from the α-helix containing *Cys*105, which is slightly shorter than the corresponding helix in members of the papain family, D-I is otherwise entirely unrelated to the corresponding domain in the typical thiol protease, because it is larger and shows no overall structural similarity *(25)*. The sequence and structural differences shown by D-I when compared with other thiol proteases may contribute to the proteolytic specificity characteristic of calpain, which usually cleaves its substrates at a limited number of sites between domains to form fragments with modified function, rather than causing complete degradation *(1)*.

Although more flexible (at least in the absence of Ca^{2+}), D-II of m-calpain is similar to that of the corresponding domain in other cysteine proteases *(24,25)* and contains two three-stranded antiparallel β-sheets. These strands position *His*262 and *Asn*286 of the catalytic triad at the interface of D-I and D-II, where the active site resides. D-II makes several contacts with D-III largely through a structurally conserved (within papain-like proteases) three-turn α-helix. D-III is an eight-stranded antiparallel β-sandwich that shares structural characteristics with the C_2-domain *(26,27)* (*see* below). This β-sandwich domain makes contacts with each domain in the enzyme, and leads into an

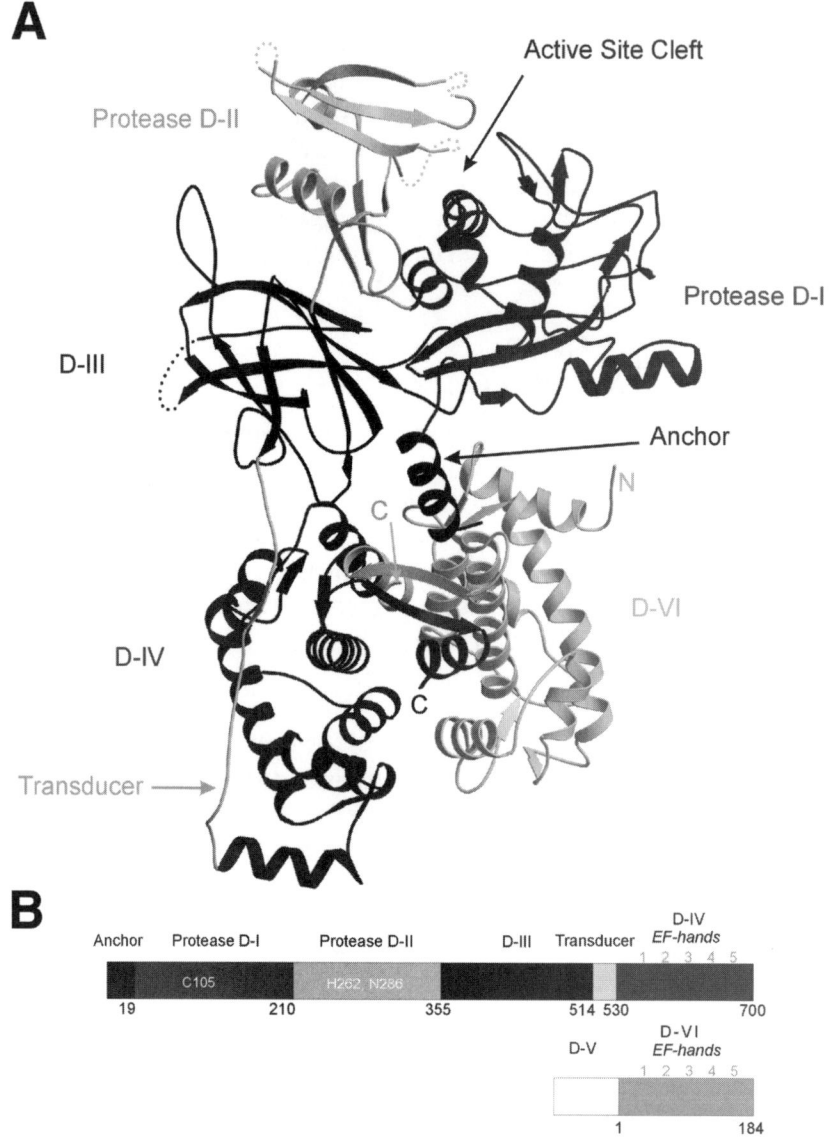

Fig. 1. **(A)** Ribbon diagram of m-calpain. The active site cleft is located at the interface of D-I and D-II. Dotted lines represent disordered regions that were omitted. **(B)** Schematic diagram illustrating the domain organization of m-calpain. The 80-kDa subunit is composed of a 19 residue anchor (*black*), protease domains I and II (*black* and *light gray*, respectively), domain III (*black*), an approx 15 residue transducer (*light gray*), and domain IV (*dark gray*). The regulatory subunit contains only domain VI (*light gray*). Catalytic triad residues are indicated, and approximate domain boundaries are indicated by residue number.

approx 15-residue extended linker (the transducer) that connects D-III to D-IV, a Ca^{2+}-binding domain with five EF-hand motifs. The transducer exists in an extended conformation, with three residues (516–518) that form a short antiparallel β-sheet with three residues (636–638) from D-IV.

D-IV and D-VI, often referred to as the calmodulin-like domains of each subunit, are approx 50% identical in sequence and are predominantly α-helical. Their structures are very similar to each other (root-mean square deviation [rmsd] of 1.7 Å on main chain atoms), as each contains five EF-hand motifs and the heterodimer between D-IV and D-VI possesses pseudo-twofold symmetry. Heterodimerization of the catalytic and regulatory subunits occurs primarily through hydrophobic interactions in the C-terminal regions of D-IV and D-VI, an observation previously predicted from the known crystal structure of D-VI (which is a homodimer) *(28,29)*. In the homodimer of D-VI, it has been shown that EF-hands 1 to 4 bind Ca^{2+} while EF-hand 5 does not bind Ca^{2+} *(28,29)*. The structure of D-VI in the heterodimer described here, is virtually identical to its structure in the homodimer (rmsd of 1.29 Å on main chain atoms).

2.2. Why is Calpain Activity Absolutely Dependent on Ca^{2+}?

One of the major objectives of this structure determination was to elucidate why calpain activity is completely dependent on Ca^{2+}, a subject that has been heavily debated over the years. The structure reveals a regulatory mechanism that is highly unusual in proteases and is unprecedented within the cysteine protease family. Although the α-helix containing *Cys*105 in D-I, and the fold of D-II, are each similar to those of other thiol proteases, the active site is not assembled. Numerous crystal structures of cysteine proteases have clearly shown that the catalytic *Cys* and *His* residues form an ion-pair in the active site. Specifically, the interatomic distance between the S atom of *Cys* and the Nδ atom of *His* is approx 3.7 Å *(25)*. In this orientation, the *His*-Nδ is at an appropriate distance to coordinate the hydrogen atom bonded to the *Cys*-S, significantly decreasing the pK_a of the thiol proton and increasing the nucleophilicity of the sulfur atom. In calpain, the catalytic *Cys*105-S in D-I is approx 10.5 Å away from *His*262-Nδ and therefore is too remote to form a competent catalytic triad with its counterparts *His*262 and *Asn*286 in D-II (**Fig. 2A**). Accordingly, the Ca^{2+}-induced conformational change must reduce this distance to approx 3.7 Å in order to assemble the triad and form an active protease.

As we have not yet been able to obtain crystals of a Ca^{2+}-bound form of calpain, the details of the required conformational change remain speculative, but it is clear that the catalytic triad must be assembled into the required geometry for calpain to be active. Molecular modeling illustrates that such a conformational change may be achieved by small rotations of D-I and D-II toward each other (approx 5°), together with a small translation of approx 1–2 Å. These

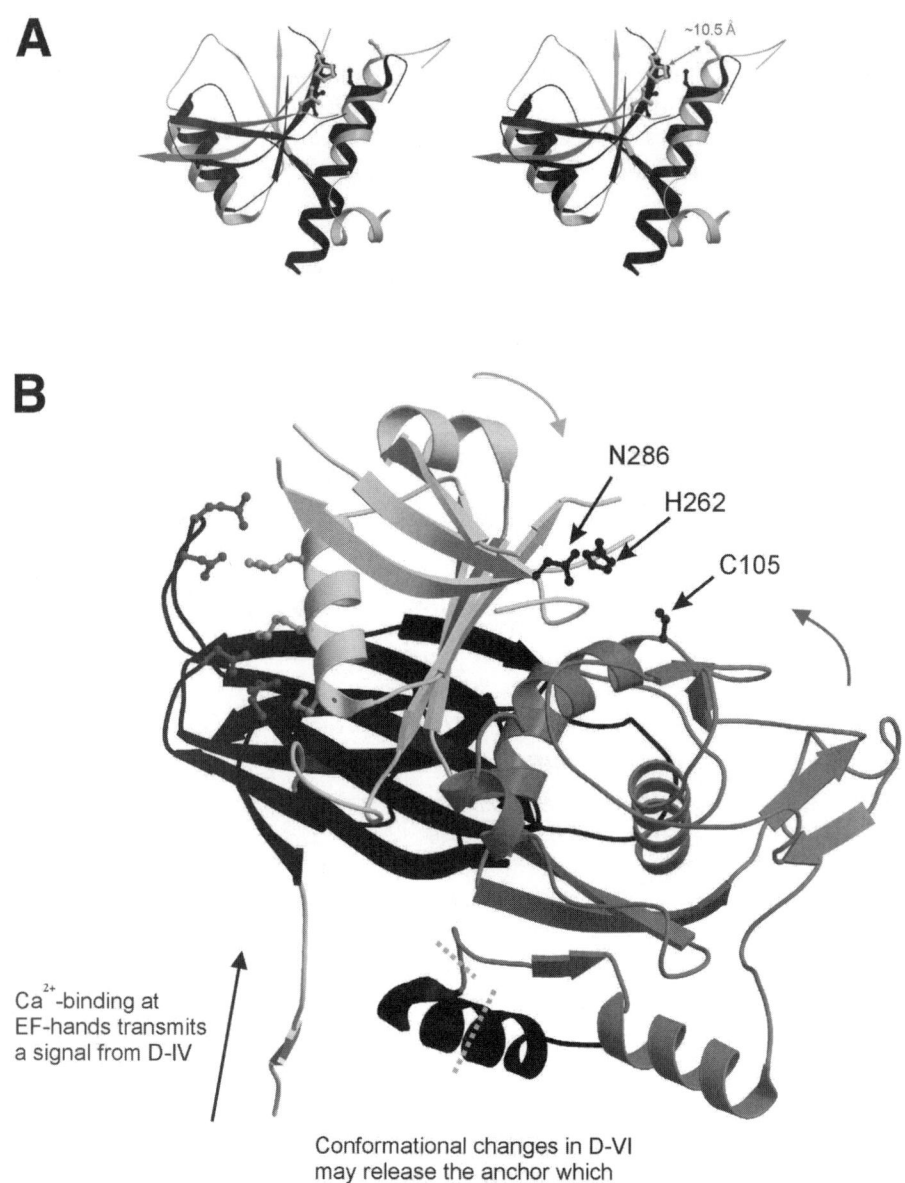

Fig. 2. **(A)** Structural basis for the inactivity of calpain. Stereoview of the active site of apo-Ca^{2+}-calpain (*light gray*) superimposed upon the active site of papain (*dark gray*) (PDB accession-code 9PAP) *(25)*. *His*262 and *Asn*286 of calpain are arranged in a similar orientation to that in papain. Inactivity of calpain in the absence of Ca^{2+} is caused by the catalytic *Cys*105 being approx 7 Å too distant from *His*262 (the distance between *His*-Nδ and *Cys*-S is approx 10.5 Å in calpain and is approx 3.7 Å in active cysteine

observations provide the first example of a cysteine protease catalytic triad that undergoes a substantial conformational rearrangement during enzyme activation. This suggests an attractive strategy for the design of inhibitory drugs specific for calpain, as it may prove possible to synthesize inhibitors that recognize the unique preactive-site, thereby preventing calpain from attaining an active conformation.

2.3. The Anchor and the Transducer Have Important Roles in the Inactivation/Activation Mechanism

It is apparent from the structure that D-I and D-II, which house the catalytic triad residues, are held apart and have their movement restricted by both N- and C-terminal constraints. The N-terminal anchor (residues 2–19 of the catalytic subunit) is an α-helix that tethers D-I to D-VI in the regulatory subunit (**Fig. 1A**), and plays a key role in the activation process. Most interestingly, this helix, which was thought to be analogous to the prosegment of typical cysteine proteases, does not occupy the active site. In all members of the papain-family of cysteine proteases, the prosegment binds to the active site in a reverse orientation to that of native substrates, thereby maintaining the protease inactive by preventing substrate-binding *(8)*. Upon activation, the prosegment is autolytically cleaved, and the active site becomes accessible.

In calpain, the anchor instead inhibits activity through its interaction with the regulatory subunit. The anchor makes several contacts with a hydrophobic pocket in D-VI, and has a strong helical dipole–dipole interaction with a helix in D-VI. This anchor is autolyzed upon Ca^{2+} activation. Activation may result in loss of the contacts between the anchor and D-VI through Ca^{2+}-induced conformational changes, with subsequent release of the anchor from the regulatory subunit. Alternatively, the anchor could be released through dissociation of the two subunits upon Ca^{2+}-binding *(18)*. Release of the anchor would increase the conformational freedom of D-I, and allow it to move toward D-II. Autolysis of the exposed anchor at *Ala9-Lys10* and subsequently at *Gly19-Ser20* is not strictly required for activity, yet in practice occurs rapidly

---

(Fig. 2. *continued*) proteases *(25)*. (**B**) Proposed activation mechanism. Ca^{2+}-binding causes conformational changes in D-IV, which may allow the transducer to release constraints on other domains, and affords increased flexibility to D-II. This may occur by change in the interactions between basic residues *(light gray)* in D-II and acidic residues *(dark)* in D-III, or by movement of D-II *(light gray)* and D-III *(black)* together. Concomitantly, Ca^{2+}-binding to D-VI causes the release of the anchor *(black)*, yielding a more flexible D-I *(dark gray)*. Release of the conformational restraints imposed on D-I and D-II would permit formation of the active site *(catalytic triad residues are indicated in black)* at the interface of D-I and D-II through rotations of D-I and D-II relative to each other.

during activation *(14,30)*. The crystal structure shows that these cleavage sites are too far away from the active site (approx 40 Å) to be autolyzed by an intramolecular mechanism, so that the cleavage must be intermolecular. In natural calpains, D-V of the regulatory subunit is also autolyzed, possibly in several steps. The details of regulatory subunit autolysis cannot be determined from this structure, but the final step must again be intermolecular because this cleavage site, which is close to the N-terminus of the recombinant regulatory subunit used here, is also approx 40 Å from the active site.

The fact that calpain remains Ca^{2+}-dependent following large subunit autolysis indicates that additional conformational changes are required to assemble the active site. Assuming that D-I is free to move after the release of the anchor, it follows that movement of D-II is also required to form the competent catalytic triad. Based on the structure, an appealing mechanism is suggested by the linker of approx 15 residues that directly connects the Ca^{2+}-binding D-IV to D-III. It appears that this extended segment is acting as an intramolecular signal transducer, communicating a conformational change resulting from Ca^{2+}-binding in D-IV, through D-III, to the protease domains. D-III is able to mediate these conformational changes because it makes several contacts with each domain in the protein, including a significant interaction at the interface with the protease D-II. At this interface, there are several electrostatic interactions involving *Lys*226, *Lys*230, and *Lys*234, which reside on an amphipathic helix in D-II, and several acidic residues including *Asp*395, *Asp*398, *Glu*399, and *Glu*504 on loops in D-III (**Fig. 2B**). These strong interactions suggest that both D-III and D-II may move in unison in assembling the active site, but the nature of these movements is presently unknown.

2.4. Does Calpain Contain Ca^{2+}-Binding Sites Distinct from EF-Hands? The Elusive D-III

The function of D-III has remained unclear, partly owing to its lack of sequence homology with any known protein. The C_2-like fold revealed by this crystal structure, however, suggests a role for this domain in membrane-association processes. D-III makes several contacts to the protease domains, and possesses an antiparallel β-sheet sandwich structure (**Fig. 3**) with similar overall dimensions to that of the C_2-domain. In typical C_2-domains, several loops contribute acidic residues to form a binding cradle at one end of the β-sandwich that is often surrounded by basic residues *(26,27)*. In addition to possessing a fold similar to a Ca^{2+}-dependent C_2-domain, D-III also has a large number of acidic residues in the loop region, flanked by basic residues. Further, primary sequence analysis indicates that a C_2-domain exists in the C-terminal region of *Tra*-3, a calpain homolog found in C. elegans *(31)*. Many C_2-domains have been described as Ca^{2+}-dependent lipid-binding domains *(26,27)*. Because

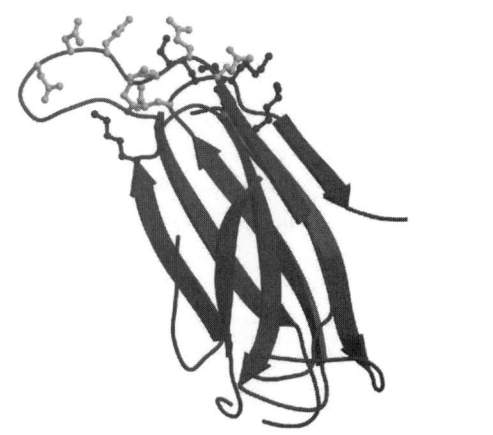

Fig. 3. Domain III shares similar characteristics with a C_2 domain. A typical C_2 domain exists as an antiparallel β-sandwich with several acidic residues at one end that form a binding cradle for Ca^{2+}. The first C_2 domain from synaptotagmin (*light gray*, PDB accession-code 1RSY) *(26)* and D-III (*dark gray*) have roughly the same overall dimensions, though slightly differing topologies. Numerous acidic residues (*light gray*) result in a highly negative potential, which is partially stabilized by adjacent basic residues (*dark gray*).

calpain activation in vivo is commonly believed to depend in part on Ca^{2+}-dependent interactions with the membrane *(13)*, D-III may target the active enzyme to the membrane where it would have access to physiologically relevant substrates associated with the cytoskeletal matrix. Also, it is possible that Ca^{2+}-induced conformational changes in D-III could enhance its affinity for the cell membrane. Although an overall similarity of the fold is observed between calpain D-III and Ca^{2+}-dependent C_2-domains, whether D-III of calpain actually binds Ca^{2+} or binds to the membrane needs to be explored.

2.5. The Anchor–Regulatory Subunit Interaction has a Key Role in Enzyme Assembly

In addition to its role in activation and in modulating Ca^{2+}-affinity, the observation that the N-terminal anchor associates exclusively with the regulatory subunit implies that their functions must be interrelated. It has been demonstrated previously that the regulatory subunit has chaperone-like effects on the refolding of denatured catalytic subunit *(17)*. Chaperonins such as *GroE* are also able to assist in the folding of the denatured catalytic subunit in the presence of ATP, though not as effectively as the regulatory subunit *(17)*. It is well documented that classical prosegments often promote folding of proenzymes *(32)*, and we suggest that the N-terminal anchor has a chaperone-like activity. In accordance with this suggestion is the finding that expression of various N-terminal large subunit-truncated constructs yielded calpains that were stable heterodimers, but lacked enzymatic activity *(30)*, presumably owing to misfolding. Taken together with the fact that the crystal structure shows a direct interaction between these structural elements, it seems likely that the anchor and the regulatory subunit act as cochaperones that promote productive folding of the catalytic subunit.

3. Conclusions

The fusion of calmodulin-like and papain-like activities suggests a unique mechanism for Ca^{2+}-dependent regulation of calpain, whose significance in vivo is indicated by the numerous pathological conditions resulting from excessive or inappropriate calpain activation. Now that a crystal structure has been determined, it has provided many insights into Ca^{2+}-dependent activation of calpain. Accordingly, we propose the following mechanism that addresses aspects of calpain regulation ranging from enzyme assembly and Ca^{2+}-induced activation, to autolytic processing and cellular localization. The regulatory subunit of calpain aids the formation of a properly folded catalytic subunit through interactions with the N-terminal anchor and D-IV. The assembled heterodimer remains inactive in the absence of Ca^{2+} as a result of a network of conformational restraints that effectively hold the active site apart. Binding of Ca^{2+} to EF-hands induces conformational changes in D-IV and D-VI, and these changes have two immediate results: the release of the catalytic subunit N-terminal anchor from the regulatory subunit, permitting movement of D-I, and movement of the transducer attached to D-IV, permitting movement of D-II and D-III. Consequently, D-I and D-II are able to move together to form a competent active site. Intermolecular autolysis of the N-terminal anchor activates calpain by permitting activity at lower Ca^{2+}-concentrations. Subunit dissociation, if it occurs, will affect these events because of the close association of the N-terminal anchor of the catalytic subunit with the regulatory subunit. Finally, the C_2-like D-III may be responsible for promoting binding of active calpain to the membrane in response to Ca^{2+}, thereby relieving inhibition from calpastatin and promoting digestion of physiologically relevant substrates.

Calpains are unique because they are the only family of enzymes that combines papain-like cysteine protease activity with a dependence on Ca^{2+}-binding to calmodulin-like EF-hand domains. The regulatory mechanisms governing calpain activity are complex, and the nature of the Ca^{2+}-induced switch between inactive and active forms has remained elusive in the absence of structural information. The crystal structure of m-calpain in the Ca^{2+}-free form has revealed the structural basis for the inactivity of calpain in the absence of Ca^{2+}. In the absence of Ca^{2+}, the protease catalytic triad is not assembled, indicating that Ca^{2+}-binding induces conformational changes that reorient the protease domains to form a functional active site. It has also revealed an unusual thiol protease fold, which is associated with Ca^{2+}-binding EF-hand domains through a C_2-like β-sandwich domain and dimerization with the regulatory subunit. The finding of a C_2-like domain is of considerable interest because it suggests that Ca^{2+} may regulate calpain activity in vivo through yet another new aspect. In contrast to other papain-like proteases, which have a N-terminal prosegment that physically blocks the active site, the α-helical N-terminal anchor of the catalytic subunit does not do so in calpain, but instead inhibits active site assembly and regulates Ca^{2+}-sensitivity through association with the regulatory subunit. The Ca^{2+}-dependent regulatory features that govern calpain activation in vivo are clearly quite different from all other proteases.

4. Methods and Materials
4.1. Protein Expression, Purification, and Storage

We have expressed a recombinant version of rat m-calpain in *Escherichia coli*, which has a few modifications. The catalytic subunit was expressed with a C-terminal histidine-tag and a mutation of the active site *Cys* 105 to *Ser*. The regulatory subunit was expressed as a truncated 21-kDa form from which D-V has been omitted, because it contains approx 30% glycine residues *(33)* and is partially degraded when expressed and purified. This led to heterogeneous preparations of calpain *(15)*, a factor that was expected to hinder crystallization. Rat m-calpain ($C105S$-m-80k-$CHis_6$/21k) was expressed in 4–8 l of *E. coli* strain *BL21*(DE3) from compatible plasmids that encode the two subunits, which associate within *E. coli* to produce fully active heterodimers *(34)*. The selenomethionine derivative was expressed in *B834*(DE3) cells *(35)* in defined medium lacking methionine, but supplemented with selenomethionine. Both mass-spectrometry and amino-acid analysis have confirmed that uptake of selenomethionine is nearly 100% complete (data not shown). Protein purification involved several chromatographic steps at 4°C on successive columns of DEAE-Sepharose, Ni-NTA-agarose, Ultra-gel AcA44, occasionally on Reactive-Red-agarose, and finally by FPLC on Q-Sepharose. In most instances the buffers contained 50 mM Tris-HCl, pH 7.6, 2 mM ethylenediaminetetracetic acid (EDTA), 10 mM 2-mercaptoethanol, and appropriate NaCl concentrations *(6,36)*. Typical yields of fully purified material were 10–15 mg of the methionine form, and 4–8 mg of the selenomethionine derivative, from 8 l of *E. coli*. The final product was concentrated to 10 mg/mL in 50 mM Tris-HCl, pH 7.6, 2 mM EDTA, 0.01% sodium azide and 10 mM dithiothreitol (DTT).

Calpain degrades relatively quickly at 4°C, as indicated by a reduction of the crystallization quality within a few days following purification and by the appearance of degradation and aggregation products on SDS gels in the presence and absence of reducing agent (data not shown). Therefore, any protein sample that was not immediately used for crystallization following purification and concentration was snap-frozen in 100 µL aliquots using liquid N_2 and stored at –70°C. On thawing, the protein retained both enzymatic activity and its ability to crystallize, indicating that the freezing procedure did not significantly disrupt the protein.

4.2. Crystallization and X-Ray Structure Determination

Sparse matrix screening was carried out using the hanging drop method at room temperature *(37)*. Several PEG-containing conditions yielded small crystals. After routine optimization, the best condition was determined to be 100 m*M* MES, pH 6.5, 10–12% PEG-6000, 50 m*M* NaCl, and approx 10 mg/mL protein. In the initial stages of the work, the only reducing agent in the drops was 2-ME remaining from protein purification. Under these conditions, most crystals grew as small multicrystals with many surface defects, and diffracted poorly using the in-house X-ray facility. In addition, reproducibility of crystallization was a major problem, because only 1% of drops contained diffraction-size crystals. These crystals were determined to belong to triclinic space group *P*1. The best *P*1 crystals diffracted to 3.0 Å resolution at 100 K using the in-house X-ray facility.

Although the process of structure determination was limited because of the poor crystal quality, adjustments were made both to the purification procedure and to the crystallization conditions to improve the reproducibility and quality of crystals. Acceleration of the purification procedure, and the presence of fresh 10 m*M* DTT in the final protein sample, gave rise to a new crystal form $P2_1$. The $P2_1$ crystals were of higher quality, as they could typically diffract to 2.8 Å resolution at 100 K at the home source. Most crystallization drops contained crystals, but the number of drops containing well-formed single crystals large enough for in-house X-ray analysis remained low (5–8%). In addition, crystals of the selenomethionine derivatives of C105S-m-calpain were obtained in both *P*1 and $P2_1$ forms *(7)*.

Both forms decayed very fast at room temperature upon exposure to X-rays, making it essential to collect data at cryogenic temperatures. For *P*1 crystals, serial transfer to crystallization solutions supplemented with 10%, 20%, and 30% glucose for 2 min at each concentration followed by immediate flash cooling in liquid propane gave the best conditions for data collection. For $P2_1$ crystals, serial transfer through solutions supplemented with 5%, 10%, 20%, 30%, and 40% ethylene glycol for approx 5 min at each concentration followed by flash cooling in liquid propane was found to be optimal.

Selenomethionine-containing crystals of the C105S form in the absence of Ca^{2+} were subjected to MAD data collection in the vicinity of the selenium absorption edge (approx 0.98 Å) using an ADSC Quantum IV-CCD at Stanford Synchrotron Radiation Laboratory beamline 1–5. Diffraction data were processed with the HKL program suite *(38)* (**Table 1**) and the CCP4 program suite *(39)*, and 17 of 19 selenium positions were determined by direct methods with the DREAR/SnB package *(40)*. Heavy-atom refinement and phasing with SHARP *(41)* gave interpretable electron density maps (figure of merit 0.63, 0.38 for $P2_1$ and $P1$ crystals, respectively) that were substantially improved upon solvent flattening with SOLOMON *(42)*. The model was traced with XFIT *(43)* independently in both space groups to reduce the possibility of tracing mistakes in disordered regions and was subjected to iterative cycles of manual fitting and maximum-likelihood refinement as implemented in the CNS package *(44)*. The final model was obtained using data collected at the Cornell High Energy Synchrotron Source on the $P1$ native crystal form *(7)*. The resultant structure was well defined with the exception of the following disordered regions that were omitted: 245–260, 273–278, 292–321, 437–459, 565–566 (indicated by dotted lines in **Fig. 1A**). The quality of the model was assessed with PROCHECK *(45)*, and displayed good stereochemistry with 83.2% of residues in the most favorable regions and no residues in the disallowed regions of a Ramachandran plot. **Figures 1A, 2,** and **3,** were prepared with the programs MOLSCRIPT *(46)*, and RASTER3D *(47)*.

Data were also collected and the structure was determined for enzymatically active calpain containing the natural *Cys*105. The data were at a lower resolution but the structure was fully consistent with the model of the *Ser*105 mutant. All attempts to crystallize calpain in the presence of Ca^{2+} were unsuccessful because of massive aggregation. Alternatively, we attempted to soak native crystals in solutions containing Ca^{2+}, however addition of Ca^{2+} immediately shattered the crystals.

Acknowledgments

We would like to thank D. Hosfield and T. Moldoveanu for insightful discussions, C. Hegadorn and Q. Ye for skilled technical assistance, H. Bellamy and other staff at SSRL for technical support at beamline 1–5. This work was funded by grants from MRC, PENCE and Warner-Lambert Canada. Z. Jia is an MRC Scholar and C. M. Hosfield is a recipient of studentships from NSERC and MRC. Coordinates have been deposited in the RCSB Protein Data Bank with accession code 1DF0.

Table 1
Crystallographic Analysis

Crystal	λ (Å)	dmin (Å)	Observed reflections	Unique reflections	Completeness (%)	I/sig I	R_{merge} (%)[a]
SeMet $P2_1$§							
Low energy remote	1.069	2.6	80,764	28,841	92.7 (66.1)	25.3	2.7 (15.6)
Anomalous peak	0.9795	2.6	155,288	57,860	93.4 (88.7)	21.5	3.1 (21.9)
Inflection point	0.9799	2.6	136,982	54,126	88.3 (81.1)	19.7	3.2 (26.0)
High energy remote	0.9252	2.6	153,100	57,581	93.2 (89.0)	21.6	3.2 (21.5)
SeMet P1¶							
Anomalous peak	0.9795	2.6	128,734	74,656	93.7 (92.1)	16.9	2.7 (37.4)
Inflecion point	0.9799	2.6	67,174	43,860	57.2 (55.1)	18.7	2.0 (23.4)

Refinement statistics	
Native P1 (25 to 2.6 Å)	
Unique/free reflections	34 794/3454
R-factor[b]/R-free[c] (%)	22.3/29.3
Protein/water atoms	6428/361
rmsd bond lengths (Å)	0.0095
rmsd bond angles (°)	1.452

[a] $R_{merge} = \sum |I(k) - \langle I \rangle| / \sum I(k)$, where $I(k)$ and $\langle I \rangle$ represent the diffraction intensity values of the individual measurements and the corresponding mean values. The summation is over all measurements.
[b] R-factor = $\sum ||F_o| - |F_c|| / \sum |F_o|$.
[c] $R_{free} = \sum ||F_{o\ (free)}| - |F_{c\ (free)}|| / \sum |F_{o\ (free)}|$.
F_o is observed structure factor, F_c is calculated structure factor based on the model.
Values given in parentheses refer to reflections in the outer resolution shell, 2.7 – 2.6 Å.
Unit cell dimensions: $P2_1$§: $a = 51.85$, $b = 156.7$, $c = 64.43$, $\beta = 95.32°$; P1¶: $a = 65.17$, $b = 79.91$, $c = 81.59$, $\alpha = 108.5°$, $\beta = 103.37°$, $\gamma = 112.95°$.
All MAD data were collected in inverse beam mode; with the exception of the low energy remote of the $P2_1$ form, which served as the "native" for phasing.
No sigma cutoff was applied to the data; 10% of reflections were excluded from refinement for calculation of R-free; rmsd, root-mean-square deviation.

References

1. Sorimachi, H., Ishiura, S., and Suzuki, K. (1997) Structure and physiological function of calpains. *Biochem. J.* **328,** 721–732.
2. Molinari, M. and Carafoli, E. (1997) Calpain: a cytosolic proteinase active at the membranes. *J. Membrane Biol.* **156,** 1–8.
3. Ono, Y., Sorimachi, H., and Suzuki, K. (1998) Structure and physiology of calpain, an enigmatic protease. *Biochem. Biophys. Res. Commun.* **245,** 289–294.
4. Carafoli, E. and Molinari, M. (1998) Calpain: a protease in search of a function. *Biochem. Biophys. Res. Commun.* **247,** 193–203.
5. Wang, K. K. W. and Yuen, P.-W. (1997) Development and therapeutic potential of calpain inhibitors. *Adv. Pharmacol.* **37,** 117–152.
6. Elce, J. S. (2000) Expression of m-Calpain in *E. coli*, in *Methods in Molecular Biology*, vol. 144: *Calpain Methods and Protocols* (Elce, J. S., ed.), Humana Press, Totowa, NJ, pp. 47–54.
7. Hosfield, C. M., Ye, Q., Arthur, J. S. C., Hegadorn, C., Croall, D. E., Elce, J. S., and Jia, Z. (1999) Crystallization and X-ray crystallographic analysis of m-calpain: a Ca^{2+}-dependent protease. *Acta Crystallogr.* **D55,** 1484–1486.
8. Khan, A. R. and James, M. N. G. (1998) Molecular mechanisms for the conversion of zymogens to active proteolytic enzymes. *Protein Sci.* **7,** 815–836.
9. Jing, H., Macon, K. J., Moore, D., DeLucas, L. J., Volanakis, J. E., and Narayana, S. V. L. (1999) Structural basis of profactor D activation: from a highly flexible zymogen to a novel self-inhibited serine protease, complement factor D. *EMBO J.* **18,** 804–814.
10. Bernstein, N. K., Cherney, M. M., Loetscher, H., Ridley, R. G., and James, M. N. G. (1999) Crystal structure of the novel aspartic proteinase zymogen proplasmepsin II from plasmodium falciparum. *Nat. Struct. Biol.* **6,** 32–37.
11. Goll, D. E., Thompson, V. F., Taylor, R. G., and Zalewska, T. (1992) Is calpain activity regulated by membranes and autolysis or by calcium and calpastatin? *BioEssays* **14,** 549–556.
12. Melloni, E., Michetti, M., Salamino, F., and Pontremoli, S. (1998) Molecular and functional properties of a calpain activator protein specific for mu-isoforms. *J. Biol. Chem.* **273,** 12,827–12,831.
13. Mellgren, R. L. (1987) Calcium-dependent proteases: an enzyme system active at cellular membranes? *FASEB J.* **1,** 110–115.
14. Molinari, M., Anagli, J., and Carafoli, E. (1994) Ca(2+)-activated neutral protease is active in the erythrocyte membrane in its nonautolyzed 80-kDa form. *J. Biol. Chem.* **269,** 27,992–27,995.
15. Elce, J. S., Davies, P. L., Hegadorn, C., Maurice, D. H., and Arthur, J. S. C. (1997b) The effects of truncations of the small subunit on m-calpain activity and heterodimer formation. *Biochem. J.* **326,** 31–38.
16. Yoshizawa, T., Sorimachi, H., Tomioka, S., Ishiura, S., and Suzuki, K. (1995a) Calpain dissociates into subunits in the presence of calcium ions. *Biochem. Biophys. Res. Commun.* **208,** 376–383.

17. Yoshizawa, T., Sorimachi, H., Tomioka, S., Ishiura, S., and Suzuki, K. (1995b) A catalytic subunit of calpain possess full proteolytic activity. *FEBS Lett.* **358,** 101–103.
18. Suzuki, K. and Sorimachi, H. (1998) A novel aspect of calpain activation. *FEBS Lett.* **433,** 1–4.
19. Zhang, W. and Mellgren, R. L. (1996) Calpain subunits remain associated during catalysis. *Biochem. Biophys. Res. Commun.* **227,** 890–896.
20. Dutt, P., Arthur, J. S. C., Croall, D. E., and Elce, J. S. (1998) m-Calpain subunits remain associated in the presence of calcium. *FEBS Lett.* **436,** 367–371.
21. Hosfield, C. M., Elce, J. S., Davies, P. L., and Jia, Z. (1999) Crystal structure of calpain reveals the structural basis for Ca^{2+}-dependent protease activity and a novel mode of enzyme activation. *EMBO J.* **18,** 6880–6889.
22. Ohno, S., Emori, Y., Imajoh, S., Kawasaki, H., Kisaragi, M., and Suzuki, K. (1984) Evolutionary origin of a calcium-dependent protease by fusion of genes for a thiol protease and a calcium-binding protein? *Nature* **312,** 566–570.
23. Berti, P. J. and Storer, A. C. (1995) Alignment/phylogeny of the papain superfamily of cysteine proteases. *J. Mol. Biol.* **246,** 273–283.
24. Groves, M. R., Coulombe, R., Jenkins, J., and Cygler, M. (1998) Structural basis for specificity of papain-like cysteine protease proregions toward their cognate enzymes. *Proteins* **32,** 504–514.
25. Kamphuis, I. G., Kalk, K. H., Swarte, M. B. A., and Drenth, J. (1984) Structure of papain refined at 1.65 Å resolution. *J. Mol. Biol.* **179,** 233–256.
26. Sutton, R. B., Davletov, B. A., Berghuis, A. M., Sudhof, T. C., and Sprang, S. R. (1995) Structure of the first C2 domain of synaptotagmin I: a novel Ca^{2+}/phospholipid-binding fold. *Cell* **80,** 929–938.
27. Rizo, J. and Sudhof, T. C. (1998) C_2-domains, structure and function of a universal Ca^{2+}-binding domain. *J. Biol. Chem.* **273,** 15,879–15,882.
28. Blanchard, H., Grochulski, P., Li, Y., Arthur, J. S. C., Davies, P. L., Elce, J. S., and Cygler, M. (1997) Structure of a calpain Ca^{2+}-binding domain reveals a novel EF-hand and Ca^{2+}-induced conformational changes. *Nat. Struct. Biol.* **4,** 532–538.
29. Lin, G. D., Chattopadhyay, D., Maki, M., Wang, K. K., Carson, M., Jin, L., et al. (1997) Crystal structure of calcium bound domain VI of calpain at 1.9 Å resolution and its role in enzyme assembly, regulation, and inhibitor binding. *Nat. Struct. Biol.* **4,** 538–547.
30. Elce, J. S., Hegadorn, C., and Arthur, J. S. C. (1997a) Autolysis, Ca^{2+} requirement, and heterodimer stability in m-calpain. *J. Biol. Chem.* **272,** 11,268–11,275.
31. Barnes, T. M. and Hodgkin, J. (1996) The tra-3 sex determination gene of Caenorhabditis elegans encodes a member of the calpain regulatory protease family. *EMBO J.* **15,** 4477–4484.
32. Baker, D., Shiau, A. K., and Agard, D. A. (1993) The role of pro regions in protein folding. *Curr. Opin. Cell Biol.* **5,** 966–970.
33. Sorimachi, H., Amano, S., Ishiura, S., and Suzuki, K. (1996) Primary sequences of rat µ-calpain large and small subunits are, respectively, moderately and highly similar to those of human. *Biochim. Biophys. Acta* **1309,** 37–41.

34. Elce, J. S., Hegadorn, C., Gauthier, S., Vince, J. W., and Davies, P. L. (1995) Recombinant calpain II: improved expression systems and production of a C105A active-site mutant for crystallography. *Protein Eng.* **8,** 843–848.
35. Hendrickson, W. A., Horton, J. R., and LeMaster, D. M. (1990) Selenomethionyl proteins produced for analysis by multiwavelength anomalous diffraction (MAD): a vehicle for direct determination of three-dimensional structure. *EMBO J.* **9,** 1665–1672.
36. Croall, D. E. (2000) Purification of calpain by affinity chromatography on reactive red-agarose or on casein-sepharose, in *Methods in Molecular Biology*, vol. 144: *Calpain Methods and Protocols* (Elce, J. S., ed.), Humana Press, Totowa, NJ, pp. 33–40.
37. Jancarik, J. R. and Kim, S. H. (1991) Sparse matrix sampling: a screening method for crystallization of proteins. *J. Appl. Crystallogr.* **24,** 409–411.
38. Otwinowski, Z. and Minor, W. (1997) Processing of X-ray diffraction data collected in oscillation mode. *Methods Enzymol.* **276,** 307–326.
39. Collaborative Computational Project No. 4. (1994) The CCP4 suite, *Acta Crystallogr.* **D50,** 760–763.
40. Weeks, C. M. and Miller, R. (1999) Optimizing Shake-and-Bake for proteins. *Acta Crystallogr.* **D55,** 492–500.
41. De la Fortelle, E. and Bricogne, G. (1997) Maximum-likelihood heavy-atom parameter refinement for multiple isomorphous replacement and multiwavelength anomalous diffraction methods. *Methods Enzymol.* **276,** 472–494.
42. Abrahams, J. P. and Leslie, A. G. W. (1996) Methods used in the structure determination of bovine mitochondrial F1 ATPase. *Acta Crystallogr.* **D52,** 30–42.
43. McRee, D. E. (1992) A visual protein crystallographic software system for X11/XView. *J. Mol. Graphics* **10,** 44–46.
44. Brünger, A. T., Adams, P. D., Clore, G. M., DeLano, W. L., Gros, P., Grosse-Kunstleve, R. W., et al. (1998) Crystallography and NMR system: a new software system for macromolecular structure determination. *Acta Crystallogr.* **D54,** 905–921.
45. Laskowski, R. A., MacArthur, M. W., Moss, D. S., and Thronton, J. M. (1993) Procheck — a program to check the stereochemical quality of protein structures. *J. Appl. Crystallogr.* **26,** 283–291.
46. Kraulis, P. J. (1991) MOLSCRIPT: a program to produce both detailed and schematic plots of protein structures. *J. Appl. Crystallogr.* **24,** 946–950.
47. Merritt, E. A. and Bacon, D. J. (1997) Raster3D: Photorealistic molecular graphics. *Methods Enzymol.* **277,** 505–524.
48. Nicholls, A., Sharp, K., and Honig, B. (1991) Protein folding and association: insights from the interfacial and thermodynamic properties of hydrocarbons. *Proteins* **11,** 281–296.

4

The Multifunctional S100 Protein Family

Claus W. Heizmann

1. Introduction

The S100 protein family has grown from originally two members (S100A1 and S100B) *(1)* to one of the largest subfamilies of EF-hand Ca^{2+}-binding proteins, consisting now of 18 members displaying a diverse pattern of cell- and tissue-specific distribution, subcellular expression, and relocation upon Ca^{2+} stimulation *(2–7)*. Their pleiotropic intra- and extracellular functions are consistent with their distinct affinities for Ca^{2+}, Zn^{2+}, and Cu^{2+}, their binding to diverse target proteins, and different post-translational modifications.

There are several indications that S100 proteins and calmodulin act in a concerted manner in some tissues and cells.

Thirteen S100 genes were found to be clustered on human chromosome $1q21$ *(8–10)*. Furthermore, a similar organization was discovered in mouse where S100 genes cluster in the syntenic region on chromosome 3, demonstrating conservation of this pattern during evolution *(11)*.

A wide range of human diseases, including cardiomyopathies, neurological diseases and chronic inflammations, and cancer was recently linked to deregulation of S100 genes expression. Thus, S100 proteins are currently being investigated in clinical diagnostics of these diseases and as potential therapeutic targets.

2. Organization of the S100 Genes and Chromosomal Localization

The structural organization of S100 genes is highly conserved (**Fig. 1**), and they generally consist of three exons. The first exon carries exclusively the 5' untranslated sequences, the second exon contains some 5' untranslated

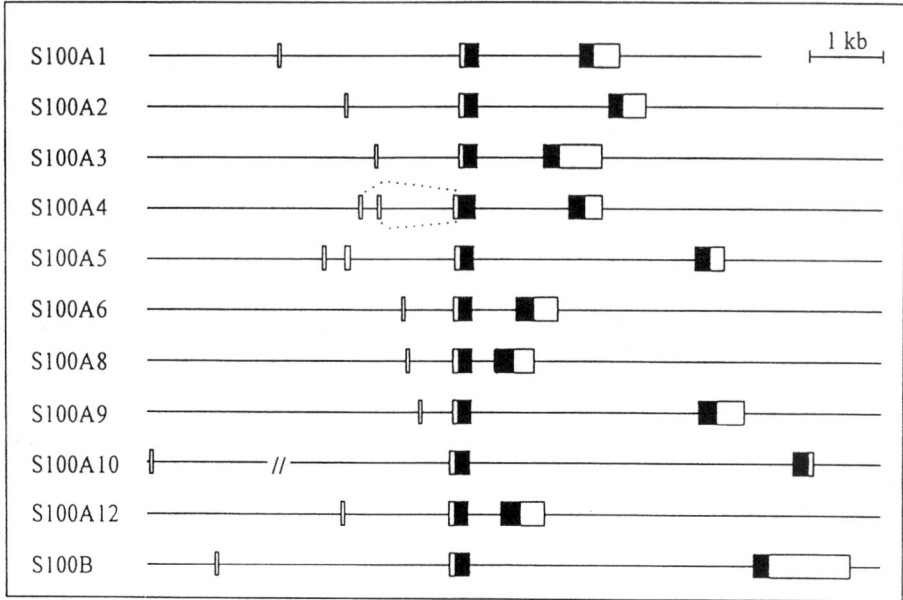

Fig. 1. Structural organization of human S100 genes. The diagram depicts the organization of the human genes. The filled boxes represent exons, the open boxes untranslated sequences. For comparison, all genes are aligned at the exon encoding the N-terminal EF-hand. The broken lines (//) represent gaps in the sequences. The dotted lines in the S100A4 gene represent alternate splicing. Modified from Wicki, R., PhD thesis (1997), University of Zurich.

sequences and the N-terminal S100-specific EF-hand, while the third exon encodes the carboxy-terminal canonical EF-hand and 3' untranslated sequences.

Exceptions are S100A4, containing two first exons that are alternatively spliced into the mature mRNA; S100A5, containing four exons each of which is present in the mature mRNA, and S100A11 (S100C), with an unique exon-intron organization *(12)*. The S100A11 gene consists of three exons and two introns. The first exon, located after the ATG translation initiation codon, is unique among S100 genes but characteristic for most calmodulin genes, suggesting a closer evolutionary relationship of S100A11 to calmodulin genes than to the other S100 gene family members.

The clustering of S100 on human chromosome 1q21 (*see* **Fig. 2**) led to the introduction of a new nomenclature used throughout this review (*see* **Table 1**). The gene density in this region is very high, increasing the likelihood that additional S100 genes will be discovered in this region. Interestingly, the epidermal differentiation genes trichohyalin and profilaggrin, containing an amino termi-

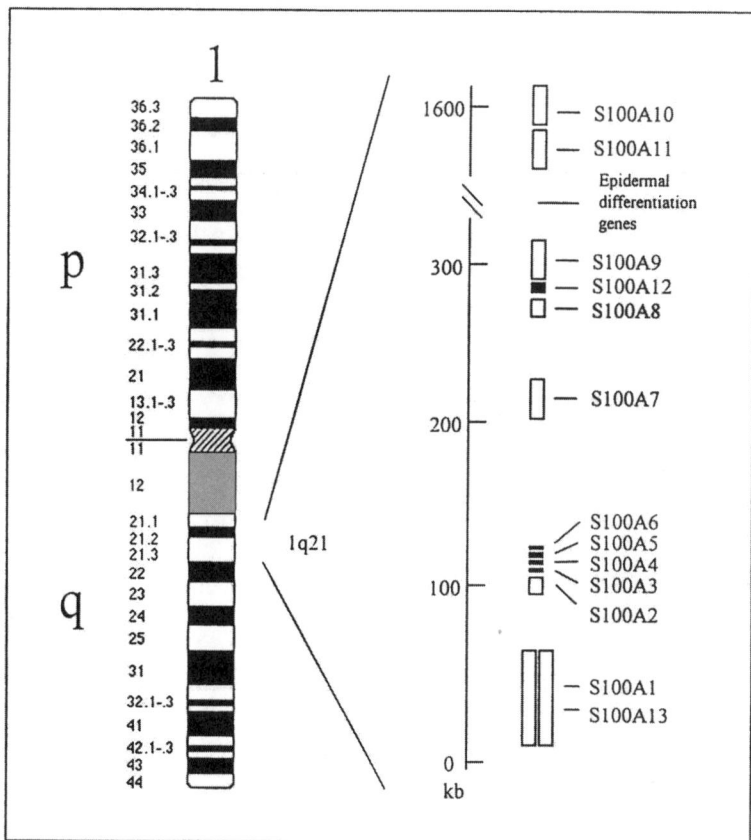

Fig. 2. Organization of S100 genes on human chromosome 1q21. A region of 1600 kb is drawn schematically with the genes and their locations depicted by open boxes. Exactly located genes are depicted as filled boxes. Modified from *TIBS* **21**, 134–140 (1996).

nal S100 domain that is cleaved off upon maturation of the protein, are also located within the cluster of S100 genes *(13)*.

A number of abnormalities in chromosome 1q21 such as deletions, rearrangements, or translocations *(14)* are often associated with human neoplasia, and further detailed analysis of this chromosomal region is therefore in progress.

The regulation of the individual genes within this cluster poses the question whether each gene is regulated by its own elements or by possible superior locus regulatory control elements as has been suggested for the epidermal differentiation genes.

**Table 1
Nomenclature for S100 Genes Clustered on Human Chromosome 1q21**

New name	Previous symbols/synonyms
S100A1	S100α
S100A2	S100L, CaN19
S100A3	S100E
S100A4	CAPL, p9ka, pEL98, mts1, metastatin, calvasculin, murine placental calcium protein
S100A5	S100D
S100A6	Calcyclin, CACY, 2A9, PRA, CaBP, 5B10
S100A7	Psoriasin, PSOR1, BDA11, CAAF2
S100A8	Calgranulin A, CAGA, CFAg, MRP8, p8, MAC387, 60B8Ag, L1Ag, CP–10, MIF, NIF
S100A9	Calgranulin B, CAGB, CFAg, MRP–14, p14, MAC 387, 60B8Ag, L1Ag, MIF, NIF
S100A10	Calpactin light chain, CAL12, CLP11, p11, p10, 42C
S100A11	Calgizzarin, S10°C
S100A12	Calgranulin C, p6, CAAF1, CGRP, corned-associated antigen
S100A13	—

Nomenclature for S100 Genes Located on Different Chromosomes

Name	Previous symbols/synonyms	Chromosome
Calbindin	9k CALB3, CaBP9k, I CaBP	Xp22
S100B	S100β, NEF	21q22
S100P	S100P	4p16

3. Biochemical Properties and Protein Structures

S100 proteins are characterized by two distinct EF-hand motifs with different Ca^{2+} affinities. Both EF-hands, flanked by hydrophobic regions, are separated by a central hinge region. The carboxy-terminal EF-hand is usually referred to as a canonical Ca^{2+}-binding loop and encompasses 12 amino acids, whereas the amino-terminal S100-specific loop is composed of 14 amino acids and has a lower affinity for Ca^{2+} *(7)*.

S100 proteins vary in their biochemical and metal-binding properties, also reflected in their different purification protocols.

For example, the purification of human recombinant S100A1 from *Escherichia coli* (BL21 *Lys*) using a pGEMEXT7 expression vector can easily be accomplished. After IPTG induction, cell lysis, centrifugation, and ammonium sulfate precipitation, S100A1 can easily be purified by hydrophobic interaction chromatography and HPLC, yielding large quantities of recombinant proteins for structural analysis *(15)*.

Table 2
Ca^{2+}- and Zn^{2+}-Binding Parameters of the S100 Proteins

	Ca^{2+} sites[a]	[Ca^{2+}]$_{0.5}$ (mM)	n$_H$	Zn^{2+} sites[a]	[Zn^{2+}]$_{0.5}$ (µM)	n$_H$	Ca^{2+}/Zn^{2+} effect
S100B	4	0.50	1.0	6–8	500	nd	+
S100A1	4	0.29	1.1				—
S100A2	4	0.47	2.04	4	4.5	1	—
S100A3[b]	nd	35	0.76	4	0.005	1.4	No
S100A4	4	0.15	1.15				
S100A5	4	0.013	1.76	2	nd		+
S100A6	4	0.32	1.33	4	2000		(–)
S100A10	no				nd		
S100A11	4	0.52	1.4				
S100A12	2	0.05	1	2	<0.1	nd	+
S100P	4	0.002/0.8	na				

[a]per dimer; na, not applicable; nd, not determined; no, no binding.
The Ca^{2+}-binding parameters of S100A7, S100A8, S100A9, and S100A13 have not been reported. Adapted from *BioMetals* **11**, 383–397 (1998).
[b]Fritz et al. (2001), in preparation.

However, purification of recombinant S100A3, using a similar procedure, resulted in very poor yield, most likely because of the unusual properties (very high cysteine content) of this protein. The yield was improved by expressing the MBP-S100A3 fusion protein in *E. coli* followed by induction, cell lysis, centrifugation, and application to an amylose resin affinity column. After elution, the fusion protein was cleaved by protease factor Xa and subjected to a second round of chromatography resulting in a low amount of purified S100A3. Unfortunately, the protein was difficult to handle because of its high tendency to aggregate *(16)*. Thus, a new protocol was developed for purification of S100A3 under anaerobic conditions on a large scale for the purpose of structural analysis. This purification of S100A3 resulted in a high yield, a 10-fold increase in Ca^{2+} affinity and high stability, allowing a detailed analysis of the spectroscopic properties of the Ca^{2+}- and Zn^{2+}-binding S100A3 protein and the proposal of a three-dimensional structural model *(17)*.

The distinct biochemical properties observed during purification of the individual S100 proteins, reflected in their different Ca^{2+}- and Zn^{2+}-binding values, are listed in **Table 2**.

Generally, the dimeric S100 proteins bind 4 Ca^{2+} per dimer, displaying a lower affinity for Ca^{2+}. This is strongly influenced by the ionic strength of the buffer system *(7)*. Several S100 proteins also bind Zn^{2+}, however, to a site different from the Ca^{2+}-binding site. In a few cases, Zn^{2+} is able to modify the Ca^{2+} affinity. Furthermore, S100B *(18)* and S100A5 *(33)* also bind Cu^{2+}.

S100B, identified as an inhibitor of the Cu^{2+}-catalyzed oxidation of L-ascorbate, binds 4 Cu^{2+} ions per dimer, which can be exchanged by Zn^{2+}, suggesting that both metal ions occupy the same binding site. The variable cation-binding sites of S100 proteins seem to be consistent with their diverse intra- and extracellular functions.

The reported post-translational modifications of several S100 proteins may interfere with the metal-binding properties, their interactions with target proteins, and their Ca^{2+}-dependent cellular translocation/secretion upon Ca^{2+} activation.

S100A8 and S100A9 were found to be phosphorylated *(19,20)*. S100A8, S100A9, and S100A7 *(21)* bind unsaturated fatty acids, important for their relocations and/or attachment to the plasma membrane, perhaps in a manner similar to the calcium-myristoyl switch of recoverin *(22)*.

Each S100 protein was found to interact in a Ca^{2+}-dependent manner with a number of target proteins, leading to a physiological response. In fact, binding of S100 proteins to their effector proteins could in turn affect their metal-binding properties as has been found for S100A10 when bound to annexin II. Furthermore, some of the S100 protein members target the same molecules as calmodulin, e.g., cytoskeletal proteins, microtubule-associated proteins, caldesmon, the enzymes adenyl cyclase and glycogen phosphorylase, nuclear kinase (Ndr), and the cell cycle-associated proteins neuromodulin and p53 (for review, *see* **refs.** *2,3* and *7*).

Perhaps some S100 proteins have similar functions as the ubiquitously expressed calmodulin but in a cell- and tissue-specific manner.

4. Biological Functions

S100 proteins are involved in a large number of cellular activities, such as signal transduction, cell differentiation, regulation of cell motility, transcription, and cell cycle progression (**Table 3**). This is achieved by modulating target proteins in a Ca^{2+}-dependent (possibly also in a Zn^{2+}- and Cu^{2+}-dependent) manner.

Recently, we observed that several S100 proteins are able to translocate within the cell in response to rises in intracellular calcium *(23,24)*. Moreover, several S100 proteins such as S100A2, S100A4, S100A7, S100A8, S100A9, S100A13, and S100B are probably secreted from the cells and appear to exert extracellular functions. For example, S100A13 is probably involved in the release of FGF–1 and p40 synaptotagmin-1 in response to temperature stress *(25)*.

The mechanism of secretion *(34)* as well as their interaction with the cell surface receptors RAGE *(35)* are presently under investigation.

The glial-derived S100B protein secreted as a neurotrophic factor *(26)* or a mitogen *(27)* undergoes oxidation on cysteine residues, leading to its conforma-

Table 3
S100 Proteins in Health and Disease

Protein	Postulated functions	Disease association
S100A1	Regulation of cell motility, muscle contraction, phosphorylation, Ca^{2+} release channel, transcription	Cardiomyopathies
S100A2	Tumor suppression, nuclear functions, chemotaxis	Cancer
S100A3	Hair shaft formation, tumor suppression, secretion, and extracellular functions	Hair damage, cancer
S100A4	Regulation of cell motility, secretion, and extracellular functions	Cancer (metastasis)
S100A5	Ca^{2+}-, Zn^{2+}-, and Cu^{2+}-binding protein in the CNS; unknown function	Not known
S100A6	Regulation of insulin release, prolactin secretion, Ca^{2+} homeostasis, tumor progression	Melanoma, cancer
S100A7	S100A7-fatty acid binding protein complex regulates differentiation of keratinocytes	Psoriasis
S100A8 (S100A8/A9 complex)	Chemotactic activities, adhesion of neutrophils, myeloid cell differentiation, apoptosis, fatty acid metabolism	Inflammation, cystic fibrosis
S100A9 (S100A8/A9 complex)	Cytostatic activities, myeloid cell differentiation, neutrophil immobilizing activity, apoptosis, fatty acid metabolism	Inflammation, cystic fibrosis
S100A10	Inhibition of phospholipase A2, neurotransmitter release, in connection with annexin II regulates membrane traffic, ion currents	Inflammation
S100A11	Organization of early endosomes, inhibition of annexin I function, regulation of phosphorylation, physiological role in keratinocyte cornified envelope	Skin diseases, ocular melanoma
S100A12	Host-parasite interaction, differentiation of squamous epithelial cells and extracellular functions	Mooren's ulcer (autoimmune disease)
S100A13	Regulation of FGF–1 and synaptotagmin-1 release	—
S100B	Cell motility, proliferation, inhibition of phosphorylation, inhibition of microtubule assembly transcription, regulation of nuclear kinase, extracellular functions, e.g., neurite extension	Alzheimer's disease, Down's syndrome, melanoma, amyotrophic lateral sclerosis, epilepsy
S100P	Function in the placenta	—
Calbindin 9k	Ca^{2+} buffer and Ca^{2+} transport	Vitamin D deficiency, abnormal mineralization

tional change. As a result, a site within the typical EF-hand Ca^{2+}-binding domain is exposed to phosphorylation by CKII. The production of these disulfide-bridged S100B monomers is enhanced by peroxynitrite anions, suggesting their possible function in neuropathology related to the production of this anion.

Another member of the S100 family, the S100A3, usually expressed in the inner root sheath cuticle of selected hair cuticles *(16)*, was found to be overexpressed in several tumor types *(28)*. S100A3 undergoes oxidation of *Cys*23 and *Cys*30, which form a disulfide bridge, distorting the Ca^{2+}-binding loop I and lowering the affinity for Ca^{2+} required for extracellular functions. However, fully reduced purified S100A3 has a higher affinity for Ca^{2+}, possibly required for intracellular regulatory functions.

To study the subcellular localizations, relocations, and secretion patterns of S100 proteins, we prepared Green Fluorescent Protein (GFP) and several S100 protein chimeras with the intention of monitoring those processes in live cells. A first study on S100A11 translocations has already been reported

Furthermore, an increasing number of animal models is available. Transgenic mice were generated to examine whether elevated S100B levels can cause brain dysfunction. In female mice overexpressing S100B, hyperactivity and lack of habituation to novelty were observed, which indicates hippocampal dysfunction *(29)*.

An association of S100A4 with metastasis was demonstrated in two transgenic mouse models *(30)*, and these and other cell transfection studies provide strong evidence that S100A4 alters cell motility, and thus, invasiveness of tumor cells *(5,36)*.

5. Association of S100 Proteins with Human Diseases

A wide range of different human diseases has been attributed to a deregulation of S100 genes expression (*see* **Table 3**). Earlier studies showed that 13 S100 genes are located in a gene cluster on human chromosome 1*q*21 (*see* **Fig. 2**). Aberrations in this chromosomal region are associated with the etiology of skin diseases, cardiovascular diseases, and cancer. Amplifications of this chromosomal region are found in various tumors and are possibly linked to metastases. Thus, understanding the underlying mechanisms would be of high clinical relevance.

The levels of expression of individual S100 proteins vary most remarkably in brain and other tumor types in respect to the progression of malignancy. In fact, specific antibodies are already in use to distinguish, e.g., low (WHO grade I and II) and high (WHO grade III and IV) grade astrocytic tumors, helping neuropathologists with diagnosis and selection of treatment *(28)*. Recently, it has also been discovered that the expression of S100 proteins is closely correlated with tumor prognosis *(37,38)*.

Another example for a disease association is S100A1, which is a major regulator in human heart function *(39)*. This protein is downregulated in human cardiomyopathies, probably affecting Ca^{2+} homeostasis in end-stage human heart failure *(31)*. Our preliminary results indicate that the measurement of S100A1 in serum might be a putative indicator of myocardial infarction *(40)*.

Differential expression of S100A6 was detected in colon adenomas compared to colorectal carcinomas, and several S100A6 variants with different p*I*s were identified by 2D-PAGE *(32)*. It appears that different post-translational modifications of S100A6 correlate with colon carcinoma progression.

6. Conclusion and Perspectives

A number of S100 proteins target molecules and several 3-D structures of S100 proteins (S100B, S100A6, S100A7, S100A8, S100A9, S100A10, and S100A12 some in their Ca^{2+}-bound form) were identified. This knowledge will help to design specific drugs to test the physiological functions of these proteins and provide therapeutic solutions.

Thirteen S100 genes together with the epidermal differentiation complex are clustered on the *q*21 sequence of human chromosome 1. Inherited or acquired aberrations in this segment are associated with cardiomyopathies, cancer, and skin diseases. An ambitious project has been started to identify all genes on the 1*q*21 segment, which most likely will lead to the discovery of novel S100 genes.

A further goal will be to extend ongoing studies to explore S100 proteins and antibodies against them in clinical diagnostics.

Acknowledgments

The author is grateful to M. Killen for help with the preparation of and to Drs. G. Fritz and G. Davey for reading the manuscript.

References

1. Moore, B. W. (1965) A soluble protein characteristic of the nervous system. *Biochem. Biophys. Res. Commun.* **19,** 739–744.
2. Donato, R. (1999) Functional roles of S100 proteins, clacium-binding proteins of the EF-hand type. *Biochem. Biophys. Acta* **1.150,** 191–231.
3. Schäfer, B. W. and Heizmann, C. W. (1996) The S100 family of EF-hand calcium-binding proteins: functions and pathology. *Trends Biochem. Sci.* **21,** 134–140.
4. Nelson, M. R. and Chazin, W. J. (1998) Structures of EF-hand Ca^{2+}-binding proteins: Diversity in the organization, packing and response to Ca^{2+} binding. *BioMetals* **11,** 297–318.
5. Barraclough, R. (1998) Calcium-binding protein S100A4 in health and disease. *Biochim. Biophys. Acta* **1448,** 190–199.

6. Kerkhoff, C., Klempt, M., and Sorg, C. (1998) Novel insights into structure and function of MRR8 (S100A8) and MRP14 (S100A9). *Biochim. Biophys. Acta* **1448**, 200–211.
7. Heizmann, C. W. and Cox, J. A. (1998) New perspectives on S100 proteins: a multifunctional Ca^{2+}-, Zn^{2+}- and Cu^{2+}-binding protein family. *BioMetals* **11**, 383–397.
8. Schäfer, B. W., Wicki, R., Engelkamp, D., Mattei, M. G., and Heizmann, C. W. (1995) Isolation of a YAC clone covering a cluster of nine S100 genes on human chromosome 1q21: rationale for a new nomenclature of the S100 calcium-binding protein family. *Genomics* **25**, 638–643.
9. Wicki, R., Schäfer, B. W., Erne, P., and Heizmann, C. W. (1996a) Characterization of the human and mouse cDNAs coding for S100A13, a new member of the S100 protein family. *Biochem. Biophys. Res. Commun.* **227**, 597–599.
10. Wicki, R., Marenholz, I., Mischke, D., Schäfer, B. W., and Heizmann, C. W. (1996b) Characterization of the human S100A12 (calgranulin C, p6, CAAF1, CGRP) gene, a new member of the S100 gene cluster on chromosome 1q21. *Cell Calcium* **20**, 459–464.
11. Ridinger, K., Ilg, E. C., Niggli, F. K., Heizmann, C. W., and Schäfer, B. W. (1998) Clustered organization of S100 genes in human and mouse. *Biochim. Biophys. Acta* **1448**, 254–263.
12. Nakamura, T., Hayashi, M., Kato, A., Sawazaki, T., Yasue, H., Nakano, T., and Tanaka, T. (1998) A unique exon-intron organization of a porcine S100C gene: close evolutionary relationship to calmodulin genes. *Biochem. Biophys. Res. Commun.* **243**, 647–652.
13. South, A. P., Cabral, A., Ives, J. H., James, C. H., Mirza, G., Marenholz, I., et al. (1999) Human epidermal differentiation complex in a single 2.5 Mbp long continuum of overlapping DNA cloned in bacteria integrating physical and transcript maps. *J. Invest. Dermatol.* **112**, 910–918.
14. Forus, A., Berner, J.-M., Meza-Zepeda, L. A., Fodstad, O., Mischke, D., and Myklebost, O. (1998) Molecular characterisation of a novel amplicon at 1q21-q22 frequently observed in human sarcomas. *Br. J. Cancer* **78**, 495–503.
15. Engelkamp, D., Schäfer, B. W., Erne, P., and Heizmann, C. W. (1992) S100a, CAPL, and CACY: molecular cloning and expression analysis of three calcium-binding proteins from human heart. *Biochemistry* **31**, 10,258–10,264.
16. Föhr, U., Heizmann, C. W., Engelkamp, D., Schäfer, B. W., and Cox, J. A. (1995) Purification and cation binding properties of the recombinant human S100 calcium-binding protein A3, an EF-hand motif protein with high affinity for zinc. *J. Biol. Chem.* **270**, 21,056–21,061.
17. Fritz, G., Heizmann, C. W., and Kroneck, P. M. H. (1998) Probing the structure of the human Ca^{2+}- and Zn^{2+}-binding protein S100A3: spectroscopic investigations of its transition metal ion complexes, and three-dimensional structural model. *Biochim. Biophys. Acta* **1448**, 264–276.
18. Nishikawa, T., Lee, I. S. M., Shiraishi, N., Ishikawa, T., Ohta, Y., and Nishikimi, M. (1997) Identification of S100b protein as copper-binding protein and its suppression of copper-induced cell damage. *J. Biol. Chem.* **272**, 23,037–23,041.

19. Edgeworth, J., Freemont, P., and Hogg, N. (1989) Ionomycin-regulated phosphorylation of the myeloid calcium-binding protein p14. *Nature* **342,** 189–192.
20. Guignard, F., Mauel, J., and Markert, M. (1996) Phosphorylation of myeloid-related proteins MRP–14 and MRP–8 during human neutrophil activation. *Eur. J. Biochem.* **241,** 265–271.
21. Siegenthaler, G., Roulin, K., Chatellard-Gruaz, D., Hotz, R., Saurat, J. H., Hellman, U., and Hagens, G. (1997) A heterocomplex formed by the calcium-binding proteins MRP8 (S100A8) and MRP14 (S100A9) bind unsaturated fatty acids with high affinity. *J. Biol. Chem.* **272,** 9371–9377.
22. Ames, J. B., Ishima, R., Tanaka, T., Gordon, J. I., Stryer, L., and Ikura, M. (1997) Molecular mechanics of calcium-myristoyl switches. *Nature* **389,** 198–202.
23. Mueller, A., Bächi, T., Höchli, M., Schäfer, B. W., and Heizmann, C. W. (1999) Subcellular distribution of S100 proteins in tumor cells and their relocation in response to calcium activation. *Histochem. Cell Biol.* **111,** 453–459.
24. Mandinova, A., Atar, D., Schäfer, B. W., Spiess, M., Aebi, U., and Heizmann, C. W. (1998) Distinct subcellular localization of calcium binding S100 proteins in human smooth muscle cells and their relocation in response to rises in intracellular calcium. *J. Cell Sci.* **111,** 2043–2054.
25. Mouta Carreira, C., LaVallee, T. M., Tarantini, F., Jackson, A., Lathrop, J. T., Hampton, B., et al. (1998) S100A13 is involved in the regulation of fibroblast growth factor-1 and p40 synaptotagmin-1 release in vitro. *J. Biol. Chem.* **273,** 22,224–22,231.
26. Barger, S. W., Wolchok, S. R., and Van Eldik, L. J. (1992) Disulfide-linked S100 beta dimers and signal transduction. *Biochim. Biophys. Acta* **1160,** 105–112.
27. Scotto, Ch., Mély, Y., Ohshima, H., Garin, J., Cochet, C., Chambaz, E., and Baudier, J. (1998) Cysteine oxidation in the mitogenic S100B protein leads to changes in phosphorylation by catalytic CKII-a subunit. *J. Biol. Chem.* **273,** 3901–3908.
28. Camby, I., Nagy, N., Lopes, M.-B., Schäfer, B. W., Maurage, C.-A., Ruchoux, M.-M., et al. (1999) Supratentorial pilocytic astrocytomas, astrocytomas, anaplastic astrocytomas and glioblastomas are characterized by a differential expression of S100 proteins. *Brain Path.* **9,** 1–19.
29. Friend, W. C., Clapoff, S., Landry, Ch., Becker, L. E., O'Hanlon, D., Allore, R. J., et al. (1992) Cell-specific expression of high levels of human S100b in transgenic mouse brain is dependent on gene dosage. *J. Neurosci.* **12,** 4337–4346.
30. Lloyd, B. H., Platt-Higgins, A., Rudland, P. S., and Barraclough, R. (1998) Human S100A4 (p9Ka) induces the metastatic phenotype upon benign tumour cells. *Oncogene* **17,** 465–473.
31. Remppis, A., Greten, T., Schäfer, B. W., Hunziker, P., Erne, P., Katus, H. A., and Heizmann, C. W. (1996) Altered expression of the Ca^{2+}-binding protein S100A1 in human cardiomyopathy. *Biochim. Biophys. Acta* **1313,** 253–257.
32. Stulik, J., Oesterreicher, J., Koupilova, K., Knicek, J., Bures, J., Jandik, P., et al. (2000) Differential expression of the Ca^{2+}-binding S100A6 protein in normal, preneoplastic and neoplastic colon mucosa. *Eur. J. Cancer* **36,** 1050–1059.

33. Schäfer, B. W., Fritschy, J.-M., Murmann, P., Troxler, H., Durussel, I., Heizmann, C. W., and Cox, J. A. (2000) Brain S100A5 is a novel calcium-, zinc-, and copper ion-binding protein of the EF-hand superfamily. *J. Biol. Chem.* **275,** 30,623–30,630.
34. Davey, E. G., Murmann, P., Hoechli, M., Tanaka, T., and Heizmann, C. W. (2000) Calcium-dependent translocation of S100A11 requires tubulin filaments. *Biochim. Biophys. Acta* **1498,** 220–232.
35. Schmidt, A. M., Yan, S. D., and Stern, D. M. (2000) The biology of the receptor for advanced glycation end products and its ligands. *Biochim. Biophys. Acta* **1498,** 99–111.
36. Yonemura, T., Endou, T., Kimura, K., Fushida, S., Bandon, E., Taniguchi, K., Kinoshita, K., Ninomiya, I., Sugiyama, K., Heizmann, C. W., Schäfer, B. W., and Sasaki, T. (2000) Inverse expression of S100A4 and E-cadherin is associated with metastatic potential in gastric cancer. *Clin. Cancer Res.* **6,** 4234–4242.
37. Lauriola, L., Michetti, F., Maggiano, N., Galli, J., Cadoni, G., Schäfer, B. W., Heizmann, C. W., and Ranelletti, F. O. (2000) Prognostic significance of the Ca^{2+}-binding protein S100A2 in laryngeal squamous-cell carcinoma. *Int. J. Cancer* **89,** 345–349.
38. Kimura, K., Endo, Y., Yonemura, Y., Heizmann, C. W., Schäfer, B. W., Watanabe, Y., and Sasaki, T. (2000) Clinical significance of S100A4 and E-cadhein-related adhesion molecules in non-small cell lung cancer. *Int. J. Oncol.* **16,** 1125–1131.
39. Kiewitz, R., Lyons, G. E., Schäfer, B. W., and Heizmann, C. W. (2000) Transcriptional regulation of S100A1 and expression during mouse heart development. *Biochim. Biophys. Acta* **1498,** 207–219.
40. Kiewitz, R., Acklin, C., Minder, E., Huber, P. R. Schäfer, B. W., and Heizmann, C. W. (2000) S100A1, a new marker for acute myocardial ischemia. *Biochem. Biophys. Res. Commun.* **274,** 865–871.

5

Ca²⁺ Binding to Proteins Containing γ-Carboxyglutamic Acid Residues

Egon Persson

1. Introduction

The family of proteins and natural peptides containing γ-carboxyglutamic acid (Gla) residues has a limited number of members found in such diverse sources as blood plasma, bone tissue, and snail venoms. The best characterized subfamily consists of the so-called vitamin K-dependent plasma proteins found in the blood and is the topic of this chapter. In the conversion of glutamate to Gla, an extra carboxyl group is added post-translationally to the γ-carbon of a restricted number of glutamate residues in the amino-terminal Gla domain, which received its name from this unusual residue, by a vitamin K-dependent carboxylase *(1)*. The carboxylase binds primarily to a propeptide that is cleaved off before secretion of the proteins, but the signal that limits the carboxylation to glutamate residues within the first 40 amino acid residues has not been identified. The Gla-containing plasma proteins have 9–12 Gla residues depending on how many Glu residues are encoded by the two exons covering the first approx 46 residues of the proteins because all Glu residues in this part of the protein become γ-carboxylated. The resulting dicarboxylic Gla residues have significantly higher affinity for Ca^{2+}, compared to the precursor Glu residues, and this ability to chelate Ca^{2+} ions under physiological conditions ($[Ca^{2+}]_{free}$ = 1–1.5 mM in the blood) is pivotal for biological activity.

Coagulation factors VII, IX, and X, and protein C are vitamin K-dependent plasma proteins. They are synthesized and circulate as zymogens which are activated to enzymes (denoted by an a, e.g., factor VIIa) by limited proteolysis, a conversion that does not affect their Ca^{2+}-binding properties. These proteins are either substrates in or, in their active forms, catalysts of reactions taking place on a membrane surface, usually in a macromolecular complex containing a membrane-associated cofactor *(2)*. The accumulation of Gla-containing proteins on phospholipid surfaces requires a functional, Ca^{2+}-loaded Gla domain. Factors VII, IX, and X, and protein C share the same domain architecture *(3)*. The N-terminal Gla domain is followed by two domains homologous to the epidermal growth factor (EGF), and the C-terminal catalytic serine protease domain is homologous to trypsin (*see* **Fig. 1**). The Gla and EGF-like domains constitute the light chain, whereas the protease domain occupies the heavy chain. Factor X and protein C exist as two-chain molecules already in their zymogen forms, but factors VII and IX molecules go from single-chain to two-chain on activation. The two chains are connected by a single disulfide bond. Two or more of these domains, the Gla domain included, are found in the other three Gla-containing plasma proteins (prothrombin: Gla and catalytic domains; protein S: Gla and four EGF-like domains; and protein Z: Gla and two EGF-like domains, and a protease domain that lacks the catalytic triad). All Gla-containing plasma proteins are presumably involved in blood coagulation, some by promoting the process and others by attenuating it.

Factors VII, IX, and X, and protein C bind Ca^{2+} not only in the Gla domain (seven Ca^{2+} ions), but, in addition, contain one site in the first EGF-like domain and one in the catalytic domain *(4)*. The affinities and requirements of the different sites, as well as the functional and structural consequences of Ca^{2+} binding, have been studied in detail and will be described in the following sections. Studies of individual Ca^{2+} sites in the intact proteins are complicated by the binding of a total of nine Ca^{2+} ions (with similar dissociation constants) to three different domains and by the fact that Ca^{2+} binding to the Gla domain is a cooperative event. Various approaches to circumvent these problems are also presented.

2. Calcium Binding to Gla Residues

Because prothrombin is the most abundant vitamin K-dependent plasma protein, the early, and majority of, studies have been conducted on this protein. Subsequent studies have shown that the whole family of vitamin K-dependent plasma proteins display similar Ca^{2+}-binding isotherms. Initially, it was observed that the electrophoretic mobility of prothrombin in agarose gels was influenced by calcium ions, indicating a measurable affinity for this metal ion. Ca^{2+} binding to prothrombin and its membrane-binding region (fragment 1; the Gla and first kringle domains) was characterized by equilibrium dialysis and was found to be cooperative with half-saturation around 0.6 mM free Ca^{2+} *(5,6)*. Moreover, Ca^{2+} binding was abolished in prothrombin isolated from cows given dicoumarol, a vitamin K antagonist. The Ca^{2+}-binding ability and the resulting membrane-interactive properties of prothrombin were attributed to the pres-

Ca²⁺ Binding to Gla-Containing Proteins

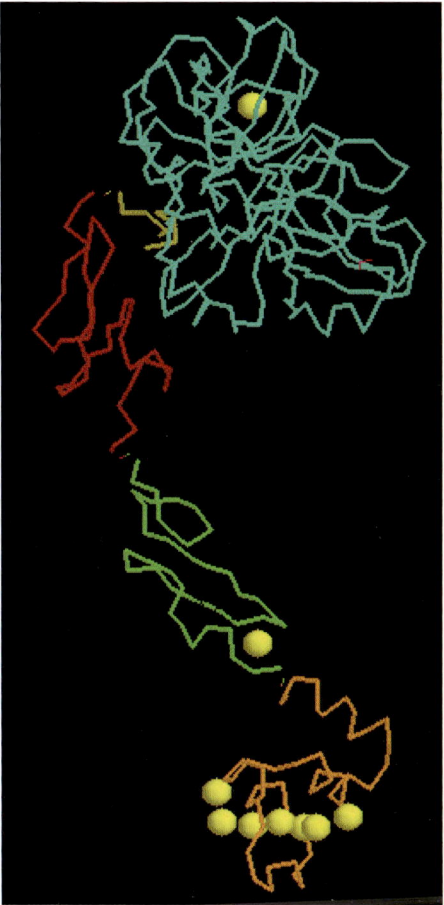

Fig. 1. The domain organization of factor VIIa (and factors IX, X, and protein C). The picture shows the Gla domain (*orange*), the first EGF-like domain (*green*), the second EGF-like domain (*red*), the C-terminal part of the light chain (*yellow*) and the protease domain (*cyan*). The bound Ca^{2+} ions are depicted as yellow spheres; seven in the Gla domain, one in the first EGF-like and protease domains. The picture was generated using RasMol version 2.6 using atomic coordinates from the structure of the complex between factor VIIa and tissue factor (PDB ID 1dan, **ref. 46**).

ence of a previously unidentified amino acid residue, Gla, in the amino-terminal region *(7,8)*. Gla-mediated Ca^{2+} binding appears to result in a structural change that precedes prothrombin-phospholipid interactions *(9)*. Using a more direct method of monitoring Ca^{2+}-induced conformational changes by means of intrinsic fluorescence measurements, the Ca^{2+} concentration at which half of the protein molecules had undergone this transition was found to be 0.4 m*M*

Table 1
Dissociation Constants for Ca^{2+} Binding to the Sites in the Different Domains of Factor X Obtained Using Different Fragments of the Protein[a]

	Dissociation constants for Ca^{2+} sites in			
Peptide (domains) used	Gla domain	*EGF*1 domain	Protease domain	Ref.
Gla	2.5 m*M*			*(12)*
*EGF*1		2 m*M*		*(38)*
Gla-*EGF*1	0.7 m*M*	0.1 m*M*		*(12)*
*EGF*1-*EGF*2-protease		1.2 m*M*	0.16 m*M*	*(40)*
*EGF*2-protease			0.15 m*M*	*(56)*
Intact factor X	0.8 m*M*	0.06 m*M*	ND	*(12,40)*

[a]The values given for the Gla domain represent the average affinities obtained from measurements of Ca^{2+}-induced quenching of the intrinsic fluorescence *(12)* or of binding to Ca^{2+}-dependent antibodies *(40)*. Ca^{2+} binding to the first EGF-like domain was measured using NMR spectroscopy *(38)*, fluorescence spectroscopy *(12)* or Ca^{2+}-dependent domain-specific antibodies *(40)* and to the serine protease domain using equilibrium dialysis *(56)* or Ca^{2+}-dependent domain-specific antibodies *(40)*.

in the absence of phospholipids and somewhat lower in their presence, indicating that membrane binding stabilizes the Ca^{2+} conformation. Circular dichroism spectroscopy was also employed to demonstrate apparent changes in secondary structural content on Ca^{2+} binding and the Ca^{2+} dependence agreed with that observed when measuring fluorescence *(10)*.

One way to study Ca^{2+} binding to the Gla domain without the influence of other Ca^{2+} binding sites is to isolate the Gla domain from factor IX or X by proteolytic cleavage *(11,12)*. The Ca^{2+}-induced intrinsic fluorescence quenching, shown to result from alterations in the environment of the tryptophan residue in the C-terminal helix of the Gla domain, was employed to monitor Ca^{2+} binding. Half-maximal quenching was observed at 2.5–4 m*M* Ca^{2+}. This value is three- to fourfold higher than that obtained using the intact proteins or fragments containing the Gla domain and one or two EGF-like domains. In another study, it was shown that both quenching of the intrinsic fluorescence and phospholipid vesicle binding of the Gla peptides from prothrombin and factor X requires about 10 times higher Ca^{2+} concentrations than required for the parent protein *(13)*. Hence, the presence of the first EGF-like domain (or the first kringle domain in prothrombin) facilitates Ca^{2+} binding to the Gla domain. A similar stabilizing effect of the first EGF-like domain was observed using fragments derived from protein C *(14)*. Whether the entire EGF-like domain, or only the N-terminal part of it, is required for optimal Ca^{2+} binding to the Gla domain has yet to be elucidated. However, experiments using a Gla-EGF1 fragment of protein Z show that Ca^{2+} binding to the first EGF-like domain is not necessary to stabilize Ca^{2+} binding to the Gla domain *(15)*. The C-terminal helix of the Gla domain appears to stabilize the Ca^{2+}- and membrane-binding structure, as shown in a study using Gla peptides from protein C of various lengths *(16)*. A summary of the affinities of the Ca^{2+}-binding sites in factor X and the influence of neighboring domains is given in **Table 1**.

An elegant approach to obtain the dissociation constants for Ca^{2+} binding to each individual Gla residue was used by Colpitts and co-workers *(17)*. Nine protein C-Gla peptides were synthesized, each containing one Gla residue enriched with ^{13}C at its two γ-carboxylate carbon atoms. Ca^{2+} titrations were carried out and the induced ^{13}C chemical shifts monitored by NMR spectroscopy. The results showed that four of the nine Gla residues are involved in the Ca^{2+} sites of relatively higher affinity. Hence, the Gla residues are not equivalent. This has also been shown in functional analyses of protein C and prothrombin mutants in which *Asp* was substituted for one Gla residue at a time *(18,19)*. Some substitutions resulted in severe reduction or complete abolishment of coagulant activity, whereas others had limited effects. The explanation for the different affinities of the sites in the Gla domain and the varying functional consequences of different Gla replacements is obvious when looking at how individual Gla residues participate in Ca^{2+} binding and how different Ca^{2+} ions are bound. The Gla residues display varying extents of involvement in Ca^{2+} coordination, binding one to three Ca^{2+} ions in mono- or divalent fashions *(20)*. For instance, Gla^{15} in prothrombin only binds one Ca^{2+} ion in a unidentate manner, whereas Gla^{17} binds three Ca^{2+} ions in a bidentate mode.

The Ca^{2+}-loaded Gla domain mediates the localization of the vitamin K-dependent plasma proteins to membrane surfaces, which is essential for the assembly into functional ternary enzyme-cofactor-substrate complexes. A deletion of the Gla domain or impaired Ca^{2+} binding to the Gla domain results in a dramatic loss of biological function *(21)*. The structural changes induced by Ca^{2+} virtually turn the Gla domain inside-out, as revealed in a comparison of the structures of this domain from factor IX determined by NMR spectroscopy in the absence and presence of the metal ion *(22,23)*. Ca^{2+} binding results in the exposure of hydrophobic residues residing in the interior of the apo form and a concomitant internalization of charged Gla residues involved in Ca^{2+} coordination, which are exposed in the apo form (*see* **Fig. 2**). Three hydrophobic residues, *Leu*^{6}, *Phe*^{9}, and *Val*^{10}, close to the N-terminus were proposed to be involved in phospholipid binding. The same Ca^{2+}-induced conformational transition was seen and a corresponding phospholipid-interactive patch (*Phe*^{4}, *Leu*^{5}, and *Val*^{8}) was inferred when comparing the structures of the apo form of the Gla domain from factor X and the Ca^{2+}-loaded form of the Gla domain from prothrombin *(20,24)*. Moreover, site-directed mutagenesis directly demonstrated contributions from the corresponding residues in protein C to phos-

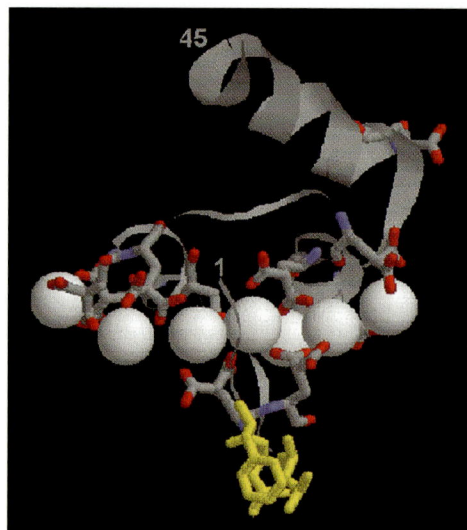

Fig. 2. The structure of the Ca^{2+}-loaded Gla domain of factor VIIa. Residues 1–45 are shown with the seven Ca^{2+} ions (*white balls*) in one plane. The *Gla* residues are shown in stick representation and all but one (Gla^{35}) are directly involved in Ca^{2+} binding. It is also striking how the *Gla* residues form two carboxylate surfaces, a minor surface below the plane of the Ca^{2+} ions comprising residues Gla^6 and Gla^7 and a major surface on top of the metal ions. The three residues postulated to be in contact with the phospholipid surface are shown in yellow at the bottom. For details of picture generation, *see* **Fig. 1**.

pholipid affinity *(25,26)*. Interestingly, although Mg^{2+} binds to the Gla domain in a similar number and manner as Ca^{2+}, this metal ion does not render the proteins membrane-interactive. This is presumably explained by the inability of the amino-teminal part of the Mg^{2+}-loaded Gla domain to attain the structure needed to properly present the three hydrophobic residues *(27)*. It has been shown that the Gla domain undergoes two sequential structural changes on Ca^{2+} binding *(28–30)*. The second transition is essential for membrane binding, but Mg^{2+} is only able to induce the first one.

3. Calcium Binding Outside the Gla Domain

Close to 10 yr after the discovery of Gla, it was demonstrated, using antibodies directed against the individual chains of factor X, that Ca^{2+} binding alters the conformation of both the light and heavy chains *(31)*. This suggested that factor X contains Ca^{2+} sites outside the Gla domain. One or two Gla-independent Ca^{2+} sites with dissociation constants of around 0.1 mM were subsequently shown to be present in factors IX and X and in protein C *(32–35)*. Two

Gla-independent sites were later confirmed to exist in these proteins and in factor VII and they are located in the first EGF-like domain and in the catalytic domain, respectively.

3.1. Calcium Binding to the First EGF-Like Domain

The solution structure of human EGF determined by NMR spectroscopy inferred that a number of negatively charged residues in the first EGF-like domain of the coagulation proteins are in close proximity and potentially could form a Ca^{2+} binding site *(36)*. One of these residues is a post-translationally modified aspartic acid, β-hydroxyaspartic acid, whose function was, and still is, elusive. The first direct demonstration of calcium binding to an EGF-like domain was provided using a fragment from protein C containing both its EGF-like domains isolated after limited proteolysis *(37)*. The fragment binds one Ca^{2+} ion with a dissociation constant of approx 100 μM as measured using a calcium-selective electrode. Owing to the lack of plausible Ca^{2+} ligands in the second EGF-like domain, it was assumed that this site was located in the first EGF-like domain. This was confirmed when the latter domain from factors X and IX was obtained after proteolytic degradation of bovine factor X and by expression in yeast, respectively *(38,39)*. However, the dissociation constant was unexpectedly high, above 1 mM, at physiological pH and ionic strength. Interestingly, when the affinity of this Ca^{2+} site was measured using a fragment containing the Gla domain and the first EGF-like domain its affinity increased to around 0.1 mM *(12)*. Selective monitoring of Ca^{2+} binding to the first EGF-like domain in the presence of the Gla domain was achieved after decarboxylation of the Gla residues to abolish their Ca^{2+} binding and subsequent measurements of the changes in intrinsic fluorescence, now solely reflecting Ca^{2+} binding to the EGF-like domain. It should be kept in mind that the average affinity of the Ca^{2+} sites in the Gla domain were similarly higher in this fragment compared to in the isolated Gla domain demonstrating the mutual stabilizing effect the Gla and first EGF-like domains exert on each other (*see* **Table 1**). Using antibodies that recognize Ca^{2+}-induced epitopes in specific domains of factor X, the affinity of the Ca^{2+} site in the first EGF-like domain of the intact protein showed the same dependence on the presence of the Gla domain, i.e., the affinity in Gla-domainless factor X was similar to that in the isolated first EGF-like domain and the affinity in intact factor X was similar to that in the Gla-EGF fragment *(40)*. Subsequent work showed that the C-terminal helix of the Gla domain is primarily responsible for increasing the affinity of Ca^{2+} binding to the first EGF-like domain *(41)*.

The first attempt to identify the Ca^{2+} ligands in the first EGF-like domain was carried out by mutagenesis *(42)*. In agreement with the hypothesis based on the human EGF structure, a number of charged residues were found to be

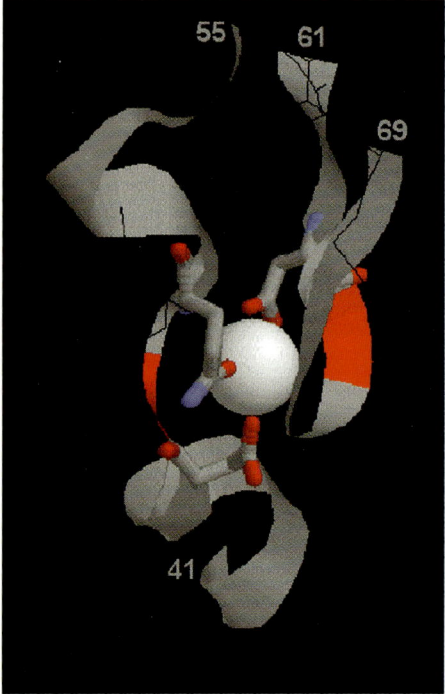

Fig. 3. The structure of the Ca^{2+} site in the first EGF-like domain of factor VIIa. Residues 41–55 plus 61–69 are shown. The site is located in the N-terminal part of the domain close to the Gla domain, whose C-terminal helix is shown at the bottom. The side-chain ligands *Asp^{46}*, *Gln^{49}*, and *Asp^{63}* are shown in stick representation and the backbone carbonyl ligands *Gly^{47}* and *Gln^{64}* are indicated by a red backbone around the central Ca^{2+} ion (*white ball*). *Asp^{63}* is β-hydroxylated in some of the vitamin K-dependent plasma proteins. For details of picture generation, *see* **Fig. 1**.

involved. This includes three aspartic acid and one glutamine residue. Even conservative point replacements with glutamic acid resulted in a 5- to 50-fold increase in the dissociation constant depending on in what position the mutation was introduced. The determinations of the solution structures of the Ca^{2+} form of the first EGF-like domain from factors X and IX confirmed three of these ligands, namely those corresponding to *Asp^{46}*, *Gln^{49}*, and *Asp^{63}* in factor X, but excluded the importance of one of the *Asp* residues (*43,44*) (*see* **Fig. 3**). *Asp^{63}* in factor X is β-hydroxylated. However, when comparing Ca^{2+} binding to the proteolytically isolated domain to that of its chemically synthesized counterpart with *Asp* in position 63, the extra hydroxyl group does not appear to influence the Ca^{2+} affinity under physiological conditions (*45*). The Ca^{2+} site in the first EGF-like domain was later found to be virtually identical in factor

VIIa *(46)*. It is noteworthy that the EGF-like domains in protein S bind Ca^{2+} with dissociation constants several orders of magnitude lower than those measured for Ca^{2+}-binding EGF-like domains in other vitamin K-dependent proteins *(47)*. The reason for the high affinities found in protein S appears to be that the affinity of each domain is influenced by neighboring domains, rather than unique primary structures *(48,49)*. The structural changes within the EGF-like domain induced by Ca^{2+} binding are minor and restricted to the Ca^{2+} ligands and neighboring residues *(43)*. The functions of Ca^{2+} binding to the first EGF-like domain include stabilization of the relative orientation of the Gla and EGF-like domains *(50)*, which, in turn, appears to be important for cofactor interactions, for instance, that between factor VIIa and tissue factor *(51,52)*.

3.2. Calcium Binding to the Catalytic Domain

Almost 25 yr ago, Bode and Schwager identified a Ca^{2+} site in trypsin *(53,54)*. Even though the homology with the protease domains of the vitamin K-dependent plasma proteins was obvious, not until many years later did data appear that suggested the presence of a corresponding Ca^{2+} site in these proteins. Rezaie et al. expressed truncated variants of protein C and factor X, containing only the second EGF-like and protease domains, to demonstrate the existence of a Ca^{2+} site C-terminal of the first EGF-like domain *(55,56)*. In both proteins, this site bound Ca^{2+} with a dissociation constant of 70–150 µM (*see* **Table 1**). The binding characteristics of antibodies specific for the calcium-induced conformation of the protease domain of factor X tentatively located the site to this domain *(40)*. In addition, the Ca^{2+}-binding property of a peptide from the protease domain of factor IX *(57)* and the effects of mutations of the putative Ca^{2+} ligand Glu^{220} in factor VIIa *(58)* completed the picture that all four homologous proteins harbor a Ca^{2+} site in the protease domain. The available crystal structures of Ca^{2+}-loaded vitamin K-dependent plasma proteins, of factor Xa *(59,60)* and of factor VIIa in complex with tissue factor *(46,61)*, demonstrate that the ligands are identical to those originally identified in trypsin (*see* **Fig. 4**). The role of this Ca^{2+} site is less clear. Currently available data show that only factor VIIa's activity (but not those of the other vitamin K-dependent enzymes) is significantly stimulated by Ca^{2+} and that there might be a weak influence of Ca^{2+} binding to the protease domain on protein–cofactor interactions.

4. Concluding Remarks

The zymogen and enzyme forms of factors VII, IX, and X, and protein C contain three types of Ca^{2+}-binding domains: Gla, EGF-like, and protease domains. The different functions of Ca^{2+} binding have been outlined, and the

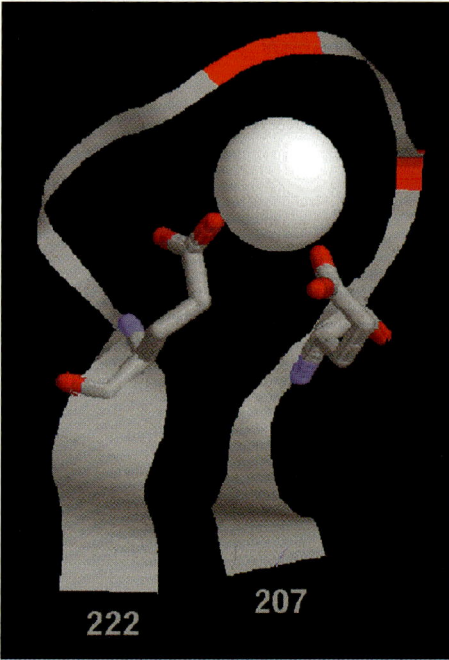

Fig. 4. The structure of the Ca^{2+}-binding loop in the serine protease domain of factor VIIa. Residues 207–222 are depicted with the side-chain ligands *Glu^{210}* and *Glu^{220}* in stick representation and the backbone carbonyl ligands *Asp^{212}* and *Glu^{215}* denoted by a red backbone. The Ca^{2+} ion is shown as a white sphere. Water molecules involved in Ca^{2+} coordination are not shown for clarity. For details of picture generation, *see* **Fig. 1**.

most fundamental and important role is to render the Gla domain phospholipid-binding. The dissociation constants of some Ca^{2+} sites in the Gla domain are close to the physiological Ca^{2+} concentration, whereas the sites in the other two domains are virtually saturated in vivo. In order to study individual Ca^{2+} sites and to investigate the effects of neighboring domains, various fragments of the proteins have been isolated, domain-specific Ca^{2+}-dependent antibodies and the fluorescent properties of strategically positioned tryptophan residues have been exploited, and isotope labeling has been employed. Based on available data, both from Ca^{2+}-binding studies and structure determinations, the proteins can be considered to contain two relatively independent units. One comprises the Gla and first EGF-like domains, which influence each other's Ca^{2+}-binding properties and relative orientation, and tethers the proteins to membrane surfaces and protein cofactors. The other comprises the second EGF-like and protease domains, which form a compact structural unit, and contributes the catalytic activity.

References

1. Furie, B., Bouchard, B. A., and Furie, B. C. (1999) Vitamin K-dependent biosynthesis of γ-carboxyglutamic acid. *Blood* **93,** 1798–1808.
2. Kalafatis, M., Swords, N. A., Rand, M. D., and Mann, K. G. (1994) Membrane-dependent reactions in blood coagulation: role of the vitamin K-dependent enzyme complexes. *Biochim. Biophys. Acta* **1227,** 113–129.
3. Furie, B. and Furie, B. C. (1988) The molecular basis of blood coagulation. *Cell* **53,** 505–518.
4. Stenflo, J., Selander, M., Persson, E., Astermark, J., Valcarce, C., and Drakenberg, T. (1993) Calcium binding properties of vitamin K-dependent clotting factors, in *Current Aspects of Blood Coagulation, Fibrinolysis, and Platelets* (Shen, M.-C. and Takada, A., eds.), Springer-Verlag, Tokyo, pp. 3–13.
5. Stenflo, J. and Ganrot, P.-O. (1973) Binding of Ca^{2+} to normal and dicoumarol-induced prothrombin. *Biochem. Biophys. Res. Commun.* **50,** 98–104.
6. Henriksen, R. A. and Jackson, C. M. (1975) Cooperative calcium binding by the phospholipid binding region of bovine prothrombin: a requirement for intact disulfide bridges. *Arch. Biochem. Biophys.* **170,** 149–159.
7. Stenflo, J., Fernlund, P., Egan, W., and Roepstorff, P. (1974) Vitamin K dependent modifications of glutamic acid residues in prothrombin. *Proc. Natl. Acad. Sci. USA* **71,** 2730–2733.
8. Nelsestuen, G. L., Zytkovicz, T. H., and Howard, J. B. (1974) The mode of action of vitamin K. Identification of γ-carboxyglutamic acid as a component of prothrombin. *J. Biol. Chem.* **249,** 6347–6350.
9. Nelsestuen, G. L. (1976) Role of γ-carboxyglutamic acid. An unusual protein transition required for the calcium-dependent binding of prothrombin to phospholipid. *J. Biol. Chem.* **251,** 5648–5656.
10. Bloom, J. W. and Mann, K. G. (1978) Metal ion induced conformational transitions of prothrombin and prothrombin fragment 1. *Biochemistry* **17,** 4430–4438.
11. Astermark, J., Björk, I., Öhlin, A.-K., and Stenflo, J. (1991) Structural requirements for Ca^{2+} binding to the γ-carboxyglutamic acid and epidermal growth factor-like regions of factor IX. Studies using intact domains isolated from controlled proteolytic digests of bovine factor IX. *J. Biol. Chem.* **266,** 2430–2437.
12. Persson, E., Björk, I., and Stenflo, J. (1991) Protein structural requirements for Ca^{2+} binding to the light chain of factor X. Studies using isolated intact fragments containing the γ-carboxyglutamic acid region and/or the epidermal growth factor-like domains. *J. Biol. Chem.* **266,** 2444–2452.
13. Schwalbe, R. A., Ryan, J., Stern, D. M., Kisiel, W., Dahlbäck, B., and Nelsestuen, G. L. (1989) Protein structural requirements and properties of membrane binding by γ-carboxyglutamic acid-containing plasma proteins and peptides. *J. Biol. Chem.* **264,** 20,288–20,296.

14. Öhlin, A.-K., Björk, I., and Stenflo, J. (1990) Proteolytic formation and properties of a fragment of protein C containing the γ-carboxyglutamic acid rich domain and the EGF-like region. *Biochemistry* **29,** 644–651.
15. Persson, E. and Stenflo, J. (1992) Comparison of the Ca^{2+} binding properties of the γ-carboxyglutamic acid-containing module of protein Z in the intact protein and in N-terminal fragments. *FEBS Lett.* **314,** 5–9.
16. Colpitts, T. L. and Castellino, F. J. (1994) Calcium and phospholipid binding properties of synthetic γ-carboxyglutamic acid-containing peptides with sequence counterparts in human protein C. *Biochemistry* **33,** 3501–3508.
17. Colpitts, T. L., Prorok, M., and Castellino, F. J. (1995) Binding of calcium to individual γ-carboxyglutamic acid residues of human protein C. *Biochemistry* **34,** 2424–2430.
18. Zhang, L., Jhingan, A., and Castellino, F. J. (1992) Role of individual γ-carboxyglutamic acid residues of activated human protein C in defining its in vitro anticoagulant activity. *Blood* **80,** 942–952.
19. Ratcliffe, J. V., Furie, B., and Furie, B. C. (1993) The importance of specific γ-carboxyglutamic acid residues in prothrombin. Evaluation by site-specific mutagenesis. *J. Biol. Chem.* **268,** 24,339–24,345.
20. Soriano-Garcia, M., Padmanabhan, K., de Vos, A. M., and Tulinsky, A. (1992) The Ca^{2+} ion and membrane binding structure of the Gla domain of Ca-prothrombin fragment 1. *Biochemistry* **31,** 2554–2566.
21. Skogen, W. F., Esmon, C. T., and Cox, A. C. (1984) Comparison of coagulation factor Xa and des-(1-44)factor Xa in the assembly of prothrombinase. *J. Biol. Chem.* **259,** 2306–2310.
22. Freedman, S. J., Furie, B. C., Furie, B., and Baleja, J. D. (1995) Structure of the metal-free γ-carboxyglutamic acid-rich membrane binding region of factor IX by two-dimensional NMR spectroscopy. *J. Biol. Chem.* **270,** 7980–7987.
23. Freedman, S. J., Furie, B. C., Furie, B. and Baleja, J. D. (1995) Structure of the calcium ion-bound γ-carboxyglutamic acid-rich domain of factor IX. *Biochemistry* **34,** 12,126–12,137.
24. Sunnerhagen, M., Forsén, S., Hoffrén, A.-M., Drakenberg, T., Teleman, O., and Stenflo, J. (1995) Structure of the Ca^{2+}-free Gla domain sheds light on membrane binding of blood coagulation proteins. *Nat. Struct. Biol.* **2,** 504–509.
25. Zhang, L. and Castellino, F. J. (1994) The binding energy of human coagulation protein C to acidic phospholipid vesicles contains a major contribution from leucine 5 in the γ-carboxyglutamic acid domain. *J. Biol. Chem.* **269,** 3590–3595.
26. Christiansen, W. T., Jalbert, L. R., Robertson, R. M., Jhingan, A., Prorok, M., and Castellino, F. J. (1995) Hydrophobic amino acid residues of human anticoagulation protein C that contribute to its functional binding to phospholipid vesicles. *Biochemistry* **34,** 10,376–10,382.
27. Freedman, S. J., Blostein, M. D., Baleja, J. D., Jacobs, M., Furie, B. C., and Furie, B. (1996) Identification of the phospholipid binding site in the vitamin K-dependent blood coagulation protein factor IX. *J. Biol. Chem.* **271,** 16,227–16,236.

28. Borowski, M., Furie, B. C., Bauminger, S., and Furie, B. (1986) Prothrombin requires two sequential metal-dependent conformational transitions to bind phospholipid. *J. Biol. Chem.* **261,** 14,969–14,975.
29. Liebman, H. A., Furie, B. C., and Furie, B. (1987) The factor IX phospholipid-binding site is required for calcium-dependent activation of factor IX by factor XIa. *J. Biol. Chem.* **262,** 7605–7612.
30. Persson, E. and Petersen, L. C. (1995) Structurally and functionally distinct Ca^{2+} binding sites in the γ-carboxyglutamic acid-containing domain of factor VIIa. *Eur. J. Biochem.* **234,** 293–300.
31. Keyt, B., Furie, B. C., and Furie, B. (1982) Structural transitions in bovine factor X associated with metal binding and zymogen activation. Studies using conformation-specific antibodies. *J. Biol. Chem.* **257,** 8687–8695.
32. Morita, T., Isaacs, B. S., Esmon, C. T., and Johnson, A. E. (1984) Derivatives of blood coagulation factor IX contain a high affinity Ca^{2+}-binding site that lacks γ-carboxyglutamic acid. *J. Biol. Chem.* **259,** 5698–5704.
33. Morita, T. and Kisiel, W. (1985) Calcium binding to a human factor IXa derivative lacking γ-carboxyglutamic acid: evidence for two high-affinity sites that do not involve β-hydroxyaspartic acid. *Biochem. Biophys. Res. Commun.* **130,** 841–847.
34. Sugo, T., Björk, I., Holmgren, A., and Stenflo, J. (1984) Calcium-binding properties of bovine factor X lacking the γ-carboxyglutamic acid-containing region. *J. Biol. Chem.* **259,** 5705–5710.
35. Johnson, A. E., Esmon, N. L., Laue, T. M., and Esmon, C. T. (1983) Structural changes required for activation of protein C are induced by Ca^{2+} binding to a high affinity site that does not contain γ-carboxyglutamic acid. *J. Biol. Chem.* **258,** 5554–5560.
36. Cooke, R. M., Wilkinson, A. J., Baron, M., Pastore, A., Tappin, M. J., Campbell, I. D., Gregory, H., and Sheard, B. (1987) The solution structure of human epidermal growth factor. *Nature* **327,** 339–341.
37. Öhlin, A.-K., Linse, S., and Stenflo, J. (1988) Calcium binding to the epidermal growth factor homology region of protein C. *J. Biol. Chem.* **263,** 7411–7417.
38. Persson, E., Selander, M., Linse, S., Drakenberg, T., Öhlin, A.-K., and Stenflo, J. (1989) Calcium binding to the isolated β-hydroxyaspartic acid-containing epidermal growth factor-like domain of bovine factor X. *J. Biol. Chem.* **264,** 16,897–16,904.
39. Handford, P. A., Baron, M., Mayhew, M., Willis, A., Beesley, T., Brownlee, G. G., and Campbell, I. D. (1990) The first EGF-like domain from human factor IX contains a high-affinity calcium binding site. *EMBO J.* **9,** 475–480.
40. Persson, E., Hogg, P. J., and Stenflo, J. (1993) Effects of Ca^{2+} binding on the protease module of factor Xa and its interaction with factor Va. Evidence for two Gla-independent Ca^{2+}-binding sites in factor Xa. *J. Biol. Chem.* **268,** 22,531–22,539.
41. Valcarce, C., Selander-Sunnerhagen, M., Tämlitz, A.-M., Drakenberg, T., Björk, I., and Stenflo, J. (1993) Calcium affinity of the NH_2-terminal epidermal growth factor-like module of factor X. Effect of the γ-carboxyglutamic acid-containing module. *J. Biol. Chem.* **268,** 26,673–26,678.

42. Handford, P. A., Mayhew, M., Baron, M., Winship, P. R., Campbell, I. D., and Brownlee, G. G. (1991) Key residues involved in calcium-binding motifs in EGF-like domains. *Nature* **351,** 164–167.
43. Selander-Sunnerhagen, M., Ullner, M., Persson, E., Teleman, O., Stenflo, J., and Drakenberg, T. (1992) How an epidermal growth factor (EGF)-like domain binds calcium. High resolution NMR structure of the calcium form of the NH_2-terminal EGF-like domain in coagulation factor X. *J. Biol. Chem.* **267,** 19,642–19,649.
44. Rao, Z., Handford, P., Mayhew, M., Knott, V., Brownlee, G. G., and Stuart, D. (1995) The structure of a Ca^{2+}-binding epidermal growth factor-like domain: Its role in protein-protein interactions. *Cell* **82,** 131–141.
45. Selander-Sunnerhagen, M., Persson, E., Dahlqvist, I., Drakenberg, T., Stenflo, J., Mayhew, M., et al. (1993) The effect of aspartate hydroxylation on calcium binding to epidermal growth factor-like modules in coagulation factors IX and X. *J. Biol. Chem.* **268,** 23,339–23,344.
46. Banner, D. W., D'Arcy, A., Chène, C., Winkler, F. K., Guha, A., Konigsberg, W. H., et al. (1996) The crystal structure of the complex of blood coagulation factor VIIa with soluble tissue factor. *Nature* **380,** 41–46.
47. Dahlbäck, B., Hildebrand, B., and Linse, S. (1990) Novel type of very high affinity calcium-binding sites in β-hydroxyasparagine-containing epidermal growth factor-like domains in vitamin K-dependent protein S. *J. Biol. Chem.* **265,** 18,481–18,489.
48. Stenberg, Y., Linse, S., Drakenberg, T., and Stenflo, J. (1997) The high affinity calcium-binding sites in the epidermal growth factor module region of vitamin K-dependent protein S. *J. Biol. Chem.* **272,** 23,255–23,260.
49. Stenberg, Y., Julenius, K., Dahlqvist, I., Drakenberg, T., and Stenflo, J. (1997) Calcium-binding properties of the third and fourth epidermal-growth-factor-like modules in vitamin-K-dependent protein S. *Eur. J. Biochem.* **248,** 163–170.
50. Sunnerhagen, M., Olah, G. A., Stenflo, J., Forsén, S., Drakenberg, T., and Trewhella, J. (1996) The relative orientation of Gla and EGF domains in coagulation factor X is altered by Ca^{2+} binding to the first EGF domain. A combined NMR-small angle x-ray scattering study. *Biochemistry* **35,** 11,547–11,559.
51. Kelly, C. R., Dickinson, C. D., and Ruf, W. (1997) Ca^{2+} binding to the first epidermal growth factor module of coagulation factor VIIa is important for cofactor interaction and proteolytic function. *J. Biol. Chem.* **272,** 17,467–17,472.
52. Persson, E., Olsen, O. H., Østergaard, A., and Nielsen, L. S. (1997) Ca^{2+} binding to the first epidermal growth factor-like domain of factor VIIa increases amidolytic activity and tissue factor affinity. *J. Biol. Chem.* **272,** 19,919–19,924.
53. Bode, W. and Schwager, P. (1975) The single calcium-binding site of crystalline β-trypsin. *FEBS Lett.* **56,** 139–143.
54. Bode, W. and Schwager, P. (1975) The refined crystal structure of bovine β-trypsin at 1.8 Å resolution. II. Crystallographic refinement, calcium binding site, benzamidine binding site and active site at pH 7.0. *J. Mol. Biol.* **98,** 693–717.
55. Rezaie, A. R., Esmon, N. L., and Esmon, C. T. (1992) The high affinity calcium-binding site involved in protein C activation is outside the first epidermal growth factor homology domain. *J. Biol. Chem.* **267,** 11,701–11,704.

56. Rezaie, A. R., Neuenschwander, P. F., Morrissey, J. H., and Esmon, C. T. (1993) Analysis of the functions of the first epidermal growth factor-like domain of factor X. *J. Biol. Chem.* **268,** 8176–8180.
57. Bajaj, S. P., Sabharwal, A. K., Gorka, J., and Birktoft, J. J. (1992) Antibody-probed conformational transitions in the protease domain of human factor IX upon calcium binding and zymogen activation: putative high-affinity Ca^{2+} binding site in the protease domain. *Proc. Natl. Acad. Sci. USA* **89,** 152–156.
58. Wildgoose, P., Foster, D., Schiødt, J., Wiberg, F. C., Birktoft, J. J., and Petersen, L. C. (1993) Identification of a calcium binding site in the protease domain of human blood coagulation factor VII: evidence for its role in factor VII-tissue factor interaction. *Biochemistry* **32,** 114–119.
59. Brandstetter, H., Kühne, A., Bode, W., Huber, R., von der Saal, W., Wirthensohn, K. and Engh, R. A. (1996) X-ray structure of active site-inhibited clotting factor Xa. Implications for drug design and substrate recognition. *J. Biol. Chem.* **271,** 29,988–29,992.
60. Kamata, K., Kawamoto, H., Honma, T., Iwama, T., and Kim, S.-H. (1998) Structural basis for chemical inhibition of human blood coagulation factor Xa. *Proc. Natl. Acad. Sci. USA* **95,** 6630–6635.
61. Zhang, E., Charles, R. S., and Tulinsky, A. (1999) Structure of extracellular tissue factor complexed with factor VIIa inhibited with a BPTI mutant. *J. Mol. Biol.* **285,** 2089–2104.

6

The Caseins of Milk as Calcium-Binding Proteins

Harold M. Farrell, Jr., Thomas F. Kumosinski, Edyth L. Malin, and Eleanor M. Brown

1. Introduction

The virtual image of milk, which would be constructed by most people, is that of a creamy white fluid rich in calcium. The lubricity and taste of milk are related to this perception and are based upon three unique biological structures: the colloidal calcium-protein complexes (the casein micelles), the milk-fat globules with their limiting membrane, and the milk sugar:lactose (*1*). The complexity of these structures is necessitated by the fact that milk is, in essence, predominantly water. It is the accommodation of these ingredients to an aqueous environment that forms the basis for the structure of milk at the molecular level. Adaptation of milk components to their ultimate aqueous environment begins during secretion. Lipid and protein synthesis are partitioned from the start. Amino acids and their metabolic precursors are actively transported into the secretory epithelial cells and assembled into proteins on the ribosomes of the highly developed rough endoplasmic reticulum. All proteins of mammary origin have conserved leader sequences which cause insertion of the nascent proteins into the lumen of the endoplasmic reticulum (*see* **Fig. 1**). The proteins are then transported through the Golgi apparatus, as shown in **Fig. 1**; presumably the globular proteins of milk are folded during this period. In the Golgi apparatus, the caseins, which are the major milk proteins in most species, are phosphorylated to begin the process of calcium transport. In general, when milks that contain > 2% protein are analyzed, the accompanying inorganic phosphate and calcium levels yield insoluble precipitates (apatite or brushite). Conversely in the absence of these salts, the casein components, as a result of their open structures, have a high viscosity. Thus, the gradual intercalation of

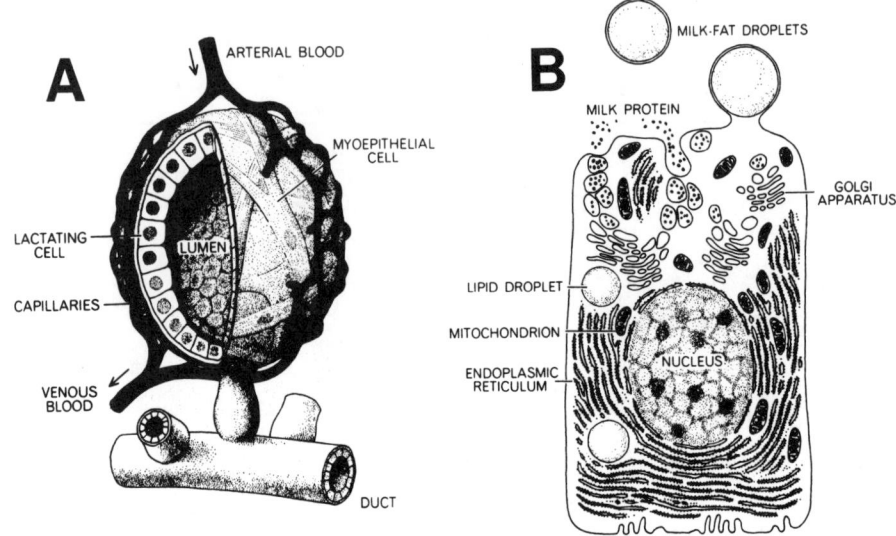

Fig. 1. Cell physiology of lactating mammary gland. (**A**) A single alveolus consisting of lactating epithelial cells surrounding the lumen. (**B**) A typical lactating cell indicating active secretion of protein and lipid by distinct mechanisms. (Reprinted with permission from S. Patton, *Sci. Am.* July 1969).

calcium, casein, and phosphate into colloidal casein micelles ensures the effective transport of these vital minerals. This process can be visualized in **Fig. 2A** where small submicellar particles are seen in the secretory vacuoles nearest the trans Golgi. Through the phosphorylation of casein *(2)*, the binding of calcium, which is actively transported by a Ca^{2+}-adenosine triphosphate (ATPase) *(3)*, and the accretion of phosphate, which is a hydrolysis product of nucleotide diphosphates *(4)*, the colloidal casein micelles are formed and finally secreted by reverse pinocytosis (*see* **Fig. 2B**).

The dominant feature of skim milk is the casein micelle (*see* **Fig. 3A**). This unique supramolecular aggregate imparts the opalescence characteristic of skim milk *(4)*. As noted above, the chief function of the micelle is to fluidize the protein and solubilize the calcium and phosphate. From research on the characterization of the caseins of cows' milk, four major casein components are recognized: α_{s1}-, α_{s2}-, β- and κ-casein. Caseins studied by protein or gene sequencing have been found to be homologous to these proteins in all species examined to date. The α_{s1}-, α_{s2}-, and β-casein are precipitated by calcium at

---

*Reference to a brand or firm name does not constitute an endorsement by the US Department of Agriculture over others of a similar nature not mentioned.

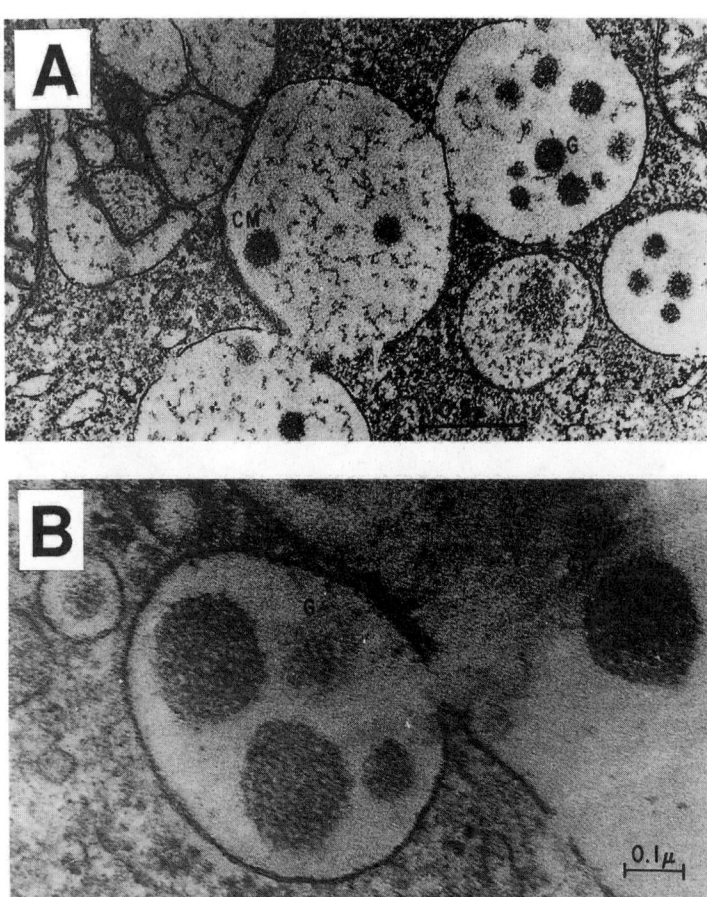

Fig. 2. Formation of casein micelles (CM) within Golgi vacuoles (G) and depicting the aggregation of small submicellar particles into larger micelles (**A**). A Golgi vacuole (G) about to discharge its contents into the alveolar lumen (**B**); a casein micelle (CM) is already present in the lumen *(10)*.

the concentrations found in most milks. However, κ-casein is not only soluble in calcium, but also interacts with and stabilizes the other calcium caseinates to initiate formation of the stable colloidal state. The casein micelle is thought to be constructed from spherical aggregates of the individual caseins (submicelles), as seen in **Fig. 2A,B**. These particles bind calcium and sequester inorganic phosphate to yield the casein micelle (*see* **Fig. 2A,B**). The protein particles are believed to be held together by calcium-phosphate linkages. κ-Casein is thought to predominate on the micellar surface (*see* **Fig. 3A**). In milk clotting in the stomach, the enzyme chymosin (rennin) specifically cleaves

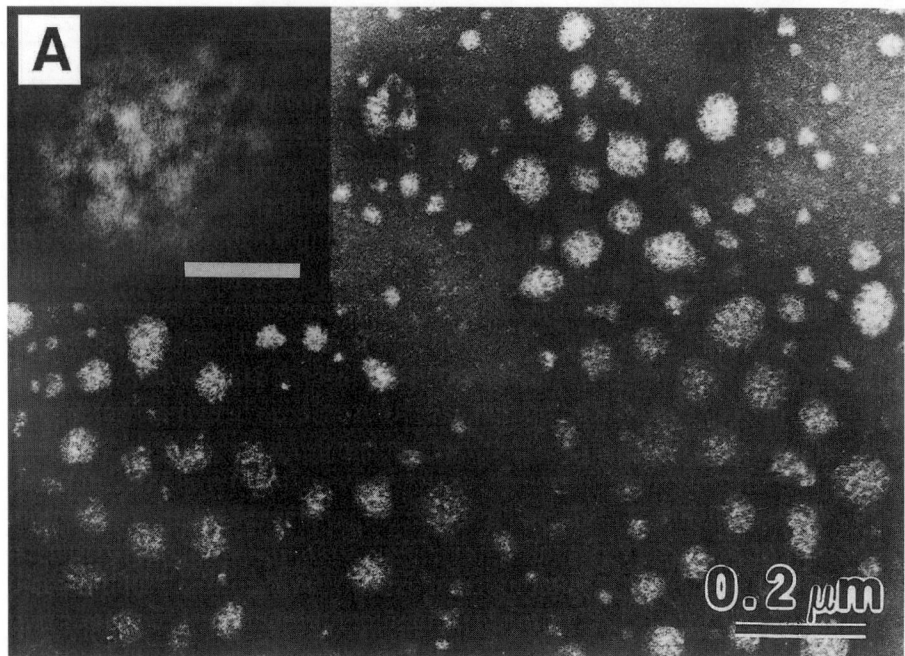

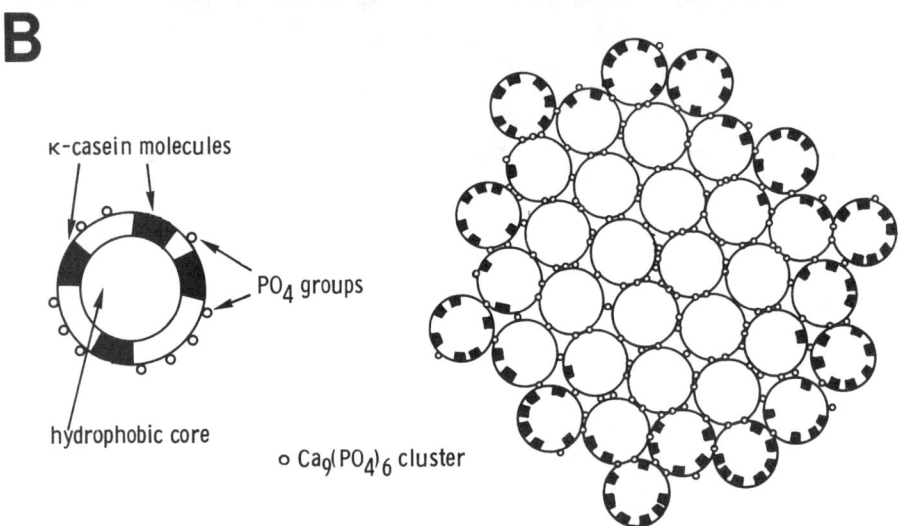

Fig. 3. Human casein micelles showing a wide range of sizes (**A**). Inset at the left shows an enlarged micelle with well defined submicellar structures (*white particles*); Bar = 30 nm. (Reprinted from Carroll et al. [1985] *Food Microstructure* **4**, 323–331). Model for casein micelle structure (bovine) showing submicellar structure and surface arrangement of κ-casein (**B**). Reprinted from **ref.** *(27)*.

Table 1
Percentage of Various Caseins in Milks

	α_{s1}-	α_{s2}-	β-	κ-
Goats[a]	5 to 17	6 to 20	50	15
Cows[a]	38	10	40	12
Human[b]	Trace	n	70	27

n = Not reported.
[a]Mora-Gutierrez et al. (1993) *J. Dairy Sci.* **76,** 3689–3710.
[b]Carroll et al. (1985) *Food Microstructure* **4,** 323–331.

one bond in κ-casein to initiate aggregation of the micelles. This step begins the digestive disassembly of the colloidal transport complex, which yields inorganic calcium and phosphate in the correct proportions for assimilation by the neonate.

At the ultrastructural level, the casein micelles of most species appear similar; however, the proportions of the various caseins vary widely. In goats, there is a high degree of variance in casein proportions among animals, which appears to be genetically controlled (*see* **Table 1**).

The casein complexes of all species contain two types of proteins: the α_{s1}-, α_{s2}-, and β-caseins, which bind calcium and aggregate or precipitate and κ-casein, which inhibits precipitation and stabilizes the former group to yield the micelle or colloid *(5)*. Because it is apparent from model studies *(6)* that the initiating reaction in the formation of casein colloids is the binding of calcium ions to casein, we have conducted a series of studies to investigate in detail calcium binding to α_{s1}-casein, as well as the changes that occur in its chemistry and physical chemistry as a result of these initial events. All of the caseins from all species studied so far contain highly conserved phosphoserine-rich sequences *(7)* that have been shown to react with Ca^{2+} by ^{31}P NMR studies *(8,9)*. The caseins of bovine milk and their naturally occurring genetic variants provide an illustration of qualitative correlations between primary structure and protein functionality *(5,10–12)*. For example, milks containing α_{s1}-casein A, rather than the more frequently occurring α_{s1}-casein B variant, yield cheeses with a softer texture and body; at the same time, these milks are more resistant to calcium-induced coagulation, i.e., they are more stable at elevated calcium concentrations *(13)*. Here the A variant is the result of the sequential deletion of 13 amino acids (residues 14 to 26) from the B variant *(12)*. However, changes in protein secondary, tertiary and quaternary structure, as well as in the thermodynamic parameters resulting from this mutation, have not been obvious. The calcium-α_{s1} system is selected for three reasons. From the point of view of the food industry, caseinate is an important commodity and milk and dairy products are widely consumed for their calcium content. Second, the interactions occurring in this important colloidal-transport system are still not well defined. Finally, a wealth of information of a qualitative nature is available in the literature on calcium-induced casein solubility *(6,13–15,16)*. In order to better understand these calcium–protein interactions, the precipitation and resolubilization of α_{s1}-casein were reinvestigated; it must be noted that the phosphate-rich region, responsible for calcium binding is highly conserved among caseins and across species *(16)*. The data were analyzed with respect to computer-generated models. Analysis of the data indicates a thermodynamic linkage between calcium binding and salting-out and salting-in reactions.

Thus, a quantitative thermodynamic mechanism could be established for the salting-in and -out of casein, and binding free energies for these ligand-induced protein solubility profiles may easily be calculated. Molecular modeling techniques, such as energy minimization and molecular dynamics were utilized to mimic protein–salt-water interactions, which lead to the salt-induced solubility profiles of α_{s1}-casein. Here, a recently developed predicted energy minimized three dimensional structure of α_{s1}-casein was employed to predict the hydrophobic sites responsible for precipitation of the protein and potential salt-binding sites responsible for the salting-in process when divalent salts are added. Studies of peptide fragments of α_{s1}-casein (1–23 and 136–196) by analytical ultracentrifugation and circular dichroism indicated that the protein-protein interactions involved in the salting-out and -in behavior are restricted to relatively select portions of α_{s1}-casein. The experimental data and the modeling predictions appear to agree well.

2. Theory and Methods
2.1. Thermodynamic Linkage

Wyman's theory of thermodynamic linkage *(17)* is based upon the concept that changes in an observable physical quantity (in this case solubility) can be linked to ligand binding. In previous studies on isolated caseins *(6,15)* it was shown that the precipitation of the caseins in the presence of calcium is indeed linked to calcium binding, and that calcium binding is the driving force in both salting-out and -in.

Here, we assume that there are essentially two classes of binding sites for ligands responsible for the sequential salting-out and -in processes and, therefore, Wyman's linked-function equations *(17)* can be used to analyze these processes with the assumption that the following equilibria occur:

$$P + nx \underset{(S_o)}{\overset{k_1^n}{\longleftrightarrow}} PXn + mX \underset{(S_1)}{\overset{k_2^m}{\longleftrightarrow}} PXnXm \quad (S_2) \tag{1}$$

where p is the unbound protein, x is the free salt, n and m are the number of X moles bound to species PX_n and $PX_n X_m$, and S_o, S_1 and S_2 are the solubilities of the species indicated. For this study, S_1 and S_2 will be relative to S_o. The mathematical relationship representing the above stoichiometry can be represented according to the following:

$$S_{app} = S_o f(p) + S_1 f(PXn) + S_2 f(PXn\,Xm) \tag{2}$$

where S_{app} is the apparent protein solubility at a given salt concentration (X_T), $f(I)$ values are the protein fractional component of species I, and the S's are species previously defined. Incorporation of the salt-binding equilibrium constants (k_1 and k_2) as defined by (1) into (2) yields the following:

$$S_{app} = \frac{S_o p}{p + k_1^n p x^n} + \frac{S_1 k_1^n p x^n}{p + k_1^n p x^n} + \frac{(S_2 - S_1) k_2^m p x^m}{p + k_2^m p x^m} \tag{3}$$

where p is the concentration in percent of the unbound protein and x is the concentration of unbound salt. Cancellation of common terms yields

$$S_{app} = \frac{S_o}{1 + k_1^n x^n} + \frac{S_1 k_1^n x^n}{1 + k_1^n x^n} + \frac{(S_2 - S_1) k_2^m x^m}{1 + k_2^m x^m} \tag{4}$$

It should be stressed here that the above expression is valid for sequential binding, i.e., $k_1 > k_2$, and n sites saturate prior to the binding of m sites on the protein and, for simplicity, that n and m do not interact. Also, for n or m greater than one, k_1 and k_2 represent an average value for each class of the n or m binding sites. In reality, n or m moles of salt will bind with only one equilibrium constant (K_1), i.e., $K_1 = k_1^n$ and $K_2 = k_2^m$.

Now, because only the total calcium concentration is known and the free and associated concentrations at each state are required, solving directly for these constants is difficult. For simplicity in these experiments, the total concentration is used in **Eq. 4** instead of the free concentration, and so the values obtained are only approximations. For unaltered purified caseins, the apparent K_a for Ca^{2+} is known, so the resultant more definitive k's can be calculated (6). For comparisons of altered and unaltered caseins, we have used total concentrations. The derivation of these equations and analysis using a Gauss–Newton nonlinear regression analysis program have been presented in detail elsewhere (18). All profiles were analyzed by fixing the values of n and m and calculating the best least-squared fit for the optimum evaluated k_1 and k_2 values. The n and m values were then fixed to new integer values and the whole procedure repeated. The n and m values, which yielded the minimum root-mean-square (rms) deviation and lowest error values for k_1 and k_2 were then reported following the procedures previously described for ligand binding and enzyme kinetics (19).

2.2. Molecular Modeling

The α_s-casein B monomer structure previously refined via energy minimization (20) was used as the basis for aggregate structures. Aggregates were constructed as previously described (20) using an interactive docking procedure with the program Sybyl (Tripos, St. Louis, MO) molecular modeling software on a Silicon Graphics IRIS computer. The docking procedure allowed for individual manipulation of the orientation of up to four molecular entities relative to one another. The desired orientations could then be frozen in space and merged into a single entity for further energy minimization calculations utilizing a molecular force field. The criterion for acceptance of reasonable structures was determined by a combination of experimentally determined information and the calculation of the lowest energy for that structure.

Studies concerned with the structures and/or energetics of molecules at the atomic level require a detailed knowledge of the potential energy surface (i.e., the potential energy as a function of the atomic coordinates). For systems with a small number of atoms, quantum mechanical methods may be used, but these methods become computationally intractable for larger systems (e.g., most systems of biological interest) because of the large number of atoms that must be considered. For these larger systems, semiempirical methods such as those of Weiner and Kollman (21) and Kollman (22) are used. These methods are based on the assumption that the true potential energy surface can be approximated with an empirical potential surface consisting of simple analytical functions of the atomic coordinates. The empirical potential energy model treats the atoms as a collection of point masses that are coupled to one another through covalent (bonded) and noncovalent (nonbonded) interactions. Energy minimization of these functions yield a relative minimum potential energy. The potential energy functions used were applied as previously described (23).

In the previous paragraph, we considered only static structures. However, the dynamic motion of molecules in solution contributes to their functionality. The molecular dynamics approach is a method of studying motion and molecular configuration as a function of time (24). All atoms in the molecule are assigned a kinetic energy through a velocity term that can be related to the local temperature, as well as to the average temperature of the system. These calculations can be performed *in vacuo* or in the presence of a desired number of solvent molecules, such as water and at a constant temperature and volume using a periodic boundary condition to confine the calculation within a prescribed volume (25). Theses concepts have been described elsewhere (23).

2.3. Analytical Ultracentrifugation Measurements

For analytical ultracentrifugation, the samples were dissolved at pH 6.75 at concentrations ranging from 1.0–3.0 mg/mL. Samples and solvents were filtered through Waters (Milford, MA) HVLP 0.45-μm membrane filters. Phast Gel electrophoresis in sodium dodecyl sulfate (SDS) showed a nearly identical pattern of protein components before and after filtration; less than 1% of the material was retained on the filter as ascertained by UV spectroscopy. Sedimentation equilibrium experiments were performed using a Beckman Optima XL-A (Palo Alto, CA) analytical ultracentrifuge at speeds ranging from 26,000 –46,000 rpm at 25°C. A 12-mm charcoal-epon six-channel centerpiece was used with quartz windows in a wide-aperture window holder. The solvent densities used in these experiments were 1.0016, 1.0060, and 1.0085 for the low-, medium-, and high-ionic strength experiments. These values were calculated from the data of Arakawa and Timasheff *(14)* using 0.564 cm^3/g as the partial specific volume of PIPES. The partial specific volume of α_{s1}-casein was taken as 0.728 *(26)* and those of the peptides were calculated from amino acid sequence data. Data were collected at 260, 280 or 292 nm using the standard XL-A procedure. The absorption vs radius plots were analyzed directly for weight average molecular weight using the program IDEAL 1, a part of the Optima XL-A data analysis software. As the absorbance offsets were not allowed to float in these analyses, weight average molecular weights were obtained. This was anticipated by the data of Schmidt *(27)* and Schmidt and Payens *(28)* who clearly demonstrated self-association for the α_{s1}-casein. The following model was employed by Schmidt and Payens *(28)* for α_{s1}-casein:

$$\alpha_i + \alpha_1 \underset{}{\overset{k_{1+i}}{\longleftrightarrow}} \alpha_{i+1} \cdots \quad (5)$$

where α is the unbound protein and $i = 1, 2, 3$ are consecutive steps of selfassociation. The model provided with the Beckman software (ASSOC4) differs in that it represents parallel equilibrium for all three polymers and the monomer. Analysis of the data was accomplished using the ASSOC4 algorithm. For the α_{s1}-casein, the best fits were obtained by fixing the sequence molecular weight at 23,600 for the monomer and floating K values at increasing integer values of n. Analysis of the self association of α_{s1}-casein was accomplished using the following equation:

$$\begin{aligned} A_r = \ & EXP[\ln(A_o) + H \times M \times (X^2-X)] \\ & + EXP[n_2 (\ln(A_o) + \ln(K_{a2}) + n_2 \times H (M \times (X^2-X_o^2)] \\ & + EXP[n_3 (\ln(A_o) + \ln(K_{a3}) + n_3 \times H (M \times (X^2-X_o^2)] \\ & + EXP[n_4 (\ln(A_o) + \ln(K_{a4}) + n_4 \times H (M \times (X^2-X_o^2)] + E \end{aligned} \quad (6)$$

where A_r = total absorbance of all species at radius x; EXP = exponent; ln = natural log; A_o = absorbance of the monomer species at reference radius x_0; H = constant $[(1 - \Delta\rho) \omega^2]/2RT$; M = apparent monomer molecular weight (monomer for α_{s1}-

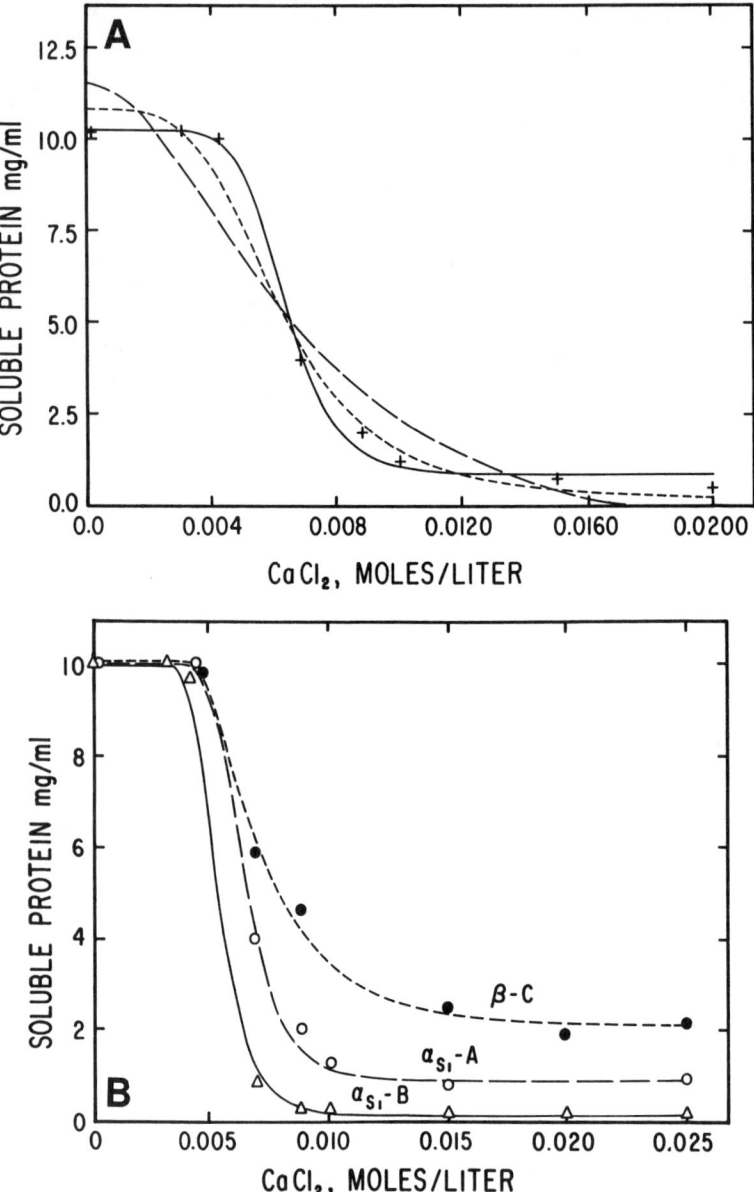

Fig. 4. Solubility at 37°C of the calcium salts of α_{s1}-caseins A and B, and β-casein C as a function of increasing $CaCl_2$ concentration. Solutions buffered at pH 7.0, 10 mM imidazole-HCl. **(A)** The experimental data for α_{s1}-A were fitted by **Eq. 4** by nonlinear regression analysis with values of 2 (– –), 4 (---), and 8 (–) assigned to n. The best fit was obtained for $n = 8$. **(B)** Similar fits for α_{s1}-B and β-casein C; results of analyses are in **Table 2**.

casein, monomer or dimer for 136–196, monomer for 1–23); X_o = reference radius; n_i = stoichiometry for species I (number of monomers); K_{ai} = association constant for the monomer-nmer equilibrium of species I; and E = baseline offset.

In these experiments, the software association constant K_{ai} is in reciprocal absorbance units. This can be converted to molar units using

$$K_{conc} = K_{abs}\,[\varepsilon l/n]^{n-1} \tag{7}$$

where K_{abs} is K_{ai} of the software, ε is the molar extinction coefficient, l is the path length, and n is the fitted degree of polymerization.

In turn the K_{con} (molar) can be converted to the weight k_n by

$$k_n = K_{conc}\,[M_1/n]^{n-1} \tag{8}$$

here, M_1 is the monomer weight or the "protomer" weight (e.g., dimer if that is the smallest fitted species) and n is the software fitted term. The k_n's reported herein were calculated in this fashion.

3. Quantitation of the Salt-Induced Solubility Profiles of the Caseins

3.1. Solubility at 37°C

Solubility determinations of α_{s1}-caseins A and B (α_{s1}-A, α_{s1}-B) and β-casein C (β-C) were performed at 37°C in 10 mM imidazole-HCl pH 7.0, 0.07 M KCl, at initial protein concentrations of 10 mg/mL *(15)*. As in the experiments of Noble and Waugh *(29)*, the proteins precipitate when added $CaCl_2$ exceeds 5 mM (**Fig. 4A,B**). Creamer and Waugh *(30)* had suggested that about 13 sites of similar calcium-binding strength exist in the α_{s1}-B at pH 6.6 and that when calcium-ion concentration exceeds this critical binding level, charge neutralization occurs and precipitation results. Comparison of the solubility profiles of α_{s1}-A and B indicates that at 37°C α_{s1}-A is more soluble than α_{s1}-B, whereas β-C is the most soluble. In order to quantify the data, nonlinear regression

Table 2
Calcium-Induced Insolubility of Casein at 37°C[a]

Casein	k_1 L/mol	n	S_1[b] mg/mL
α_{s1}-A	157 ± 3	8	0.9 ± 0.2
α_{s1}-A	186 ± 3	8	0.1 ± 0.1
β-C	156 ± 12	4	2.0 ± 0.3

[a] Solutions buffered at pH 7.0, 10 mM imidazole-HCl, 0.07 M KCl.
[b] s_1 = denotes the maximum value for soluble protein at elevated Ca^{2+} concentrations.

analyses were performed. The data of **Fig. 4A** were fitted by **Eq. 4**. Values of k_1, were obtained at fixed integer values of n; the correct value of n was taken to be the fit with the minimum rms. **Figure 4A** shows the fit to $n = 2, 4$, and 8 for α_{s1}-A; values for $n = 8$ gave the minimum rms with the lowest error in k_1. Analysis of the solubility profiles of α_{s1}-A, α_{s1}-B, and β-C at 37°C, where hydrophobic interactions are maximized, showed no salting-in behavior so that k_2 and m were essentially zero. Values obtained for k_1 (salting-out) and n are given in **Table 2**. The values of k are in terms of total calcium, which can be recalculated to free for the caseins as noted in **Subheading 2.1.** *(6)*. The interpretation of the k_i values obtained from the nonlinear regression programs used in all of these experiments is thus straight forward. In contrast, the interpretation of the n and m values is cloudy. They are nominally derived as stoichiometric values (*see* **Eqs. 3** and **4**), but even for simple enzyme substrate interactions or ligand binding *(19)*, these values may represent some degree of cooperativity. They could also represent a delay in the concentration dependent onset of a transition, or be no more than fitting constants representing a range of possible values *(19)*.

3.2. Solubility at 1°C

β-casein C is not precipitated at 1°C by Ca^{2+} at concentrations of up to 400 mM *(15,16)*. It is known that hydrophobic forces are dominant in the association reactions of β-caseins *(27)*. The solubility of β-casein C clearly distinguishes it from α_{s1}-B; it is known that β-casein binds Ca^{2+} at 1°C *(10,11)*, but in this case, binding is not linked to changes in solubility and so no analysis by thermodynamic linkage is possible. Single-aliquot additions of calcium chloride solutions to α_{s1}-casein results in a rapid decrease in solubility from 8–50 mM, where the protein is almost totally precipitated. When the calcium-chloride concentration exceeds 100 mM, a gradual salting-in of the protein ensues at 1°C. The data for α_{s1}-B were fitted by **Eq. 4** and the salting-out parameters k_1 and n, as well as the salting-in parameters k_2 and m were determined (*see* **Table 3**). The α_{s1}-A, genetic variant, in contrast to α_{s1}-B

Table 3
Cation-Induced Insolubility and Solubility of Casein at 1°C[a]

Casein	k_1[b]	n	k_2[b]	m
α_{s1}-B	123 ± 5	8	2.5 ± 0.2	4
α_{s1}-A	68 ± 3	8	10.6 ± 0.3	8
β-C	Totally soluble			

[a]Conditions as in **Table 2**.
[b]L/mol.

Calcium Binding by Caseins

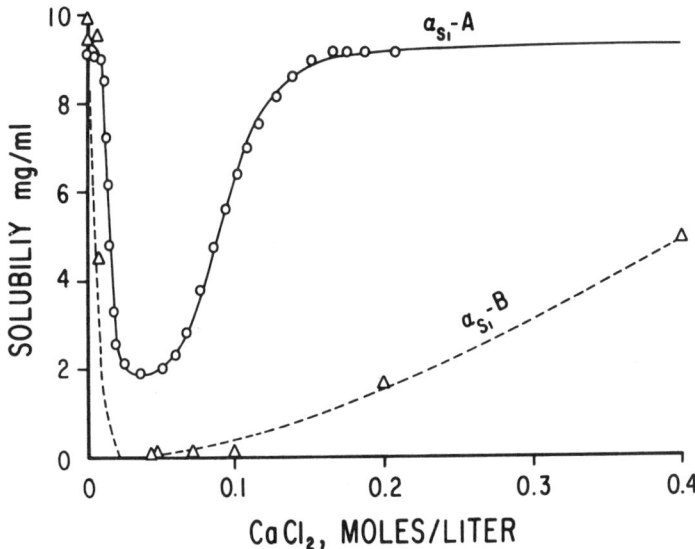

Fig. 5. Solubility at 1°C of calcium α_{s1}-B caseinate and calcium α_{s1}-A caseinate as a function of increasing $CaCl_2$ concentration. Data were fitted by **Eq. 4**; results of analyses are in **Table 3**.

exhibits extraordinary solubility behavior over a broad range of calcium-chloride concentrations. At 1°C (*see* **Fig. 5**) α_{s1}-A, like α_{s1}-B, is precipitated with calcium at about 8 mM, whereupon the net electrical charge on the Ca-protein complex may be close to zero. In the absence of electrolyte (KCl) or buffer, and after aliquot addition of $CaCl_2$, the protein is driven into solution at 90 mM $CaCl_2$. The Ca-protein complex is now positively charged and is acting as a cation. This conclusion was verified by free-boundary electrophoresis at pH 7.0, 10 mM imidazole, 150 mM $CaCl_2$, where the protein is soluble at 1°C; it migrates (+1.36 cm^2 V/s 10^{-5}) toward the cathode *(13)*. The evidence to this point favors direct salt-protein interactions as being responsible for the salting-in and -out of caseins by calcium rather than a salt-solvent interaction, as previously proposed by Melander and Horvath *(31)* following the cavity theory model of Sinanoglu *(32)*. The concept of salt binding is supported by the earlier work of Robinson and Jencks *(33)* who studied salting-out of model compounds and concluded that binding was a factor.

Previous studies of the effects of KCl on these solubility isotherms *(15)* suggest a competitive effect of elevated potassium ion for calcium-binding sites and argue for direct salt-binding to the caseins as the driving force for salting-out. It had previously been postulated *(31)* that such calcium–protein–salt interactions occur primarily through salt–solvent interactions rather than direct salt

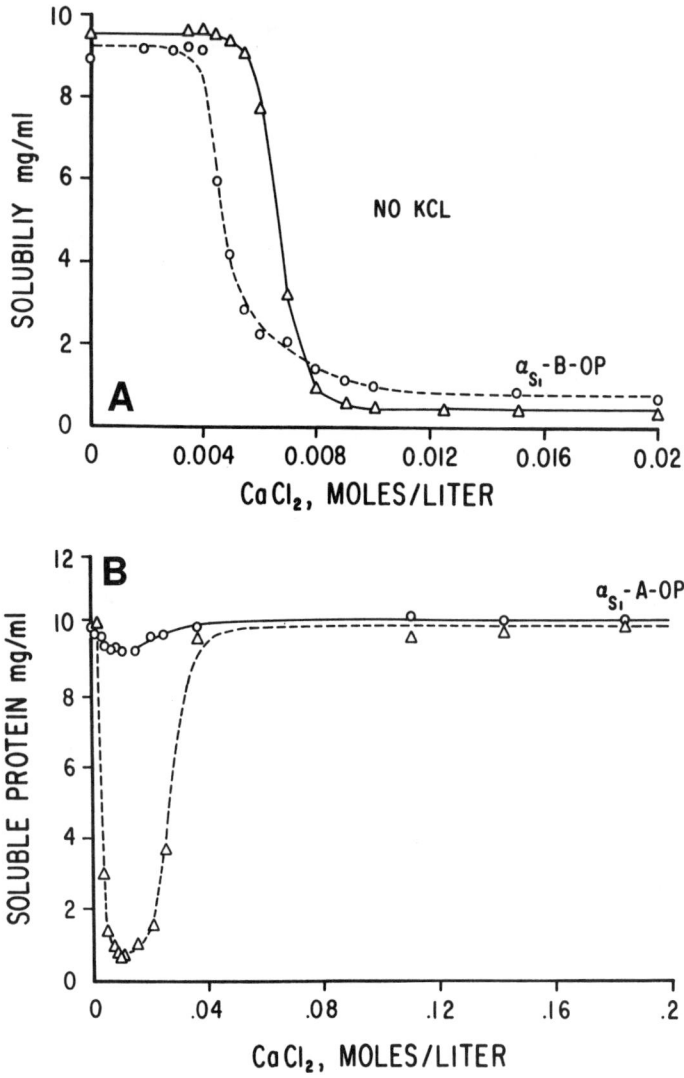

Fig. 6. **(A)** Solubility at 37°C of native and dephosphorylated α_{s1}-B (α_{s1}-B-OP) as a function of CaCl$_2$ concentrations at 10 mg/mL. **(B)** Solubility of α_{s1}-A O-P as a function of calcium ion concentrations at 1°C (—) and at 36°C (- - -). Data were fitted by **Eq. 4**. Results of analyses are given in **Table 4**.

binding to protein. The overall difference for the behavior of α_{s1}-A relative to α_{s1}-B, however, must reside in the structural differences due to the deletion mutation in the A variant. In all of the above analysis, the constants obtained are approximate binding constants, but they may not be identical with average

Table 4
Calcium-Induced Solubility of Native (N) and Dephosphorylated (O-P) α_{s1}-Casein B (B) and α_{s1}-Casein A (A)

Protein	Temp, °C	k_1, L/mol	k_2, L/mol	S_1	S_2	n	m
NB	37	151 ± 1		0.49 ± 0.04		16	
BO-P	37	219 ± 2	135 ± 12	2.6 ± 0.3	0.8 ± 0.06	16	8
NA	36	140 ± 3		0.9 ± 0.2		8	
AO-P	36	326 ± 7	36 ± 1	0.6 ± 0.3	10.0 ± 0.6	8	8
NA	1	130 ± 4	13 ± 1	0.4 ± 0.2		8	
AO-P	1	223 ± 59	46 ± 4	8.8 ± 1.8	10.0 ± 0.6	2	4

association constants, because only that binding that leads to changes in solubility is disclosed by this analysis. Bear in mind that for β-casein, binding occurs, but no change in solubility accompanies this and, thus, no analysis is possible.

3.3. Influence of Phosphate Groups on Salting-Out and Salting-In

As shown in **Fig. 4**, α_{s1}-A and B and β-C readily precipitate at 37°C. Although under these conditions, n was correlated with the number of phosphate residues in the native casein, the importance of these residues in the precipitation reaction could be tested. In previous research, the phosphate groups of α_{s1}-B were removed enzymatically *(34)* and the effects of KCl on the precipitation of native (N) and dephosphorylated (O-P) caseins by Ca^{2+} were compared, but not quantitated (**Fig. 6A** shows a typical curve). Analysis of these Ca^{2+} precipitation data by use of **Eq. 4** is summarized in **Table 4**. With no KCl present, dephosphorylation increases k_1 and some salting-in by Ca^{2+} occurs for the O-P-form; surprisingly, for both proteins (N and O-P) n = 16. Also, in the absence of KCl, salting-in occurs only for the dephosphorylated α_{s1}-B.

When α_{s1}-A is dephosphorylated, it becomes nearly completely soluble at 1°C and is salted-in even at 36°C (**Fig. 6B**). In contrast (*see* **Fig. 4**), the native α_{s1}-A is not appreciably salted-in at 37°C. Results are compared in **Table 4**. For dephosphorylated α_{s1}-A without added KCl n = 8, mirroring the numbers found for the native protcin, but for the O-P form at 1°C n = 2 and m = 4. The k_2 values observed for both native and O-P α_{s1}-A at both 1 and 36°C are similar to each other and to that of α_{s1}-B at 1°C. However, the small degree of salting-in, which occurs for O-P of α_{s1}-B does so with an elevated k_2 (*see* **Table 4**), showing another significant difference between the A and B variants.

3.4. Salt-Induced Protein Self-Association Model

Finally, it would be advantageous to test the salt-induced protein self-association to determine whether or not the monomer of α_{s1}-casein is the molecular

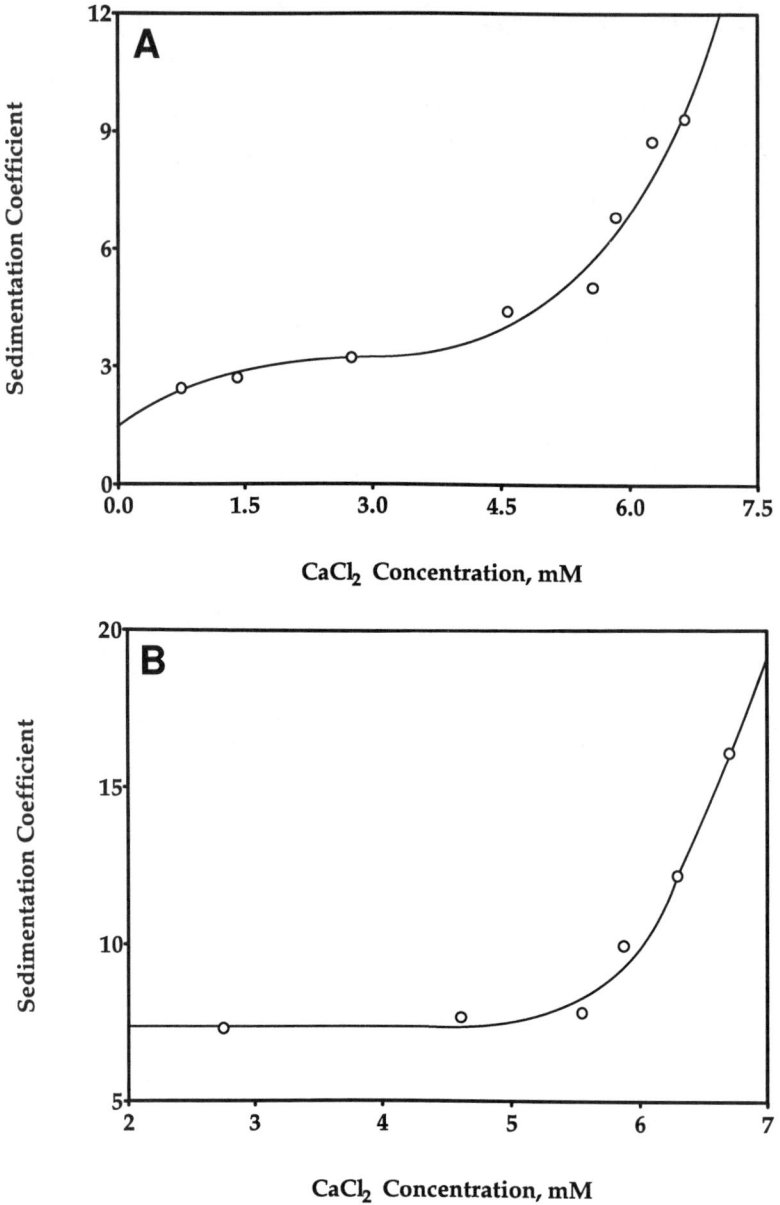

Fig. 7. Sedimentation velocity of α_{s1}-B at 10 mg/mL as a function of added $CaCl_2$. Two peaks were observed a slow one (**A**) and a fast one (**B**); analysis was by use of **Eq. 4** and given in **Table 5**; data are replotted from Waugh et al. (1971) *(35)*.

unit responsible for the salting-out process. Here, we used the sedimentation velocity data of Waugh et al. *(35)* at low $CaCl_2$ concentrations, 37°C and a protein concentration of 10 mg/mL. Two distinct peaks were observed and

Table 5
Thermodynamic Linkage Results of Sedimentation Coefficient of α_{s1}-Casein B vs Concentration of $CaCl_2$[a]

Peaks	k_1 L/mol	k_2 L/mol	S_0	S_1	S_2
Slow[b]	300 ± 50	166 ± 34	1.7 ± .3	4.6 ± .5	12.6 ± 1.1
Fast[c]	152 ± 32		7.4 ± .5	22.5 ± 1.5	

[a]At 37°C with 10 mg/mL of protein; here S denotes the sedimentation constant (*see* **Fig. 5**).
[b]$n = 1$, $m = 8$.
[c]$n = 16$.

Fig. 7 shows the variation of sedimentation coefficient $S_{20,W}$ for the slow peak (*see* **Fig. 7A**), and fast peak (*see* **Fig. 7B**) with added $CaCl_2$ up to a concentration of 0.007 M. Above this concentration, the protein precipitates. The appearance of a bimodal pattern in the sedimentation profile as the $CaCl_2$ is increased is indicative of a ligand-induced protein self-association system under equilibrium conditions *(36)*. Thermodynamic linkage and nonlinear regression analysis of the data were carried out using **Eq. 4**, but substituting the sedimentation constant(s) for S_{app}. For the slow peak, **Fig. 7A**, a biphasic binding process was determined, whereas analysis of the fast peak, **Fig. 7B**, yields only one cooperative binding mechanism leading to protein self-association. The results of this analysis of **Fig. 7** are given in **Table 5**. Here, for the slow peak the k_1 and k_2 values are 300 ± 50 and 166 ± 34 l/mol, respectively with $n = 1$ and $m = 8$. The value of S_0 of 1.7 ± 0.3 for the slow peak is of the order of magnitude expected for α_{s1}-casein monomer with a molecular mass of 24,000. The S_1 and S_2 values were 4.6 ± 0.5 and 12.6 ± 1.1 S, respectively. If we assume that S_0 is the sedimentation coefficient of monomeric α_{s1}-B, the degree of self-association can be approximated by dividing the S_1 and S_2 values by S_0 and raising that quotient to the 1.5 power which yields values of 4.3 and 18.8 for the relative size of the calcium-induced aggregates of the slow peak *(36)*. Using the same algorithm, relative degrees of self-association of 8.18 and 46.5 can be calculated from S_0 and S_1 values of 7.4 ± 0.5 and 22.5 ± 1.5 S, respectively, obtained from thermodynamic linkage analysis of the fast component (*see* **Table 5**). Thus, from the slow peak the apparent binding of one mole of $CaCl_2$ to α_{s1}-casein induces tetramer formation which, in turn, leads to an 18-mer when 8 more moles of $CaCl_2$ bind cooperatively. (Note: thermodynamic linkage reveals only those binding events which correlate with the observed physical changes). The fast peak with an S_0 value of 7.4 (an octamer), which then appears to bind 16 more moles of $CaCl_2$ cooperatively to yield a 46-mer aggregate with an S value of 22.5 ± 1.5. Subsequent increases in $CaCl_2$ concentration yields precipitation. The analysis clearly demonstrates that α_{s1}-B becomes insoluble as an aggregate and not as a monomer unit. It is also interesting that the k_2 value of the slow peak and the k_1 of the fast peak (*see* **Table 5**) are 166 and 152 L/mol respectively and these values are comparable to the salting-out binding parameters ($k_1 = 186$) seen in **Table 1**. In contrast k_1 of the slow peak approaches the average K_A for binding of Ca^{2+} to α_{s1}-casein (380 L/mol *(37)*).

In the absence of Ca^{2+}, the association of α_{s1}-casein depends strongly on ionic strength. Light-scattering studies by Schmidt and Payens *(28)* have shown that at pH 6.6, α_{s1}-casein is present predominately in monomeric form at low ionic strength (0.01 *M*), with a monomer × dimer equilibrium, which shifts to dimer at moderate ionic strength (0.05 *M*), and at high ionic strength (0.2 *M*) there exists an extended, concentration-dependent association to limiting polymer size of approx $n = 8$–10. This latter species is comparable in size to the S_o value of the Ca^{2+}-induced fast peak of **Fig. 7B**. Finally, at 0.5 *M* KCl or NaCl the protein precipitates from solution. Sedimentation equilibrium measurements made in this laboratory on purified α_{s1}-casein, at several ionic strengths, yield calculated association constants and apparent molecular weights in PIPES KCl buffer. **Table 6** compares association constants and apparent molecular weight data for α_{s1}-casein obtained in PIPES KCl in our laboratory with those obtained from light scattering *(28)*.

The low-salt sample ($\mu = 0.0175$) can best be described as a dimer in equilibrium with its component monomer; here only one association constant k_2 was needed to describe the system. The value of 0.88 L/g is in good agreement with the value obtained by Schmidt in phosphate at pH 6.0 and $\mu = 0.01$. At medium ionic strength ($\mu = 0.0889$), fits to the data were best described by adding two additional constants k_3 and k_4. The values obtained with a parallel association model in PIPES KCl for k_2 and k_3 are not in agreement with those from light scattering, but k_4 is. It may be noted that $n = 2, 4, 8$ were better fitted to the data, rather than $n = 2, 3, 4$ as used by Schmidt. It is possible then that some higher-order polymers exist. On the other hand, these numbers could represent cooperativity or statistical weighting factors. At high ionic strength, our values for k_2 and k_3 obtained at 1.0 g/L, differ significantly from light scattering values, which were carried out over a wide range of concentrations; k_4 is in reasonable agreement. The experiments here were designed to show that under these selected conditions (concentrations and solvents) there is a pronounced self-association that responds to changes in salt. The rigorous data of Schmidt are, in essence, the definitive association constants for the system. However, the similarity between self-association caused by monovalent cations and that because of Ca^{2+} allows comparison of data from all salt-induced associations to be made. Thus, spectral data, obtained in a high-KCl concentration, may be correlated with events occurring in lower concentrations, of $CaCl_2$.

Table 6
Comparison of Association Constants (L/g) and Apparent Molecular Weights for α_{s1}-Casein by Ultracentrifuge and Light Scattering Analysis

		Sedimentation equilibrium			Light scattering[a]		
	MW_w[b]	$k_2(n_2)$[c]	$k_3(n_3)$[c]	$k_4(n_4)$[c]	k_2	k_3	k_4
	Daltons	——————— L/g ———————			——— L/g ———		
Low salt $\mu = 0.0175$							
α_{s1}-Casein	34,700	0.88 ± 0.22 (2)[d]			0.44		
Medium salt $\mu = 0.0889$							
α_{s1}-casein	56,200	0.0088 ± 0.0048 (2)[d]	0.0199 ± 0.0098 (3)	6.82 ± 0.27 (4)	2.83	3.13	4.25
High salt $\mu = 0.224$							
α_{s1}-casein	68,000	0.021 ± 0.004 (2)[d]	0.32 ± 0.08 (3)	12.1 ± 0.2 (4)	15.9	20.8	14.7

[a]Schmidt (1992) following notation of **Eq. 5**.
[b]Weight average molecular weight; three determinations ± 5%.
[c]K_{ai} and n_i, as described in **Eq. 6**, and converted to L/g via **Eqs. 7** and **8**; the sequence molecular weight of α_{s1}-casein is 23,600. The values represent the average of three determinations each experiment contains three samples, one at each ionic strength.
[d]The monomer here was 23,600, the sequence weight; the values of n_i represents integer values of (4 and 19).

4. Molecular Modeling of α_{s1}-Casein, and Support from Experimental Data

In the previous subheadings, we formed a quantitative thermodynamic mechanism for the salt-induced solubility profiles of α_{s1}-casein as a function of temperature, genetic variation, dephosphorylation (post-translational modification), and types of mono and divalent cations employed. Here, the salting-out portion was described as a cooperative binding of n moles of salt with a constant of k_1 leading from a protein solubility of S_0 at zero salt concentration to a value of S_1. The salting-in process was quantitatively described by m moles of salt cooperatively binding to the protein with a constant of k_2 and yielding a limiting protein solubility of S_2. Studies of α_{s1}-caseins A and B at low-salt concentrations using electrophoresis, sedimentation velocity and sedimentation equilibrium techniques suggest the following model in terms of the molecular interactions occurring in this system:

1. Salt binding to phosphate or other groups leads to decreased charge repulsions.
2. Following salt binding, the net charge is neutralized and hydrophobic interactions yield dimers, tetramers, and octamers.
3. Oligomeric species then further aggregate to form a precipitate.
4. As the salt concentration is raised, further binding leads to resolubilization, which is easy to accomplish at 1°C for α_{S1}-A but not for α_{S1}-B.

In this subsection, we will use dimer and tetramer models as well as peptides constructed from an energy minimized, predicted three-dimensional structure of α_{s1}-casein *(20)* to simulate salt-induced interactions. Predictions from these theoretical models will be compared to experimental data. It is hoped that the presentation of such predicted structures will aid researchers in designing new chemical and biochemical modification experiments, for testing these working models.

The energy-minimized structure, generated for α_{s1}-casein B, as described above, is shown in **Fig. 8A** where it is displayed from carboxy- to amino-terminal (*left to right*). Analysis of this structure shows the molecule to be composed (*right to left*) of a short hydrophilic amino-terminal portion, a segment of rather hydrophobic β-sheet, the phosphopeptide region-represented by large loops, and a short portion of α-helix, which connects this N-terminal portion to the very hydrophobic carboxyl-terminal domain containing extended antiparallel β-strands: interspersed by proline directed turns. For clarity the backbone without side chains is shown in **Fig. 8A** with prolines (P) and serine phosphates (SP) indicated, and an accompanying chain trace stereo view (*see* **Fig. 8B**) is given.

From the overall shape of the α_{s1}-B model (*see* **Fig. 8**), it is apparent that it is impossible to approximate its structure with either a prolate or an oblate ellipsoid of revolution, as was done in the case of the β-casein refined structure

Calcium Binding by Caseins 117

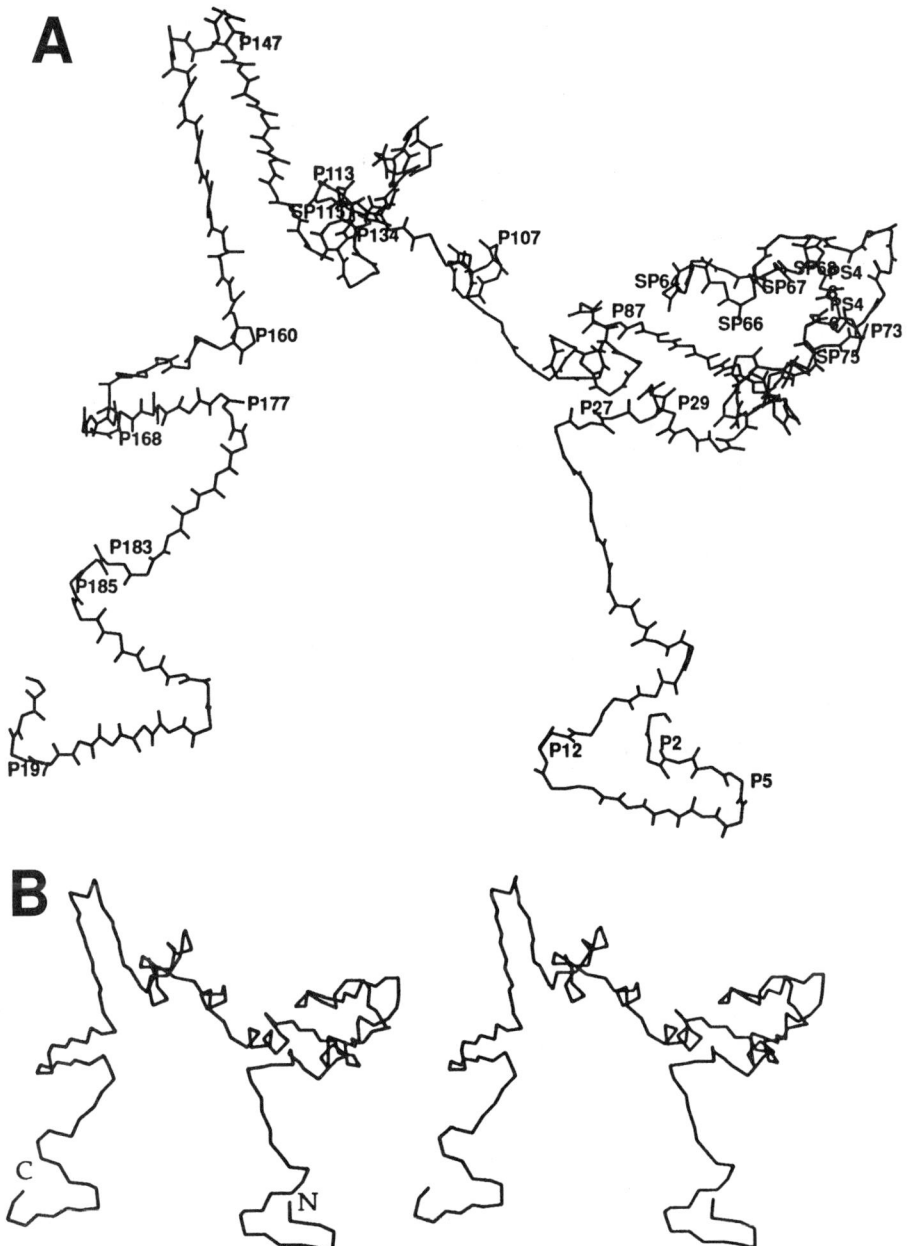

Fig. 8. (**A**) Backbone of the refined model of α_{s1}-casein B, without side chains, prolines (P) and serine phosphates (SP) indicated. (**B**) Stereo view of the refined three-dimensional molecular model of α_{s1}-casein; the N- and C-terminal ends of the molecule are labeled.

(38). Indeed, a rather large degree of asymmetry is observed. As noted above, the hydrophilic and hydrophobic domains are joined by structures whose central feature is an α-helix with its pitch perpendicular to the two domains. It may be speculated that this α-helix would be important for preserving the integrity of the two domains as these features are preserved across species *(20)*.

4.1. Step 1: Salt Binding: Molecular Dynamics of CaCl$_2$ and MgCl$_2$ in Water with the Hydrophilic Domain of α$_{s1}$-Casein-A

In this subheading, we shall attempt to form a structural basis for the thermodynamics of the salting-out/salting-in process. Here, we have chosen the hydrophilic N-terminal domain of the monomeric α$_{s1}$-casein A structure. This model was extracted from the α$_{s1}$-casein B model of **Fig. 8A** by deleting residues 14–26 *(26)* then excising residues 100 to 199 of α$_{s1}$-B. The resulting hydrophilic domain (H), i.e., residues 1–99 of the α$_{s1}$-casein B minus 14–26 (a total of 86 residues), was then energy minimized. Eleven hundred water molecules were added and the resulting structure in solution was again energy minimized with a cutoff for nonbonded interaction of 5 Å utilizing a periodic boundary condition. The resulting structure with water was subjected to molecular dynamics (MD) calculation for 20 ps above equilibrium conditions, which were determined by the stabilization of potential energy, radius of gyration of the protein backbone, root-mean-square (rms) fluctuations of the backbone atoms, and change in second moment. Such an equilibrated dynamic structure should approximate the structure, energetics, and dynamic motion of this protein domain in solution.

To mimic the salt-binding mechanism, 22 atoms of calcium or magnesium and 44 atoms of chloride with appropriate ionic charges were added to the above system in a pseudorandom fashion, energy minimized and subjected to MD for a full 40 ps. Equilibrium was easily established once again in 15–20 ps. The resulting structure for the native half of α$_{s1}$-A in CaCl$_2$ is shown in **Fig. 9A,B**. The amount of salt added was chosen to comply with a condition that would result in saturation of the Ca^{2+} binding sites. Using a value of 380 M^{-1} for K_A and 8 sites *(37)*, 22 molecules of CaCl$_2$ in 1100 water molecules with one molecule of protein is equivalent to greater than 99% occupancy of these calcium-binding sites and greater than 80% occupancy for eight additional putative salting-in sites derived from k_2 of **Table 3**. A similar structure for the dephosphorylated half of α$_{s1}$-A was also studied and is shown in **Fig. 10**. In total, seven MD calculations were performed on the hydrophilic domain (H) of α$_{s1}$-A in the presence of 1100 water molecules to 40 ps; two in the absence of salt for native (H) and dephosphorylated (HO-P); two in the presence of CaCl$_2$ for H and HO-P; one with added MgCl$_2$ for H; and one each for MgCl$_2$ and CaCl$_2$ with no protein. Each calculation specified a cutoff of 5 Å for nonbonded

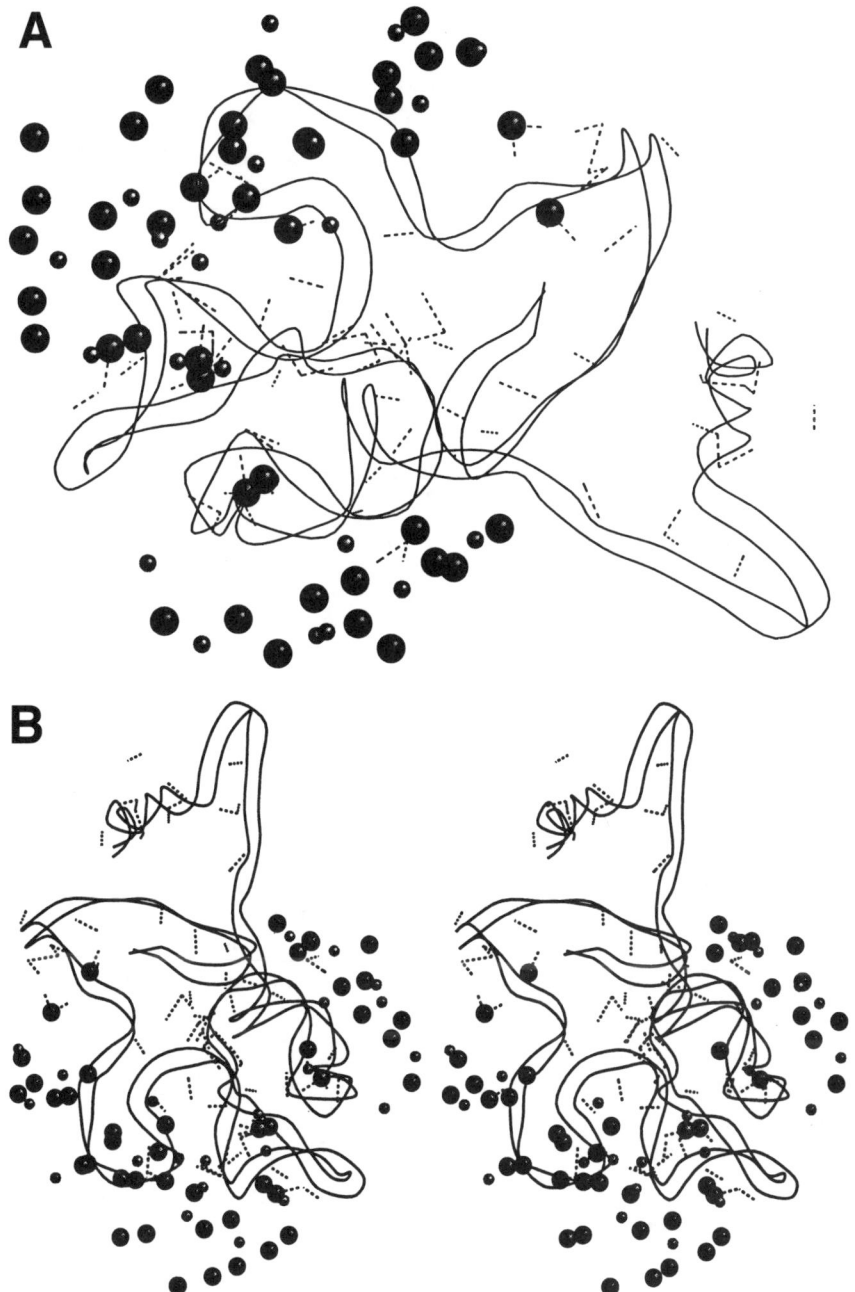

Fig. 9. **(A)** Backbone ribbon structure of the hydrophilic half of α_{s1}-casein A (residue 1–86) after molecular dynamics at 40 ps with 22 molecules of $CaCl_2$ in the presence of 1100 water molecules. Ca and Cl atoms are shown as ball models of radii equal to 0.15 times their known van der Waals radius. Dashed lines represent hydrogen bond formation. **(B)** Stereo view (relaxed view) of (A).

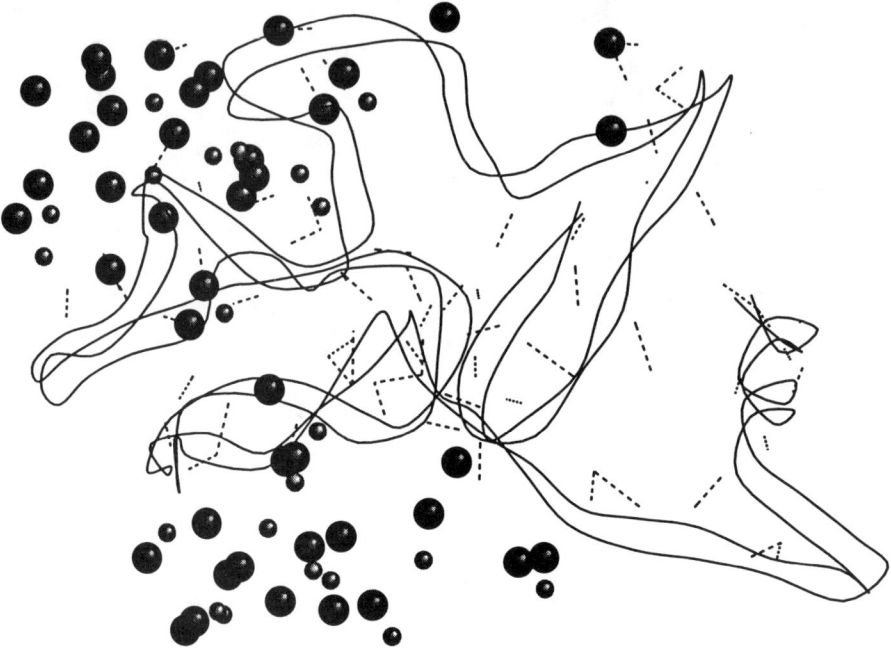

Fig. 10. Same as in **Fig. 9A**, but for dephosphorylated α_{s1}-A. All serine phosphates mutated to serine with appropriate partial charges.

interactions, the Tripos force field, and a "bump" factor of 0.7 for simulating hydrogen bond formation (i.e., a distance of >0.7 Å means no bonding). The final distribution of $CaCl_2$ with the H and HO-P forms are given in **Figs. 9** and **10**; here, the water molecules are deleted for simplicity. Qualitatively these figures show strong salt–peptide interactions.

The average quantitative results of several calculated geometric parameters with their corresponding errors are presented in **Table 7**. Here, the subscript 1 denotes the salt components, both Ca and Cl, whereas the subscript 2 describes the protein. R is the calculated dynamic radius of gyration, x is the average center of mass of the component atoms, and a is the rms fluctuation of component atoms from the center of mass. Here, a can be thought of as a dynamic Stokes radius of the chosen atoms, and x a spherical center of mass.

The distribution of the salt atoms at the end of MD calculations can be approximated by inspection of the a and x values in **Table 7** for both $CaCl_2$ and $MgCl_2$ alone, and in the presence of the hydrophilic half of α_{s1}-A (H) and its dephosphorylated form HO-P. Little significant change is seen in the a_1 values of the salts on going from the two component (salt, water) system to the three component (salt, water, peptide) system. This is because the mass of the pep-

Table 7. Molecular Dynamics of Hydrophilic Half (H) of α_{s1}-Casein A and its Dephosphorylated form (HO-P) in Water with Salt; Geometric Parameters

Peptide	Salt	a_1, Å²	x_1, Å	R_2, Å	a_2, Å²	x_2, Å
None	CaCl$_2$	9.10 ± 0.90	2.20 ± 0.10	—	—	—
None	MgCl$_2$	6.80 ± 1.10	1.5 ± 0.40	—	—	—
H	None	14.6 ± 0.2	8.2 ± 0.5	0.33 ± 0.04	—	—
HO-P	None	14.2 ± 0.04	7.0 ± 0.2	0.33 ± 0.03	—	—
H	CaCl$_2$	11.50 ± 0.70	0.80 ± 0.10	14.6 ± 0.2	8.7 ± 0.4	0.38 ± 0.04
HO-P	CaCl$_2$	9.38 ± 0.67	0.69 ± 0.06	15.5 ± 0.2	10.1 ± 0.4	0.51 ± 0.03
H	MgCl$_2$	7.80 ± 0.60	0.67 ± 0.07	14.6 ± 0.2	7.7 ± 0.5	0.37 ± 0.04

R is radius of gyration.
a is the rms fluctuation of all atoms from center of mass; a dynamic Stokes radius.
x is the calculated spherical center of mass.
Subscript 1 denotes salt, whereas 2 denotes protein atoms.

tide is far greater than the masses of the salts. The x_1 values are the best descriptors of the binding of salt atoms to the protein in these MD calculations. In all cases, the x_1 values have decreased in the presence of either peptide component. Such a decrease in the average spherical center of mass of the salt atoms is a clear indication of protein–salt interactions, because the center of mass for salt atoms alone would be larger as they move randomly about; x_1 would, of necessity, be smaller for salt bound to protein, where movement is restricted. It is assumed that the x_1 of the $MgCl_2$ was lower because of the lower value of the van der Waals radius for Mg^{+2}, 0.66 Å than the Ca^{+2} ion, 0.99 Å.

To observe the effect of salt binding on the dynamic structure of the hydrophilic half of the protein, we have calculated its radius of gyration R_2, as well as the a_2 and x_2 values in the presence or absence of salt atoms, these are presented as columns 5, 6, and 7 of **Table 7**. The H and HO-P peptides are nearly identical in the absence of salt. Virtually no changes within the calculated error are observed for the R_2, a_2, or x_2 values for the native H either in the absence or presence of $CaCl_2$ or $MgCl_2$. However, large increases in these descriptors are observed in the dephosphorylated HO-P form when $CaCl_2$ is added in the MD calculations. This change in configuration may reflect a general swelling of the HO-P structure when $CaCl_2$ is added which can also be observed by inspection of **Fig. 11A,B**. Here, the two structures are compared using representations of the protein backbones only, with no side chains displayed. The H form is represented by a ribbon trace of the backbone, whereas the HO-P form is represented by a backbone-atomic stick model. The backbone of the HO-P model is much more swollen than the ribbon (native) model. For easier observation of this conclusion in all dimensions, a stereo view is shown in **Fig. 11B**. The reason for this phenomenon is most likely caused by the hydrogen bonding of chloride ions to the serine side chains, as well as to N-H atoms of the backbone. Such anion–protein hydrogen bonding can impose important dynamic structural changes on the protein component. In the H form, these interactions may be "screened-out" by the negatively charged phosphate groups. Whether this swelling phenomenon causes increased solubility of the protein, i.e., salting-in, cannot be established at this time, because the deleted residues could play a role in keeping the α_{s1}-B variant insoluble. More MD studies in conjunction with solution structural physical chemical experiments must be performed in the future to test this hypothesis.

4.2. Step 2: Hydrophobic Dimers and Oligomers

Schmidt *(27)* and Schmidt and Payens *(28)* have summarized the light scattering studies on variants of α_{s1}-casein under a variety of environmental conditions from which a stoichiometry of the α_{s1}-casein self-association is obtained at selected temperatures and ionic strengths. From the results, it can

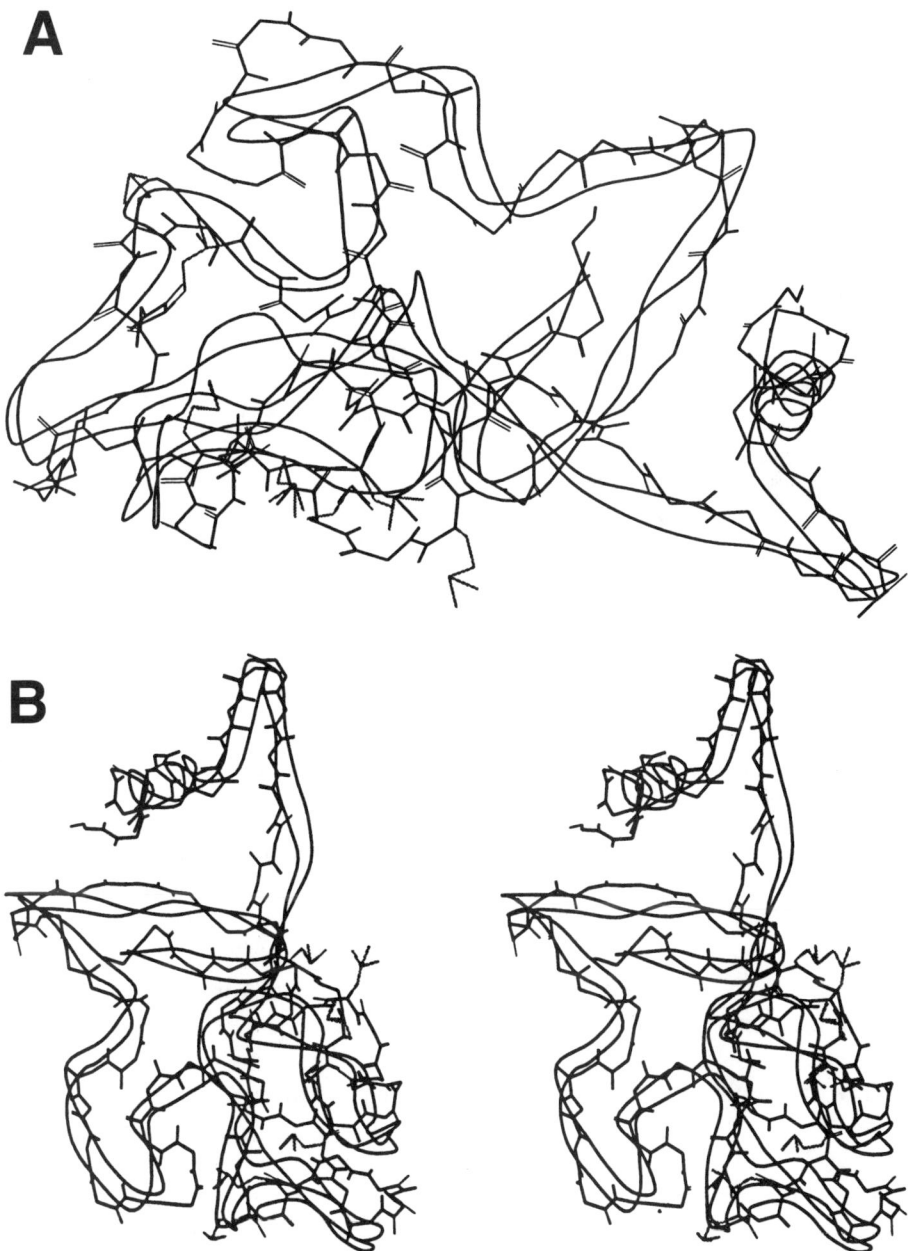

Fig. 11. (A) Comparison of hydrophilic domain of α_{s1}-A (shown with ribboned backbone) with dephosphorylated α_{s1}-A after molecular dynamics with 22 molecules of $CaCl_2$ at 40 ps in the presence of 1100 water molecules dashed lines represent hydrogen bond formations. (B) Stereo view (relaxed) of (A).

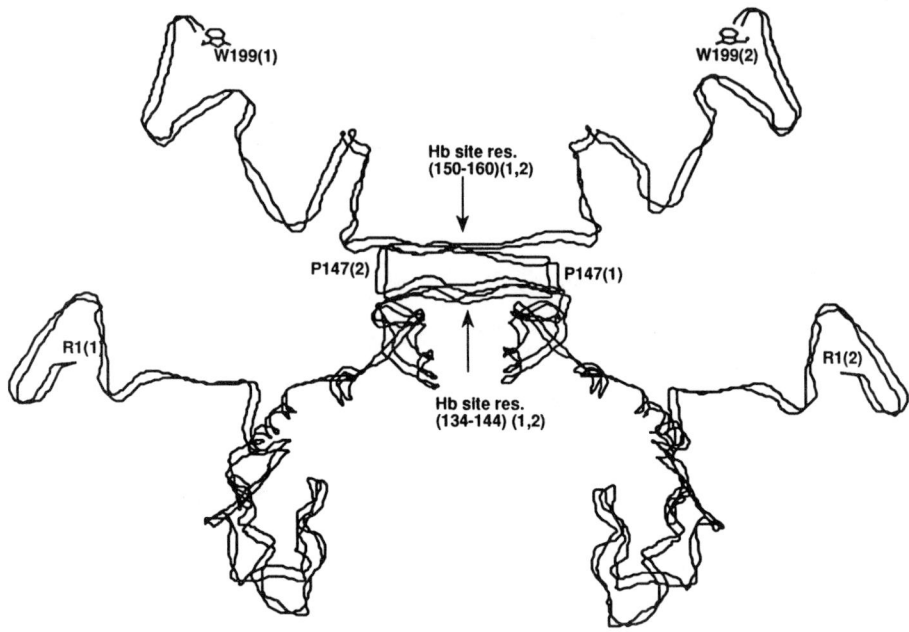

Fig. 12. α-Carbon chain trace of backbone without side chains of hydrophobic stabilized (Hb) antiparallel sheet dimer of α_{s1}-casein B; the large sheets centering on proline 147 are docked. (R) represents the N-terminal residues of molecules (1) and (2), while (W) represents the C-terminal residues of molecules (1) and (2).

be concluded that α_{s1}-casein undergoes a concentration-dependent reversible hydrophobically controlled association from monomer to dimer then tetramer, hexamer, octamer and even higher if the ionic strength is increased, especially as shown above (*see* **Tables 5, 6,** and **Fig. 7**) in the case of added Na/KCl or CaCl$_2$. To mimic hydrophobic salt-induced self-association mechanisms, we constructed energy minimized dimer, tetramer, and octamer structures using hydrophobic and hydrophilic sites.

The first step is to create a dimer using the large stranded β-sheets that occur at residues 136–158. The side chains are predominately hydrophobic and the hydrogen bonding of the sheet secondary structure adds rigidity to this site. A dimer can easily be formed if two of these stranded sheets are docked in an antiparallel fashion (*see* **Fig. 12**). Such an asymmetric arrangement minimizes the dipole–dipole interactions of the backbones while allowing the hydrophobic side chains to interact freely. In fact, this structure, following minimization (*see* **Fig. 12**), displays a lower energy of –520 kcal/mol/residue, over monomer, i.e., $-520 = E_2 - 2 \cdot E_m$, where E_2 is the energy of the dimer and E_m is the monomer energy *(20)*.

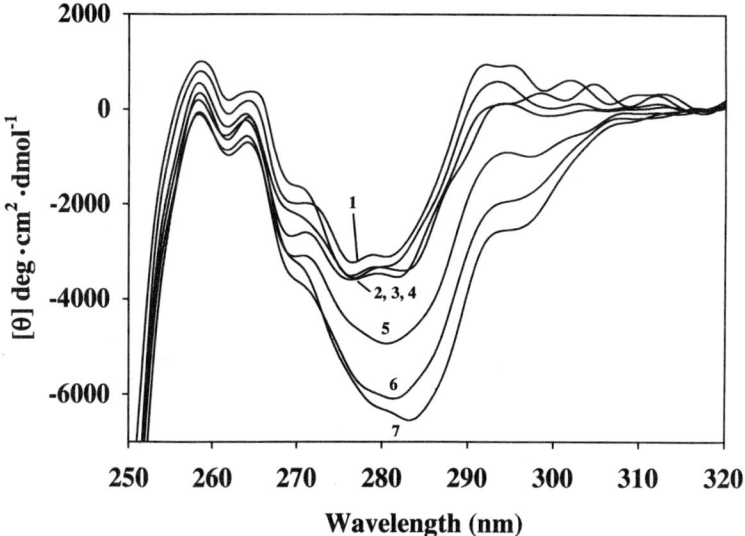

Fig. 13. Changes in near-UV CD spectra of α_{s1}-casein as a function of protein concentration at pH 6.75, 25°C, and ionic strength = 0.0175. Protein concentrations were 0.05, 0.075, 0.15, 0.50, 1.0, 2.0, and 3.0 mg/mL, lines 1–7, respectively. Molar ellipiticity [θ] is expressed in degrees · cm²/dmol.

4.2.1. Studies on α_{s1}-Casein

Previous investigations of α_{s1}-casein association have not focused on the influence of specific amino acid side chains in protein and peptide self-association. However, near-UV circular dichroism (CD) spectropolarimetry can be a useful technique in determining the role of aromatic side chain groups in polypeptide aggregation (39). In particular, the near-UV CD spectra of tyrosine and tryptophan residues are dependent upon side chain conformation and environment and can be used to monitor changes in structure during the self-association of caseins. The α_{s1}-casein monomer contains two tryptophan residues at positions 164 and 199, respectively. *Trp*164 is located in a primarily hydrophobic portion of the molecule, near *Pro*168, which is located in a predicted turn structure in the putative molecular model (*see* **Fig. 8**), and may be an important site in the formation of dimers in hydrophobically driven α_{s1}-casein aggregates (**Fig. 12**).

Changes in the near-UV CD spectrum for native α_{s1}-casein as a function of concentration are shown in **Fig. 13**. The spectra in **Fig. 13** were all obtained at pH 6.75, 25°C and low ionic strength (10 m*M* dipotassium PIPES buffer, μ = 0.0175 *M*). Prior to CD analysis, the identity and position of side chain absorption bands were determined from second-derivative UV-Vis spectra. A strong absorption band at $\lambda = 283$ nm was tentatively ascribed to tyrosine. Tryptophan was tentatively assigned to an absorption band at $\lambda = 292$ nm. However, tryptophan side chains may also contribute somewhat to the strong band at $\lambda = 283$ nm. Enzymatic cleavage by carboxypeptidase A allows for removal of two carboxy terminal residues, including $Trp199$, thus, studies of the carboxypeptidase A treated peptide α_{s1}-casein(1–197) can help determine the possible role of $Trp164$ in peptide aggregation. The second derivative band at 292 was diminished by half relative to the native in the carboxypeptidase treated sample where 1 of 2 tryptophan residues is deleted, but the 283 band was little changed in value. The vibronic CD bands for these side chains may occur at the same wavelengths as the corresponding absorption bands. Tyrosine and tryptophan side chains, while of themselves are not dichroic, may become so when their rotation is restricted because of protein aggregation. There is a gradual increase in molar ellipticity and a small red shift observed for the putative tyrosine band ($\lambda = 283$ nm) on going from 0.05 mg/mL to 3.0 mg/mL α_{s1}-casein (e.g., **Fig. 13**). There is also a change in sign and similar increase in ellipticity for the putative tryptophan band ($\lambda = 292$ nm) as a function of concentration. This indicates that both tyrosine and tryptophan side chains may play a role in the self-association of α_{s1}-casein at this ionic strength (0.0175 M). With θ being substituted for S, analysis with **Eq. 4** of the change in ellipticity in **Fig. 8** at $\lambda = 283$ nm with change in concentration yields the apparent CD derived association constant for α_{s1}-casein at low ionic strength of 0.85 L/g; the fit is given in **Fig. 14**. Under these conditions the protein is primarily monomer, but the weight average molecular weight was found to be 34,000, indicating a percentage of dimer. Sedimentation equilibrium experiments allowed for the calculation of a k_2 of 0.88 L/g (*see* **Table 8**), the apparent association constant of 0.85 as calculated from the change in θ at 283 nm with protein concentration (*see* **Fig. 14**) is in good agreement with this value.

Near-UV CD spectra were also obtained for α_{s1}-casein at medium (25 mM dipotassium PIPES/40 mM KCl buffer, $\mu = 0.0889$ M) and high ionic strength (25 mM dipotassium PIPES/180 mM KCl buffer, $\mu = 0.224$ M). All other conditions were kept the same as for the near-UV CD experiments on α_{s1}-casein at low ionic strength. Although the CD curves are not shown here, apparent association constants calculated from the CD data are given in **Table 8**. Experiments at higher ionic strengths reveal stronger casein associations. The near-UV CD spectra obtained for α_{s1}-casein at medium and high ionic strength look similar to the low ionic strength curves in **Fig. 13**, with proportionally larger band intensities at each protein concentration, but with somewhat increased 292 nm dichroism, as a function of increasing ionic strength. These results show that both tyrosine and tryptophan side chains are involved in progressive association steps of α_{s1}-casein. At medium ionic strength, the association con-

Table 8
Comparison of Association Constants (L/g) and Apparent Molecular Weights for α_{s1}-Casein and its Cleavage Fragments 136–196 and 1–197 Derived from Ultracentrifuge, and Circular Dichroism Analysis

	$MW_w{}^a$	Analytical ultracentrifuge			Circular dichroism	
		$k_2(n_2)^{b,d,e}$	$k_3(n_3)^b$	$k_4(n_4)^b$	$k_1(n)^c$	$k_2(m)^c$
	Daltons	L/g			L/g	
Low salt $\mu = 0.0175$						
α_{s1}-Casein	34,700	$0.88 (2)^b$			$0.85 \pm 0.13 (2)$	
α_{s1}-1 to 197					$0.78 \pm 0.11 (2)$	
α_{s1}-136 to 196	14,600	$1.34 \pm 0.18 (2)^d$	$2530 \pm 830 (6)$	—	$3.94 \pm 0.53 (3)$	
Medium salt $\mu = 0.0889$						
α_{s1}-Casein	56,200	$0.0088 (2)^b$	$0.0199 (3)$	$6.82 (4)$	$5.95 \pm 0.42 (1)$	
High salt $\mu = 0.224$						
α_{s1}-Casein	68,000	$0.021 (2)^b$	$0.32 (3)$	$12.1 (4)$	$0.63 \pm 0.03 (16)$	$8.1 \pm 2.5 (1)$
α_{s1}-1 to 197					$0.61 \pm 0.18 (16)$	$5.2 \pm 0.38 (4)$
α_{s1}-136 to 196	37,900	$0.0486 \pm 0.070 (2)^e$	$2220 \pm 420 (5)^e$	—	$8.80 \pm 0.65 (10)$	—

[a]Weight average molecular weight, three determinations $\pm 5\%$.
[b]k_n and n_i as described in **Table 6**.
[c]k_i and n or m as described in **Eqs. 1** and **4**, the error is the error of the fit and n and m values represent the integers at which the lowest rms is obtained for the fit [3, 16] of the model to the data.
[d]The sequence weight of α_{s1}-(136–196) is 6,995, which was used in **Eq. 6** to determine k_2 and k_3.
[e]The peptide "protomer" molecular weight here was 14,000 for **Eq. 6** to determine k_2 and k_3; $n_3 = \pm 1$.

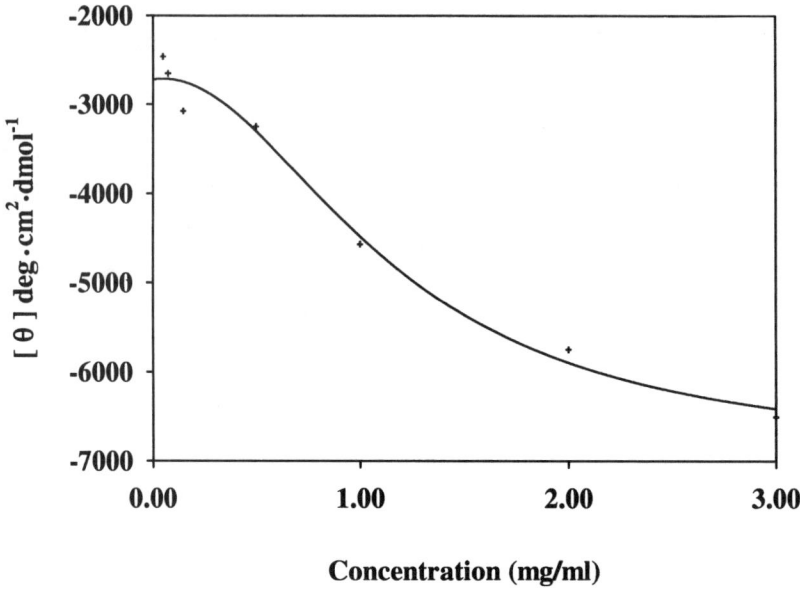

Fig. 14. Plot of change in [θ] at 283 nm as a function of protein concentration for α_{s1}-casein. The data were derived from **Fig. 13** and fitted with **Eq. 4**; the calculated apparent constants are given in **Table 8**.

stant for k_4 calculated from sedimentation equilibrium (*see* **Table 6**) shows good agreement with that calculated by Schmidt and Payens *(28)*. Those at higher ionic strength deviate more because they are extrapolations to higher concentration from ultracentrifuge studies at 1 mg/mL. Also note that the CD-derived apparent association constants are in reasonable agreement where the associations are relatively simple, but fail to detect more complicated changes at higher ionic strengths.

Changes in the near-UV CD spectra for the carboxypeptidase A-treated peptide α_{s1}-casein(1–197) as a function of concentration at pH 6.75, 25°C, at low and high ionic strength (10 m*M* dipotassium PIPES buffer) were analyzed as for the parent with **Eq. 4**. This data yields apparent association constants calculated from the CD data at both low and high ionic strengths (*see* **Table 7**), which are comparable to those of the parent protein. This comparison indicates the influence of *Trp*164 in progressive association steps, but not *Trp*199. The lack of influence of *Trp*199 in affecting associations of whole α_{s1}-casein compares favorably with casein micelle digestion with carboxypeptidase A by Ribadeau-Dumas and Garnier *(40)*. In the latter study, carboxypeptidase A was shown to quantitatively remove the C-terminal tryptophan of α_{s1}-casein in native and reconstituted micelles. The general availability of *Trp*199 to proteolytic digestion is in accord with results presented here, illustrating little participation of the two C-terminal residues in associations of α_{s1}-casein with itself or in casein colloids. These experiments offer good support for the putative model given in **Fig. 8**.

4.2.2. Studies on α_{S1}-Casein (136–196)

Although the self-association of α_{s1}-casein is governed to a large extent by hydrophobic attractions, electrostatic interactions cannot be neglected. The degree of ionization of the protein and the screening effect of the supporting electrolyte have been shown to influence associations. Much of the hydrophilicity of native α_{s1}-casein arises from the N-terminal half of the molecule. The particular significance of tyrosine and tryptophan side chains in forming α_{s1}-casein aggregates may be better interpreted by investigation of the hydrophobic portion of the protein, namely the α_{s1}-casein(136–196) peptide. Isolation of this sizable portion of the C-terminal half of the protein was achieved by chemical cleavage with cyanogen bromide and purification on DEAE-Sepharose. HPLC on a C4 reverse phase column was used to monitor the separation of the CNBr fragments; four major peaks were observed. The peptide of interest (136–196) was identified by collection of the HPLC peak, amino acid analysis, and N-terminal sequencing. In the sequencing procedure, starting with 700 pmol of peptide, the first 36 N-terminal residues were sequenced with statistical confidence. This peptide has an HPLC retention time of 16.3 ± 0.2 min, just slightly less than α_{s1}-casein(1–197). The retention time indicates the strong binding of this portion of α_{s1}-casein to the C4 reversed phase media. In fact, the late elution of the relatively low-charged peptide from DEAE-Sepharose also points to the extreme hydrophobicity of the peptide. The molecular weight calculated from sequence for this peptide is 6995.

The association properties of this peptide as a function of ionic strength give insight into the nature of the properties of its parent, α_{s1}-casein. Sedimentation equilibrium experiments on α_{s1}-casein(136–196) at 25°C, pH 6.75 at low ionic strength (0.0175 M) in 10 mM dipotassium PIPES buffer yielded a weight average molecular weight of 14,600, indicating predominantly dimer but with some associative species. An improved statistical fit was given by a three-component model, a monomer of 7000 a dimer, and an oligomer, where n_3 for the oligomer is 5 ± 1. The average effective molecular weights and association constants calculated from six such experiments are given in **Table 8**. It should be noted that there is some statistical uncertainty in the calculation of $n = 5$ for the oligomeric aggregation state. Thus, at low ionic strength for α_{s1}-casein (136–196), there is a very low population of monomer, a notable population of dimer, and a small population of associated species which may include higher-order ($n = 4$–6) aggregates.

Sedimentation equilibrium measurements for α_{s1}-casein(136–196) at high ionic strength (0.224 M) showed a much higher degree of peptide association. Association constants and effective molecular weights are given in **Table 8**. The weight average molecular weight here is 37,900. The best nonlinear regression fit to the data was obtained by using a three-component model, beginning with the 14,000 dimer and extending to tetramer and higher oligomers. Here, no fits could be made using 7000 as the starting point for fitting. At 0.224 M ionic strength, there appears to be no monomeric forms of α_{s1}-casein(136–196), indicating the predominance of hydrophobically driven *(41)* indefinite self-association under these conditions.

Experiments on the near-UV CD spectra for various concentrations of α_{s1}-casein(136–196) at low ionic strength ($\mu = 0.0175\ M$) showed a strong increase in molar ellipticity with increasing peptide concentration for tyrosine side chains monitored at $\lambda = 283$ nm as seen for the parent (*see* **Fig. 13**). This change in band intensity, along with an observed red shift, indicate that these residues are in a less-polar environment. The magnitude of change in band intensity and red shift is not quite as large for tryptophan side chains ($\lambda = 292$ nm). The results of a similar near-UV CD study of α_{s1}-casein(136–196) at high ionic strength ($\mu = 0.224\ M$) are shown in **Fig. 15**. The spectra were obtained in a 0.2 M KCl solution in dipotassium PIPES at pH 6.75 and $T = 25°C$. Perturbations in the tyrosine and tryptophan CD spectra are considerably more noticeable. The ellipticity change for tyrosine at $\lambda = 283$ nm as a function of peptide concentration is quite large, with a noteworthy jump in band intensity and red shift between 0.059–0.13 mg/mL. The most remarkable feature of the near-UV CD behavior of α_{s1}-casein(136–196) at high ionic strength is the exceptional change in the tryptophan spectrum at $\lambda = 292$ nm. Between 0.030–0.059 mg/mL peptide there is little change in the tryptophan spectrum. Above 0.059 mg/mL, the sign of the tryptophan band changes and the magnitude of the tryptophan band intensity increases substantially. Although the tryptophan band position does not appear to change to any great extent, the large jump in molar ellipticity between 0.059 and 0.13 mg/mL peptide concentration indicates the contribution of *Trp*164 to the formation of peptide aggregates at high ionic strengths. This correlates well with the ultracentrifuge data, which show the peptide to be a dimer at low ionic strengths and highly associated at high ionic strength.

The results reveal that aromatic side chains are involved in hydrophobically driven aggregation of α_{s1}-casein(136–196), with tyrosine appearing to be the predominant associating aromatic species at low ionic strength (*see* **Table 7**). In the putative molecular model for native α_{s1}-casein shown in **Fig. 8**, several proline-based turn structures were hypothesized to be effective sites of dimerization. There are three tyrosines (*Tyr*144, 146, and 153) in proximity to the *Pro*147-based turn, making this a probable site of dimerization. Additionally,

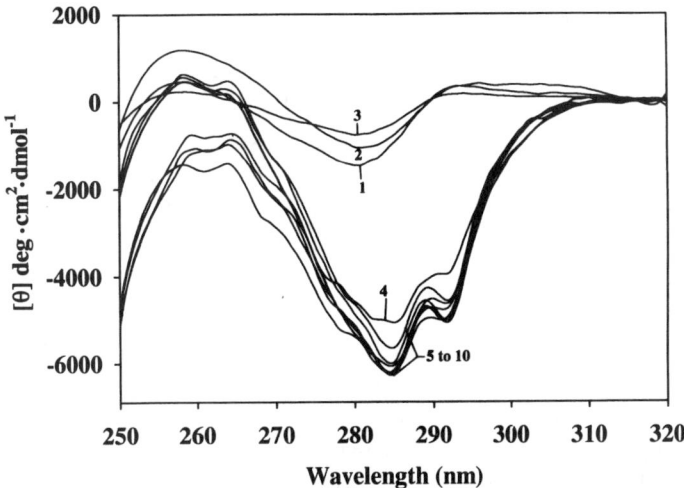

Fig. 15. Changes in the near-UV CD spectra of α_{s1}-casein (136–196) as a function of peptide concentration in PIPES at pH 6.75, 25°C, and $\mu = 0.224$. Molar elipiticy $[\theta]$ is expressed in degrees · cm^2/dmol. Peptide concentrations were 0.030, 0.044, 0.059, 0.13, 0.21, 0.26, 0.30, 0.44, 0.59, and 0.87 mg/mL, lines 1–10, respectively. The concentrations of peptide were chosen to cover the molar concentration range used for the intact protein.

there are three phenylalanine residues (*Phe*145, 150, and 152) in this region. These hydrophobic residues may also help to stabilize peptide–peptide interactions, although there is no direct spectral evidence from this study to support this latter speculation, because of the lower absorptivity of the *Phe* residues. There are three tyrosines (*Tyr*159, 165, and 166) and one tryptophan (*Trp*164) in proximity to the Pro168-based turn in the putative model. Based on the near-UV CD results, this region may also be a site for the formation of dimers and small oligomers ($n = 3$–5). However, in view of results from the near-UV CD study at high ionic strength (*see* **Fig. 15**), it is clear that *Trp*164 plays a role in the formation of high-order aggregates at high ionic strength and higher (>0.06 mg/mL) peptide concentrations. Based on comparison with association-reaction stoichiometry information obtained from the sedimentation experiment, we propose that the turn region about *Pro*147 is the initial site of hydrophobic aggregation (dimerization), with the turn region about *Pro*168 more involved in the formation of larger oligomers. As the tryptophan CD spectrum is perturbed to a much greater extent at high ionic strength for α_{s1}-casein(136–196) than it is for native α_{s1}-casein and α_{s1}-casein(1–197) under similar conditions, it is concluded that the region centering on *Pro*168 as reported upon by the CD of *Trp*164, is important in the stabilization of hydrophobic α_{s1} aggregates.

4.3. Step 3: Hydrophobic-Hydrophilic Dimerization

Another possible interaction site for hydrophobic dimerization resides in the deletion peptide of α_{s1}-casein A (i.e., the peptide which is deleted from α_{s1}-casein B to form α_{s1}-casein A). Closer inspection of the residues of this peptide within α_{s1}-casein B show β-sheet secondary structures with hydrophobic, as well as acidic (*Glu*14) and basic (*Arg*21) side chains. Thus, by docking two molecules in an antiparallel fashion, a hydrophobically stabilized intermolecular ion pair can be formed upon the construction of a dimer. Note that in this dimer, the basic residues (1–12) of the N-terminal sections of molecules 1 and 2 are at the largest distance possible from each other as are the phosphopeptide regions (59 to 79). Docking in parallel fashion produces much greater repulsions. The dimer was then energy minimized and is shown in **Fig. 16**. The formation of this hydrophobic ion pair has been used to explain the difference in the calcium-induced solubility and colloidal stability between α_{s1}-caseins B and A *(6,15)*. Small-angle X-ray scattering of micelles reconstituted from whole caseins containing α_{s1}-B or A shows large differences with respect to submicellar packing density within the micelle structure, i.e., 3:1 vs 6:1 for B and A, respectively. This difference in packing density may be a result of destructive interference in the scattered intensity due to an asymmetric structure with a center of inversion *(42)*. It was speculated that this asymmetric structure was due to the formation of a hydrophobically stabilized intermolecular ion pair in α_{s1}-B at this deletion peptide site.

4.3.1. Studies on α_{s1}-Casein (1–23)

In the second dimer model (**Fig. 16**), two electrostatic bonds could be formed between *Arg*21 and *Glu*14 on the two adjacent monomers. These bonds are in a relatively hydrophobic environment and thus have somewhat added bond strength *(41)*. The deletion of residues 14 to 26 allows for the ready solubility at 1°C (*see* **Fig. 5**) of α_{s1}-A whereas α_{s1}-B is not readily salted in (*see* **Fig. 2**). There is a fivefold difference in k_2 as seen in **Table 3** between the two genetic variants. If such bonds are formed in the protein, they would be expected to occur in the chymosin generated peptide α_{s1}-(1–23). Studies of this peptide at 20°C and low ionic strength (0.02 M) showed the peptide to be a dimer with a weight average molecular weight of 5110 (*see* **Table 9**). The K_a for this association was found to be 6.39 mL/g with $n = 4$ indicating a high degree of self-association in line with the weight average molecular weight. The degree of self association was not decreased at 4°C and so the self-association is not totally hydrophobic in nature. If *Glu*14 and *Arg*21 do form salt bridges as predicted by the model, then increased ionic strength ought to lead to dissociation. When analytical ultracentrifugation was conducted at 25°C and $\mu = 0.100\ M$,

Calcium Binding by Caseins

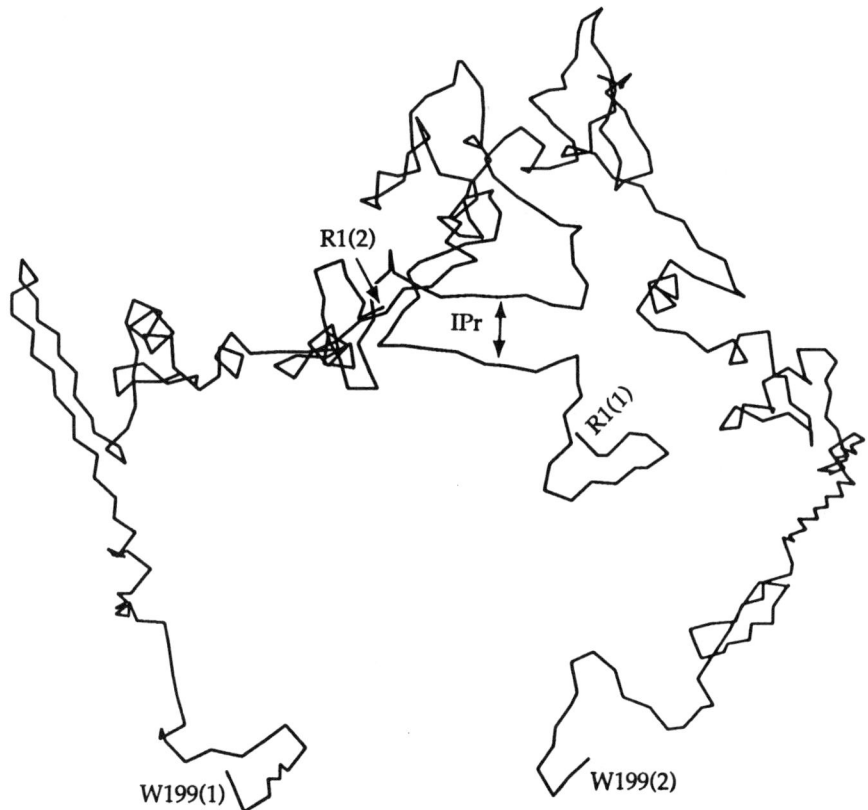

Fig. 16. Backbone structure of hydrophobic ion-pair dimer (IPr) from α_{s1}-casein B; this area contains the α_{s1}-A deletion peptide. Key for labels: $R_1(1)$, W199(1) represent N- and C-terminals of molecule 1 (arginine 1 and tryptophan 199 of molecule 1; the 2 in parentheses refers to molecule 2). Both dimeric structures energy minimized.

the peptide showed only traces of self-association and the data were readily fit to an ideal molecule with a molecular weight of 2770 (*see* **Table 9**). Thus, the salt-induced self association properties of the peptide supports the prediction of the putative molecular model *(20)*.

4.4. Step 4: Construction of High Molecular Weight Polymer Models

With these two possible dimer structures noted above (*see* **Figs. 12** and **16**), a tetramer can easily be modeled starting with the dimer formed by the intermolecular hydrophobic ion pair (*see* **Fig. 16**). To the two ends of this structure, two molecules of α_{s1}-B are added via the hydrophobic sheet–sheet interaction

Table 9
Effect of Increased Salt on the Association Constants (L/g) and Apparent Molecular Weights for α_{s1}-Casein (1–23) by Ultracentrifuge Analysis

	Sedimentation equilibrium	
	$MW_w{}^b$	$k_{a2}(n_2)^c$
Low salt $\mu = 0.02$	Daltons	mL/g
α_{s1}-Casein (1–23)	6180^a	$6.39 \pm 0.16\ (4)^c$
Medium salt $\mu = 0.100$		
α_{s1}-Casein (1–23)	2770^d	

$^a K_{ai}$ and n_i as described in **Eq. 6**; k_2 conversion as in **Table 6**. The sequence molecular weight of α_{s1}-casein (1–23) is 2770; the values represent the average of four determinations each experiment contains two samples.

^{b}Weight average molecular weight; six determinations ± 8%; two at 4°C and four at 20°C.

^{c}The monomer here was calculated as 2740 ± 77 close to the sequence weight; the values of n_i represents integer fitting with a repeatability of ± 1.

^{d}The data represent six determinations, 3 of 2 samples each at 25°C; the coefficient of variation was 6% for an ideal monomer.

shown in **Fig. 12**. Such a structure is presented in **Fig. 17**. This tetramer structure is highly asymmetric and also contains two possible hydrophobic ion pair sites, one at each end of the molecule; these sites at either end of the tetramer could lead to further aggregation resulting in a worm like rod with a large axial ratio and dipole moment, as described by SAXS data of Thurn and co-workers *(43)*. It is noteworthy that the hydrophobic stranded sheets of residues 163–174, 175–185, and ion the pair IPr site (1–26) are still open for further aggregation on the ends of the tetramers (*see* **Fig. 17**).

A very interesting feature of all the α_{s1}-caseins is the preservation of the C-terminal tryptophan. Ribabeau-Dumas and Garnier *(40)* showed that carboxypeptidase A could quantitatively remove the C-terminal tryptophan of α_{s1}-casein alone, and in native and reconstituted micelles. This was interpreted as a demonstration of the open network of the casein micelles, which allowed penetration of the protease into the micelle. This is in accord with the model shown in **Figs. 12** and **16**, where the C-terminal tryptophans are extended in space at the sides of the model. Thus, although residues 134 to 185 participate in hydrophobic interactions, a hydrophilic turn then intervenes and the C-terminal tryptophan can still be exposed in monomeric and polymeric structures, making it available to digestion with carboxypeptidase A.

Finally, it can easily be seen that deletion of residues 14–26 of the α_{s1}-B octamer structure to form α_{s1}-casein A would result in the elimination of the central hydrophobic ion pair interaction site. This mutation should result in

Calcium Binding by Caseins

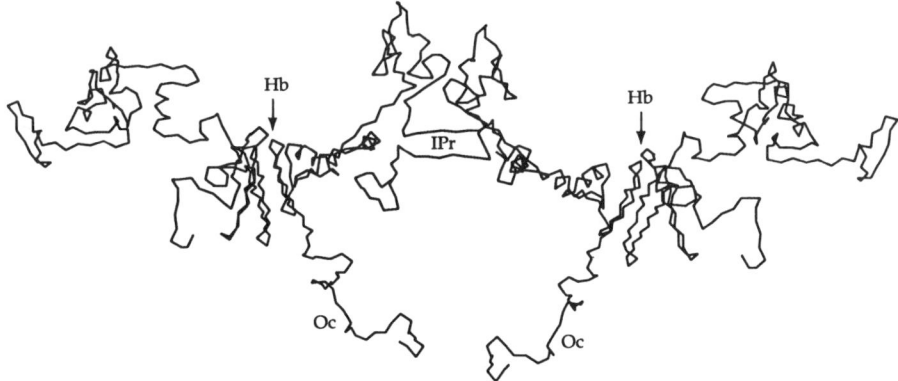

Fig. 17. α-Carbon chain trace of α_{s1}-casein B tetramer resulting from the docking of two molecules (left and right sides of the figure) of α_{s1}-B through the hydrophobic (Hb) scheme shown in **Fig. 12** with the (IPr) ion pair dimer given in **Fig. 16**. Sites for octamer formation (Oc) are noted.

only a tetramer structure arising from only the left- or right-hand portions of the tetramer α_{s1}-B structure of **Fig. 17** by Oc–Oc interactions.

Thus, we have now established a possible mechanism, as well as structures, for the calcium induced solubility profiles of α_{s1}-casein. Here, divalent cations can bind to serine phosphate or other negatively charged side chains that are mostly in the hydrophilic domain of α_{s1}-casein (*see* **Fig. 9A**) causing the hydrophobic domains to interact and self-associate in a cooperative fashion to a limiting octamer structure in the case of α_{s1}-B or a tetramer for α_{s1}-A, respectively. These limiting oligomers are thermodynamically unstable with respect to their interaction with solvents; because of their low net charge and relatively high exposure to water, noncooperative aggregation occurs leading to precipitation of the proteins. Further cation binding reverses the net charge (breaking the central IPr of **Fig. 17**) and leads to salting-in.

5. Summary Correlations and Conclusions

From all the above, it can be concluded that thermodynamic linkage in conjunction with experimental physical measurement and molecular modeling techniques can provide a powerful multifaceted approach for understanding structure–function relationships such as the salt-induced solubility profiles of this study. Here, the apparent stoichiometry and thermodynamics of the salt-induced solubility can be quantitated using thermodynamic linkage in conjunction with nonlinear regression analysis. The dynamic changes in the protein structure responsible for the salting-out and -in processes can be established using energy minimization and molecular dynamics calculations of the protein or its peptides in water and in the presence and absence of salt atoms.

For example, we have structurally defined the salting-out process in terms of a hydrophobically controlled polymerization for α_{s1}-B in compliance with the experimental results of Schmidt and Payens *(28)*, Waugh et al. *(35)* and our own studies of α_{s1}-peptides. This polymerization is most likely caused by binding of the positive divalent cations to phosphoserine and other negatively charged side chains on the hydrophilic domain of α_{s1}-casein (*see* **Fig. 9**). The geometric parameters (R_2, a_2, and x_2) for the phosphopeptide from MD calculations of the HO-P change dramatically over the H form in the presence of $CaCl_2$ (*see* **Table 7**). This phenomenon was interpreted as a swelling of the dephosphorylated form with added $CaCl_2$, which is shown in **Fig. 11A,B**. It is apparent that this structural change can also be labeled as a conformational change because many internal hydrogen bonds are disrupted. FTIR studies on model casein micelle formation *(44)* demonstrate clearly such conformational changes. In addition small-angle X-ray studies of the same model system also detected swelling when Ca^{2+} was bound *(45)* to whole native casein.

Following salt binding, resulting in the minimization of protein charge and the protein-protein electrostatic repulsion term, the first step in this self-association is most probably caused by the dimerization of both α_{s1}-B and α_{s1}-A which proceeds by the docking of two large hydrophobic stranded antiparallel sheets in an asymmetric fashion, as seen in **Fig. 12**. Support for this first step in protein precipitation is found in studies of salt-induced self-association of a_{s1}-casein and its peptide fragments 136–196. In the salting-out studies of α_{s1}-only, the next steps in the process would be aggregation and precipitation. However, in the bioassembly of casein micelles, κ-casein intervenes and colloidal complexes form (*see* **Fig. 2**). From the 3-D model studies *(45–47)*, the hydrophobic areas surrounding *Pro*168 have been proposed as possible sites for interactions with κ-casein, rather than sites that lead to precipitation for α_{s1}-alone. Further studies are needed to confirm this hypothesis.

In the events of casein micelle bioassembly, a series of complex interactions must occur. In our studies, the region of α_{s1}-casein 1–26 participates in salting out, but this region could be crucial in casein micelle formation. The difference in the calcium-induced salting-in between the α_{s1}-B variant and the α_{s1}-A variant was structurally defined in terms of an intermolecular hydrophobically stabilized ion-pair (*see* **Fig. 16**), which is present in α_{s1}-B and which delays the resolubilization process until higher ionic strength is attained. The deletion of residues 14–26 of α_{s1}-B results in the α_{s1}-A variant, which is salted-in at lower concentrations of calcium. Studies on the salt-induced self-association of the peptide fragment 1–23, indicate that the self-association of this region is sensitive to disruption at higher ionic strength. The absence of a portion of this segment in α_{s1}-A dramatically affects the overall properties of milk *(13)*. Similarly, SAXS studies of its casein micelle showed the colloids to have a very different internal packing *(42)*. It may be speculated that this positively charged region acts as a spacer arm to prevent too close packing in the colloidal casein micelles.

Overall then, the initial protein–protein interactions predicted by molecular modeling are well supported by new experimental observations. Studies such as these and their obvious extensions can lead to a better understanding of the molecular basis for casein micellar bioassembly. Once these mechanisms are known, we may have the ability to produce designer proteins by chimeric hybrids of sections of two or more proteins *(48)*. Such research on new protein forms, most likely will be initially limited to molecular biological expression, but as techniques are developed, we may someday see altered gene products that may lead to improved production of calcium containing colloids. The future of casein chemistry appears to be open ended and to yield exciting new opportunities; the best is yet to be.

References

1. Farrell, H. M., Jr. (1999) Milk synthesis and composition, in *Encyclopedia of Reproduction*, vol. 3. Academic, London, New York, pp. 256–263.
2. Bingham, E. W., Farrell, H. M., Jr., and Basch, J. J. (1972) Phosphorylation of casein: role of the Golgi apparatus. *J. Biol. Chem.* **247,** 8193–8194.
3. Bingham, E. W., McGranaghan, M. B., Wickham, E. D., Leung, C. T., and Farrell, H. M., Jr. (1993) Properties of [Ca^{2+} + Mg^{2+}]-adenosine triphosphatases in the Golgi apparatus and microsomes of the lactating mammary glands of cows. *J. Dairy Sci.* **76,** 393–400.
4. Farrell, H. M., Jr., Leung, C. T., and Wickham, E. D. (1992) Distribution of ADPase activity in the lactating rat mammary gland and its possible relationship to an ATP cycle in the Golgi apparatus. *Arch. Biochem. Biophys.* **292,** 365–375.
5. Farrell, H. M., Jr. and Thompson, M. P. (1988) The caseins of milk as calcium binding proteins in *Calcium Binding Proteins* (Thompson, M. P., ed.), CRC, Boca Raton, FL, pp. 117–137.
6. Kumosinski, T. F. and Farrell, H. M., Jr. (1991) Calcium-induced associations of the caseins: thermodynamic linkage of colloidal stability of casein micelles to calcium binding. *J. Protein Chem.* **10,** 3–16.
7. Holt, C. and Sawyer, L. (1988) Primary and predicted secondary structures of the caseins in relation to their biological functions. *Protein Eng.* **2,** 251–280.
8. Alaimo, M. H., Kumosinski, T. F., and Farrell, H. M., Jr. (1996) High resolution solid-state NMR of milk products. *J. Magnetic Res. Anal.* **2,** 267–274.
9. Kakalis, L. T., Kumosinski, T. F., and Farrell, H. M., Jr. (1990) A multinuclear, high-resolution NMR study of bovine casein micelles and submicelles. *Biophys. Chem.* **38,** 87–98.
10. Farrell, H. M., Jr. (1988) Physical equilibria: proteins, in *Fundamentals in Dairy Chemistry* (Wong, N. B., ed.), 3rd ed., Van Nostrand, Reinhold, NY.

11. Holt, C. (1992) Structure and stability of bovine casein micelles. *Adv. Protein Chem.* **43,** 63–151.
12. Whitney, R. McL. (1988) Milk proteins: composition, in *Fundamentals of Dairy Chemistry* (Wong, N. B., ed.), 3rd ed., Van Nostrand, Reinhold, NY.
13. Thompson, M. P., Gordon, W. G., Boswell, R. T., and Farrell, H. M., Jr. (1969) Solubility, solvation and stabilization of α_{s1}- and β-caseins. *J. Dairy Sci.* **52,** 1166–1173.
14. Arakawa, T. and Timasheff, S. N. (1984) Mechanisms of protein salting-in and salting-out by divalent cation salts: balance between hydration and salt binding. *Biochemistry* **23,** 5912–5923.
15. Farrell, H. M., Jr., Kumosinski, T. F., Pulaski, P., and Thompson, M. P. (1988) Calcium-induced associations of the caseins: a thermodynamic linkage approach to precipitation and resolubilization. *Arch. Biochem. Biophys.* **265,** 146–158.
16. Farrell, H. M., Jr. and Kumosinski, T. F. (1988) Modeling of calcium-induced solubility profiles of casein for biotechnology. *J. Indust. Micro.* **3,** 61–71.
17. Wyman, J., Jr. (1964) Linked functions and reciprocal effects in hemoglobin: a second look. *Adv. Protein Chem.* **19,** 223–286.
18. Rusling, J. F. and Kumosinski, T. F. (1996) in *Nonlinear Computer Modeling of Chemical and Biochemical Data*, Academic, San Diego, CA.
19. Farrell, H. M., Jr., Deeney, J. T., Hild, E. K., and Kumosinski, T. F. (1990) Stopped flow and steady state kinetics of NADP$^+$: isocitrate dehydrogenase. *J. Biol. Chem.* **265,** 17,637–17,643.
20. Kumosinski, T. F., Brown, E. M. and Farrell, H. M., Jr. (1994) Predicted energy minimized α_{s1}-casein working model in *Molecular Modeling from Virtual Tools to Real Problems* (Kumosinski, T. F. and Liebman, M. N., eds.), ACS Symposium Series 576. Am. Chem. Soc., Washington, DC, pp. 368–390.
21. Weiner, S. J., Kollman, P. A., Nguyen, D. T., and Case, D. A. (1986) Force field calculations in computational chemistry. *J. Comput. Chem.* **7,** 200–230.
22. Kollman, P. A. (1987) Application of force fields to molecular models. *Annu. Rev. Phys. Chem.* **38,** 303–333.
23. Kumosinski, T. F. and Farrell, H. M., Jr. (1994) Solubility of proteins:protein-salt-water interactions, in *Protein Functionality in Food Systems* (Hettiarachchy, N. S. and Zeigler, G. R., eds.), Marcel Dekker, New York, pp. 39–77.
24. Andersen, H. C. (1980) Molecular dynamics simulations at constant pressure and/or temperature. *J. Chem. Phys.* **72,** 2384–2394.
25. van Gunsteren, W. F. and Berendsen, H. J. C. (1977) Algorithms for molecular dynamics and constraint dynamics. *Mol. Phys.* **34,** 1311–1327.
26. Eigel, W. N., Butler, J. E., Ernstrom, C. A., Farrell, H. M., Jr., Harwalkar, V. R., Jenness, R., and Whitney, R. McL. (1984) Nomenclature of the proteins of cows' milk: 5th Revision. *J. Dairy Sci.* **67,** 1599–1631.
27. Schmidt, D. G. (1982) Association of caseins and casein micelle structure in *Developments in Dairy Chemistry* (Fox, P. F., ed.), pp. 61–86. Applied Science, London, UK.
28. Schmidt, D. G. and Payens, T. A. J. (1976) in *Surface and Colloid Science* (Matijevic, E., ed.), Wiley, New York, pp. 165–229.

29a. Noble, R. W. and Waugh, D. F. (1965) Casein micelles, formation and structure I. *J. Am. Chem. Soc.* **87,** 2236–2245.
29b. Waugh, D. F. and Noble, R. W. (1965) Casein micelles, formation and structure II. *J. Am. Chem. Soc.* **84,** 2246–2257.
30. Creamer, L. K. and Waugh, D. F. (1965) Calcium binding and precipitate solvation of Ca-α_s-caseinates. *J. Dairy Sci.* **49,** 706.
31. Melander, W. and Horvath C. (1977) Salt effects on hydrophobic interactions in precipitation and chromatography of proteins: an interpretation of the lyotropic series. *Arch. Biochem. Biophys.* **183,** 200–215.
32. Sinanoglu, O. (1968) Solvent effects on molecular association, in *Molecular Associations in Biology* (Pullman, B., ed.), Academic, New York, NY, pp. 429–445.
33. Robinson, D. R. and Jencks, W. P. (1965) Effects of concentrated salt solutions on the activity coefficient of acetyltetraglycine ethylester. *J. Am. Chem. Soc.* **87,** 2470–2479.
34. Bingham, E. W., Farrell, H. M., Jr., and Carroll, R. J. (1972) Properties of dephosphorylated α_{s1}-casein. Precipitation by calcium ions and micelle formation. *Biochemistry* **11,** 2450–2454.
35. Waugh, D. F., Slattery, C. W., and Creamer, L. K. (1971) Binding of calcium to caseins. *Biochemistry* **10,** 817–823.
36. Cann, J. R. (1978) Measurements of protein interactions mediated by small molecules using sedimentation velocity, in *Methods in Enzymology XLVIII* (Hirs, C. H. W. and Timasheff, S. N., eds.), Academic Press, New York, NY, pp. 242–248.
37. Dickson, I. R. and Perkins, J. D. (1971) Studies on the interactions between purified bovine caseins and alkaline earth metal ions. *Biochem. J.* **124,** 235–240.
38. Kumosinski, T. F., Brown, E. M., and Farrell, H. M., Jr. (1993) Three dimensional molecular modeling of bovine caseins: an energy-minimized β-casein structure. *J. Dairy Sci.* **76,** 931–945.
39. Alaimo, M. H., Wickham, E. D., and Farrell, H. M., Jr. (1999) Effect of self-association of α_{s1}-casein and its cleavage fractions α_{s1}-casein (136–196) and α_{s1}-casein (1–197), on aromatic circular dichroic spectra: comparison with predicted models. *Biochim. Biophys. Acta* **1431,** 395–409.
40. Ribadeau-Dumas, B. and Garnier, J. (1970) Structure of casein micelle. The accessability of subunits to various reagents. *J. Dairy Res.* **37,** 269–278.
41. Tanford, C. (1967) *Physical Chemistry of Macromolecules.* Wiley, New York.
42. Pessen H., Kumosinski, T. F., Farrell, H. M., Jr., and Brumberger, H. (1991) Tertiary and quaternary structural differences between two genetic variants of bovine casein by small-angle X-ray scattering. *Arch. Biochem. Biophys.* **284,** 133–142.
43. Thurn, A., Burchard, W., and Niki, R. (1987) Studies of α_{s1}-casein by SAXS, and static and dynamic light scattering. *Colloid Polymer Sci.* **265,** 897–903.
44. Curley, D. C., Kumosinski, T. F., Unruh, J. J., and Farrell, H. M., Jr. (1998) Changes in the secondary structure of bovine casein by FTIR. *J. Dairy Sci.* **81,** 3154–3162.
45. Kumosinski, T. F., Pessen, H., Farrell, H. M., Jr., and Brumberger, H. (1988) Determination of the quaternary structural states of bovine casein by small-angle X-ray scattering. *Arch. Biochem. Biophys.* **266,** 548–561.

46. Kumosinski, T. F., King, G., and Farrell, H. M., Jr. (1994) An energy-minimized casein submicelle working model. *J. Prot. Chem.* **13,** 681–700.
47. Kumosinski, T. F., King, G., and Farrell, H. M., Jr. (1994) Comparison of the three-dimensional molecular models of bovine submicellar caseins with small-angle X-ray scattering. Influence of hydration. *J. Prot. Chem.* **13,** 701–714.
48. Richardson, T., Jiminez-Flores, R., Kumosinski, T. F., Oh, S., Brown, E. M., and Farrell, H. M., Jr. (1992) Molecular modeling and genetic engineering of milk proteins, in *Advanced Dairy Chemistry 1: Proteins* (Fox, P. F., ed.), Elsevier, Essex, UK, pp. 545–578.

II

CALCIUM-BINDING PROTEINS: CASE STUDIES

7
Preparation of Recombinant Plant Calmodulin Isoforms

Raymond E. Zielinski

1. Introduction

Calmodulin and the calmodulin-like domain protein kinases (CDPKs) are the most widely distributed EF-hand Ca^{2+}-binding proteins in plants, and serve as the primary receptors for second messenger Ca^{2+} signals *(1)*. Calmodulin is the primary nonenzymatic Ca^{2+}-binding protein in plants, and is the subject of interest because a growing number of the targets of calmodulin regulation in plants are not identical to those found in animals and yeast *(2)*. Two features that complicate the study of calmodulins extracted directly from higher plant tissues, however, is the relatively low yield of protein per unit mass of tissue compared with yields obtained from most animal tissues, and the overlapping patterns of expression of multiple isoforms of the protein. For example, *Arabidopsis thaliana*, the genetically simplest higher plant, possesses at least 10 calmodulin genes encoding a family of seven protein isoforms *(2,3)*.

An additional potential pitfall in purifying plant calmodulins that is not as serious a consideration when working with calmodulin extracted from animal sources is that the plant proteins typically possess a *Cys* residue at position 27. Under the conditions typically employed and times necessary for extracting and purifying sufficient amounts of proteins directly from plant materials, sufficient oxidation can occur to result in the formation of "calmodulin dimers" consisting of two calmodulins linked by a disulfide bond *(4)*. On the other hand, a potential benefit of this change in primary structure found in plant calmodulins is that it facilitates modification of the molecules using sulfhydryl-directed fluorescent *(4,5)* or enzymatic *(6)* reporter moieties. This property makes plant calmodulins particularly useful in studies of protein–protein interaction and for screening cDNA expression libraries for clones encoding calmodulin-binding proteins.

Because the physiological responses to Ca^{2+} depend on perception and binding of the second messenger by complexes of calmodulin and its target proteins *(2,7)*, it is important to be able to examine the interaction of these proteins derived from a single, homologous system to understand how a particular organizm transduces Ca^{2+}-based signals. The availability of cloned cDNA sequences encoding calmodulin isoforms has made this goal readily attainable. This subsection describes a simple method designed to express, extract, and isolate recombinant calmodulin isoforms from *Escherichia coli* with high yield and purity. It has been used successfully to recover five of the seven isoforms of the protein that are expressed in *A. thaliana*.

2. Materials
2.1. Vector Plasmids, Host Strains, and Growth Media
1. Plasmids suitable for expressing the Arabidopsis calmodulin isoforms, Cam-2, -4, and -6, *(8)* are available from the Arabidopsis Biological Resource Center, and can be ordered via the World Wide Web (http://aims.cps.msu.edu/aims/). These plasmids are based on the vector pET5a, which drives the expression of cloned genes by the phage T7 promoter.
2. *E. coli* BL21(DE3) (F- *ompT gal [dcm] [lon] hsdS$_B$* (r_B^- m_B^-; an *E. coli* B strain) with DE3, a λ prophage carrying the T7 RNA polymerase gene) and BL21(DE3) p*Lys*S (same genotype, but with the addition of [p*Lys*S Cam^r] *(9)* are available commercially from Novagen (Madison, WI), Stratagene (La Jolla, CA), or Amersham Pharmacia Biotech (Piscataway, NJ).
3. *E. coli* strains harboring the pETCam expression plasmids are propagated in LB medium (10 g/L bactotryptone, 5 g/L yeast extract, and 10 g/L of NaCl pH adjusted to 7.0 with NaOH) containing 50 µg/mL of ampicillin or carbenicillin; 15 g/L of bactoagar are added for solid growth medium. One-mililiter overnight cultures of bacteria are mixed with equal volumes of sterile, 80% (v/v) glycerol, quick-frozen in liquid N_2, and stored at –80°C for long-term (months to years) preservation.

2.2. Protein Extraction and Initial Purification
1. Lysis Buffer: Bacterial Protein Extraction Reagent (B-PER, Pierce, Rockford, IL) containing 1 m*M* dithiothreitol (DTT), which should be added just prior to use (*see* **Note 1**).
2. Commercial bleach (10%, v/v) should be used to decontaminate glassware used to culture, harvest, and initially process bacteria harboring expression plasmids encoding nontoxic proteins in accordance with the NIH Guidelines.

2.3. Ca^{2+}-Dependent Hydrophobic Affinity Chromatography
1. Hydrophobic chromatography medium and buffers: Phenyl-Sepharose 6 Fast Flow (Amersham Pharmacia Biotech, 17-0973) has been used successfully.

Plant Calmodulin Isoforms

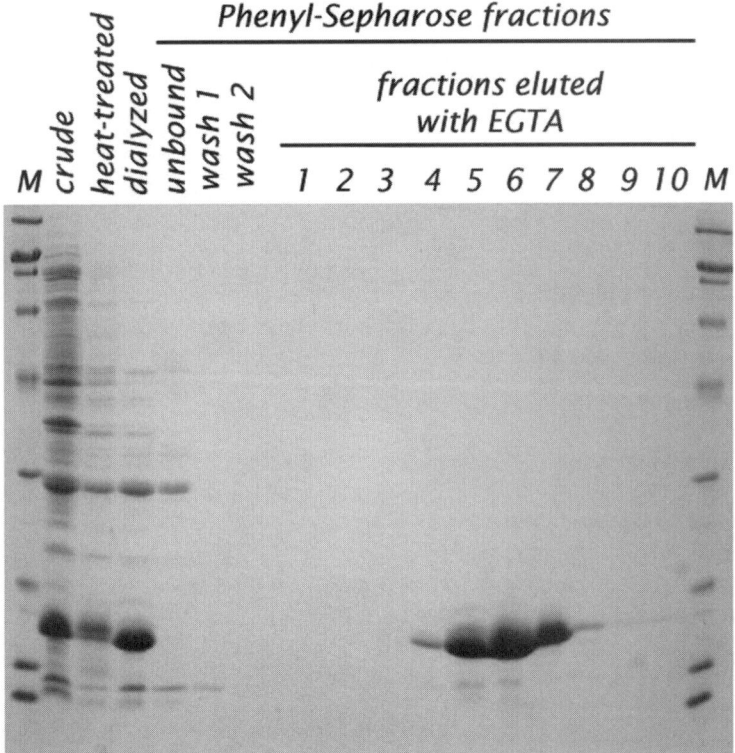

Fig. 1. SDS-gel electrophoretic separation of proteins extracted from induced BL21(DE3) harboring the expression plasmid pET-Cam2 and purified through one cycle of phenyl-Sepharose chromatography.

2. Dialysis buffer: 50 mM Tris-HCl, pH 7.5 containing 100 mM NaCl, and 1 mM DTT. Spectra/Por1 membrane tubing (Fisher Scientific) is suitable for dialyzing solutions of calmodulin. Binding buffer is 50 mM Tris-HCl, pH 7.5 containing 0.5 mM CaCl$_2$, and 0.5 mM DTT.
3. Wash buffer: has the same composition as binding buffer with the addition of 200 mM NaCl.
4. Elution buffer: 50 mM Tris-HCl, pH 7.5 containing 1 mM ethylene glycol-*bis* N,N,N',N'-tetraacetic acid (EGTA) and 0.5 mM DTT (*see* **Note 2**). All buffers in **items 2–4** can be made well in advance as long as DTT is added just prior to use.
5. A vertical minigel apparatus (0.5 to 1-mm thick gels, capable of analyzing 15–20 samples) is convenient for assaying fractions eluted from the hydrophobic affinity column (*see* **Fig. 1**).
6. Stirred ultrafiltration cells that can be driven by compressed O$_2$-free N$_2$, such as the Millipore/Amicon 8000 series (Fisher Scientific), equipped with YM-10 membranes, are useful for concentrating calmodulin protein solutions.

3. Methods
3.1. Host-Vector Systems and Protein Induction

1. Inoculate 1000 mL of fresh LB-Amp$_{50}$ with 10 mL of a fresh overnight culture and grow at 37°C with vigorous shaking (300 rpm) until the cells reach midlog phase (A_{600} = 0.6 to 0.8) (*see* **Note 3**).
2. Add IPTG to a final concentration of 0.4 mM (0.1 g/L) and grow an additional 3 to 4 h at 37°C.
3. Collect the induced cells by centrifuging at 5000g for 10 min in a sealed container to meet the NIH Guidelines for expressing nontoxic recombinant proteins.
4. Wash the pelleted cells once by suspending them in 200 mL of 50 mM Tris-HCl, pH 7.5, and collect the bacteria by centrifugation.

3.2. Protein Extraction and Initial Purification

1. Resuspend the cells in 50 mL of lysis buffer (*see* **Note 4**).
2. Incubate and lyse the bacteria by gently rocking the suspension at room temperature for 30 min (*see* **Note 5**).
3. Clarify the homogenate by centrifugation at 27,000g for 30 min at 4°C.
4. Heat the protein extract at 85°C for 3 min, then immediately chill the solution in an ice-water slurry (*see* **Note 6**).
5. Clarify the homogenate at 27,000g for 30 min at 4°C.
6. Dialyze 1 h vs 4 to 5 L of double-distilled (dd) H$_2$O, then 1 h vs 4 to 5 L of dialysis buffer.
7. Remove any particulate matter by centrifuging the dialysate at 27,000g for 15 min at 4°C.

3.3. Phenyl-Sepharose Chromatography

1. Add CaCl$_2$ to the dialyzed extract to a final concentration of 5 mM with mixing from a 100 mM solution. Apply the sample to a phenyl-Sepharose column (10- to 15-mL bed volume for protein from 1 L of induced cells) equilibrated with binding buffer.
2. Wash the column with binding buffer until the A_{280} < 0.01. Usually five column volumes of buffer are sufficient.
3. Wash again with wash buffer until the A_{280} approx 0. As in the previous step, five column volumes are normally sufficient.
4. Elute calmodulin from the column with elution buffer and collect fractions equal to about 25% of the total volume of the column.
5. Assay fractions eluted from the column by SDS-PAGE to identify the fractions containing the highest levels of calmodulin (*see* **Fig. 1** and **Note 7**). Pool the peak fractions and dialyze vs 1000 vol of 20 mM NH$_4$HCO$_3$ or double-deionized H$_2$O.
6. Lyophilize or store as dialyzed liquid (after quick-freezing in liquid N$_2$) at –80°C. If the concentrations are lower than desired, ultrafiltration in a stirred cell driven by O$_2$-free N$_2$ works well to concentrate the protein with minimal loss and minimal oxidation. To prepare solutions of lyophilized calmodulin, use either dilute

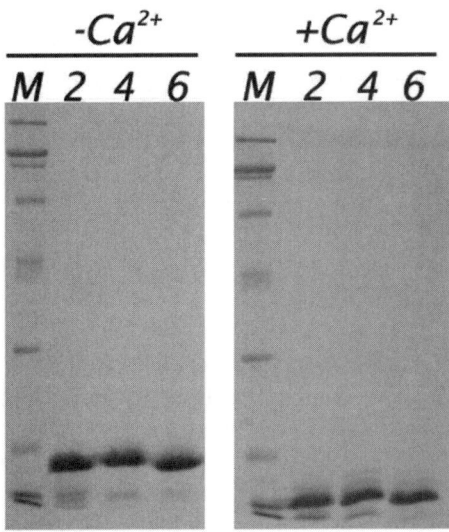

Fig. 2. SDS-gel electrophoretic separation of 5-µg portions of Arabidopsis Cam isoforms 2, 4, and 6 purified according to the method described illustrating their Ca^{2+}-induced mobility shift.

buffer at neutral pH or double-deionized H_2O to which "a few" microliters of 1 M Tris base are added to neutralize the acidic protein and dissolve the dried residue.

7. Measure protein concentration by Lowry assay or by titration of the single, free sulfhydryl in plant CaM proteins with Ellman's reagent (no DTT or 2-mercaptoethanol can be used for storage in this case). These methods normally agree within +/– 5–10%. Typical yields of *purified* protein have ranged from 15 to 50 mg/L of induced bacteria that are 90 to 95% pure as judged by SDS-PAGE (*see* **Fig. 2**).

4. Notes

1. Pierce B-PER cannot be used in conjunction with acid precipitation, which has been used previously in several methods for preparing CaM (e.g., **ref. 8**), as a massive precipitate results when the pH of the solution is lowered to 4.0.
2. In order to retard potential oxidation of plant calmodulins, buffers for dialysis and hydrophobic chromatography are typically deaerated for 5 min (a water aspirator is sufficient) and gassed for 5 to 10 min with a stream of O_2-free N_2 just prior to use. Calmodulin eluted from phenyl-Sepharose is stable for several days at 4°C, particularly if the buffers used in chromatographic separation were gassed with O_2-free N_2.
3. Bacterial strains harboring expression plasmids should be tested for stability when they are first constructed, just prior to inducing recombinant protein production (i.e., when they have grown to an A_{600} of 0.6–0.8). This involves plating cells

diluted 10^5-fold on LB-agar plates containing (a) 0.4 mM IPTG and antibiotic or (b) 0.4 mM IPTG, and plating cells diluted 2×10^6-fold on LB-agar plates containing (c) antibiotic or (d) no additives. All viable cells will grow on plate (d), and cells that have retained the expression plasmid will grow on plate (c). In stable strains harboring plasmids expressing proteins from the T7 promoter, such as pETCam-2, -4, and -6, less than 2% of the cells will grow on plate (b), and less than 0.01% of the cells will grow on plate (a).
4. Freezing the induced bacterial suspension in lysis buffer is a convenient point at which to stop the procedure. Overnight storage at –20°C causes no apparent harm to the protein; freezing at –80°C is used for longer-term storage (up to at least 1 mo).
5. This incubation permits bacterial lysis, but must be done gently to avoid aerating the suspension and oxidizing the free sulfhydryl in plant calmodulin. Slow rotation on an orbital shaker or on a rotating wheel in a tightly sealed container work well. Lysis should also mildly shear the bacterial DNA, which will permit the homogenate to be clarified by centrifugation. This can be determined by testing whether the suspension will form drops at the end of a Pasteur pipet after incubation in lysis buffer. If drops do not form, pass the solution once or twice through a 60-cm^3 syringe equipped with an 18- or 20-gage needle by positive pressure to avoid aerating the extract.
6. Heat treatment should be carried out in small portions in thin-walled glass or plastic tubes that can be rapidly transferred into a water bath held at 85°C to ensure rapid and efficient heat distribution. Alternatively, the extract can be heated in a flask sufficiently large to hold the protein solution in a layer less than 1 cm deep. Rapid chilling is accomplished by pouring the heat-treated extract into centrifuge tubes held in an ice-water slurry.
7. Absorbance measurements of plant calmodulin concentration are insensitive because of their extremely low content of aromatic amino acids. Furthermore, a common method for calculating molar extinction coefficients based on amino acid composition *(10)* is inaccurate for quantifying the concentrations of plant calmodulins compared with Lowry and Bradford methods and titration with Ellman's reagent for protein estimation. A convenient and sensitive alternative is to assay 10-µL aliquots of each fraction by SDS-PAGE in a 15% acrylamide minigel and detecting the protein by Coomassie blue staining (*see* **Fig. 1**).

References

1. Roberts, D. M. and Harmon, A. C. (1992) Calcium-modulated proteins: targets of intracellular calcium signals in higher plants. *Annu. Rev. Plant Physiol. Plant Mol. Biol.* **43,** 375–414.
2. Zielinski, R. E. (1998) Calmodulin and calmodulin-binding proteins in plants. *Annu. Rev. Plant Physiol. Plant Mol. Biol.* **49,** 697–725.
3. Zielinski, R. E. (2001) Characterization of three new members of the *Arabiodopsis thaliana* calmodulin gene family: conserved and highly diverged members of the gene family functionally complement a yeast calmodulin null (submitted).

4. Zot, H. G., Aden, R., Samy, S., and Puett, D. (1990) Fluorescent adducts of wheat calmodulin implicate the amino-terminal region in the activation of skeletal muscle myosin light chain kinase. *J. Biol. Chem.* **265,** 14,796–14,801.
5. Mills, J. S., Walsh, M. P., Nemcek, K., and Johnson, J. D. (1988) Biologically active fluorescent derivatives of spinach calmodulin that report calmodulin target protein binding. *Biochemistry* **27,** 991–996.
6. Liao, B., and Zielinski, R. E. (1995) Production of recombinant plant calmodulin and its use to detect calmodulin-binding proteins. *Methods Cell Biol.* **49,** 487–500.
7. Vogel, H. J. (1994) Calmodulin: a versatile calcium mediator protein. *Biochem. Cell Biol.* **72,** 357–376.
8. Liao, B., Gawienowski, M. C., and Zielinski, R. E. (1996) Differential stimulation of NAD kinase and binding of peptide substrates by wild-type and mutant plant calmodulin isoforms. *Arch. Biochem. Biophys.* **327,** 53–60.
9. Studier, F. W., Rosenberg, A. H., Dunn, J. J., and Dubendorf, J. W. (1990) Use of T7 RNA polymerase to direct expression of cloned genes. *Methods Enzymol.* **185,** 60–89.
10. Gill, S. C., and von Hippel, P. H. (1989) Calculation of protein extinction coefficients from amino acid sequence data. *Anal. Biochem.* **182,** 319–326.

8

Isolation of Recombinant Cardiac Troponin C

John A. Putkey and Wen Liu

1. Introduction

Troponin C (TnC) is a founding member of the EF-hand family of Ca^{2+} binding proteins. It functions as the receptor for Ca^{2+} on the thin filament of striated muscle. Association of Ca^{2+}, with the regulatory Ca^{2+} binding sites in TnC, induces muscle contraction. Subsequent release of Ca^{2+} allows the muscle to relax. There are two forms of TnC; one is found in fast skeletal muscle (sTnC), whereas the other is found in cardiac and slow skeletal muscle (cTnC).

Both isoforms of TnC are present in striated muscle as subunits of a stable heterotrimeric troponin complex, which is bound to the thin filament of myofibrils. Isolation of TnC from tissue typically involves preparation of a crude troponin complex and final isolation of the three individual troponin subunits (sTnC, sTnI, and sTnT). A variety of procedures have been reported for the isolation of sTnC *(1–6)*. Similar procedures have been adapted for cTnC *(2,7,8)*. The proliferation of recombinant expression systems provides a convenient alternative for the easy purification of large quantities of cTnC necessary for structural characterization and other studies. This chapter will provide a detailed procedure for the isolation of recombinant cTnC from bacteria. Bacterially synthesized cTnC isolated using this procedure has been used extensively for functional, biophysical, and structural studies *(9–12)*. A bacterial expression plasmid that encodes chicken cTnC is available from the authors upon request. The primary sequences of human and chicken cTnC differ by only three conservative amino acids.

Like many EF-hand Ca^{2+} binding proteins, cTnC is small and acidic. It binds three moles Ca^{2+} per mole protein and thus has a significant decrease in net negative charge upon binding Ca^{2+} at neutral pH. However, cTnC does not expose large hydrophobic surfaces upon binding Ca^{2+} as does calmodulin and skeletal troponin C. This minimizes the usefulness of Ca^{2+}-dependent hydrophobic interaction chromatography for the isolation of cTnC. The procedure described here will allow the purification of at least 200–300 mg of cTnC, and has proved effective for a variety of cTnC proteins with different mutations. It involves sequential anion exchange chromatography in the absence of Ca^{2+}, ammonium sulfate precipitation, and high-performance liquid chromatography (HPLC) anion exchange chromatography in the presence of Ca^{2+}. A useful application of hydrophobic interaction chromatography is also discussed.

2. Materials
2.1. Cell culture

1. Incubator for bacterial culture plates.
2. New Brunswick floor-model environmental shaker capable of holding at least four 4-L flasks.
3. Sorvall RC-3 centrifuge and Sorvall HG-4L rotor.
4. L-broth: 10 gm/L Bacto-tryptone, 5 gm/L Bacto-yeast extract, 7.5 gm/L NaCl.
5. Ampicillin 50 mg/mL in water, and chloramphenicol 34 mg/mL in ethanol.
6. Isopropyl thio-β-D-galactoside (IPTG), 400 mM in water.

2.2. Cell Lysis and Preparation of a Soluble Fraction

1. –70°C freezer.
2. Branson Sonifier 450 sonicator.
3. Beckman L5-50 ultracentrifuge and Beckman Ti45 rotor.
4. Spectrum Fleaker filtration unit and spectra/mesh macroporous filter paper.

2.3. Chromatography and Ammonium Sulfate Precipitation

1. Perkin Elmer Biocompatible binary pump 250 or comparable system.
2. Waters HPLC system with two or three 510 pumps and automated gradient controller or comparable system.
3. Kontes shielded glass column (2.5 × 30 cm) packed with Bio-Rad Macro-Q anion exchange resin.
4. Waters protein-Pak semipreparative diethylaminoethyl (DEAE) HPLC column (21.5 mm × 15 cm).
5. Kontes shielded glass column (2.5 × 30 cm) packed with Pharmacia/LKB high-sub phenyl-Sepharose.
6. Hoefer SE 250 Mighty Small II minigel caster and running units.
7. Sorvall RC-5B refrigerated super-speed centrifuge and Sorvall GSA rotor.
8. Low-salt buffer: 50 mM Tris-HCl, pH 7.5, 0.2 mM ethylenediaminetetracetic acid (EDTA), 14 mM 2-mercaptoethanol (add just before use). All buffers should be prepared with Milli-Q water and filtered either with Spectramesh Macroporous filter paper for low-pressure columns, or 0.45-μm filters for HPLC use.
9. Low-salt/Ca^{2+} buffer: Low-salt buffer made 5 mM in $CaCl_2$.

Cardiac Troponin C

10. High-salt buffer: Low-salt buffer made 1.0 M in KCl.
11. High-salt/Ca^{2+} buffer: High-salt buffer made 5 mM in CaCl$_2$.
12. HIC high-salt buffer: Low-salt buffer made 500 mM in (NH$_4$)$_2$SO$_4$.

3. Methods
3.1. Expression Plasmids

Our initial attempts to express cTnC in bacteria using an unmodified chicken cDNA failed to yield significant levels of protein. Expression was greatly enhanced by conversion of the first 11 codons of the cTnC cDNA to bacterial codon usage, and the second codon from GAT (*Asp*) to GCT (*Ala*) *(9)*. Similar to the bacterial expression of calmodulin *(13)*, *Ala* in this second amino acid position promotes post-translational removal of the N-terminal *Met* residue. Thus, bacterially synthesized chicken cTnC, which we call cTnC3, is missing the N-terminal *Met* found in tissue derived cTnC.

The original cTnC expression plasmid, pTnCPL3, utilized a λ phage P_L promoter. More recently, we constructed a highly efficient expression vector, *pTnC3T7*, which uses the T7 promoter system, by subcloning the small *Nco*I/*Hin*dIII fragment from pTnCPL3 into the large *Nco*I/*Hin*dIII fragment from *pET23D*(+) (Novagen). This parent expression plasmid has proved effective for the expression of numerous mutated derivatives of cTnC. Of particular interest is an expression plasmid that encodes cTnC(A-*Cys*), in which both *Cys* 35 and 84 are converted to *Ser* (*see* **Note 1**). All culture conditions described below are designed for the *pTnC3T7* vector. This plasmid is available from the authors upon request.

3.2. Small Scale Test Expressions

The *Escherichia coli* strain XL1 Blue (Stratagene) is used as a host for routine maintenance of *pTnC3T7*. For overexpression of cTnC3, *pTnC3T7* is transferred to *E. coli* strain BL21(DE3) *pLysS* (Novagen). In general, we find that the highest levels of cTnC3 expression are obtained with cells that have been freshly transfected with the expression plasmid. This subheading describes the procedure to identify transformed bacteria that efficiently express cTnC3.

1. *pTnC3T7* is transferred to transfection-competent BL21(DE3) *pLysS* cells. The cells are spread on L-plates supplemented with 100 µg/mL of ampicillin and 34 µg/mL chloramphenicol, and allowed to incubate overnight at 37°C.
2. Five to ten individual antibiotic-resistant colonies are picked and used to inoculate 1- to 5-mL aliquots of L-broth supplemented with 100 µg/mL of ampicillin and 34 µg/mL chloramphenicol. The cultures are allowed to incubate overnight at 37°C.
3. The following day 0.1 mL of each culture is used to inoculate 5 mL volumes of L-broth with antibiotics. These are allowed to grow for 3–4 h at which time two

2-mL aliquots are removed. One aliquot is made 0.4 mM in IPTG to induce protein production. The induced and control samples are incubated for an additional 2 h, whereas the remainder of the 5-mL culture is placed on ice.
4. One microliter of the induced and control cultures are transferred to separate microfuge tubes. The cells are collected by centrifugation and lysed by addition of 0.1 mL of 1X sodium dodecyl sulfate (SDS) sample buffer followed by heating for 5 min at 95°–100°C.
5. Approximately 0.02 mL of lysate is analyzed by sodium dodecyl sulfate-polyacrylamide gel electrophoresis (SDS-PAGE) to compare the presence of cTnC in induced and control cells (*see* **Note 2**). The culture that exhibits the highest levels of cTnC production is selected for large-scale expression of cTnC.

3.3. Large Scale Expression

1. Approximately 0.1 mL of culture selected in **Subheading 3.2.** is used to inoculate 50–250 mL of L-broth with antibiotics. This is allowed to incubate with shaking overnight at 37°C.
2. The overnight culture is used to inoculate 2 L of L-broth, with antibiotics, in a 4-L flask. Use 20–25 mL of overnight culture per liter. A typical preparation will use a total of 4 L of L-broth (*see* **Note 3**).
3. The cultures are allowed to incubate at 37°C with adequate shaking and aeration until the absorbance at 550 nm reaches about 0.8–1.0. IPTG is then added to a concentration 0.4 mM for induction of protein expression. The cultures are allowed to incubate with shaking for an additional 2–3 h.
4. Collect cells by centrifugation using 1-L bottles and a Sorvall RC-3 with a HG-4L rotor. It is most convenient to use only two bottles and repetitive centrifugation runs to collect the cells. The final cell pastes in the two bottles are then transferred to one bottle. The cells can be frozen or processed immediately.

3.4. Cell Lysis and Preparation of Soluble Fraction

1. The cell pellet from **Subheading 3.3.** above is subjected to 2–3 freeze/thaw cycles to facilitate cell breakage (*see* **Note 4**). This can be performed in the 1-L centrifugation bottle.
2. About 30 mL of low-salt buffer is added per liter of original culture. The cell suspension is sonicated on ice by five 2-min bursts at 50% duty cycle.
3. The cell lysate is transferred to 70-mL polycarbonate ultracentrifuge tubes and centrifuged for 60 min at 142,000g in a Ti45 rotor held at 4°C. The soluble fraction is filtered through a Spectramesh macroporous filter paper to remove residual insoluble matter.

3.5. Macro-Q Anion Exchange Chromatography

This first chromatography step uses Bio-Rad Macro-Q anion exchange resin packed into a Kontes 2.5 × 30 cm column equipped with an upper-flow adapter. We have found that this resin provides the best combination of reproducibility, ruggedness and high flow rates. The resin should be transferred to low-salt buffer, chilled and degassed prior to packing the column. All buffers used for the Macro-Q column contain EDTA. The procedure described here was optimized for the pump system and column described below and in **Subheading 2**. The flow rates have been determined to be maximal without exceeding the pressure limits of the columns. Sample loading, solvent switching, and fraction collector operation is controlled by TTL signals from the pump's microprocessor. Column buffers run from reservoirs inside a chromatography cabinet to the Perkin Elmer binary pump, located just outside the cabinet, and back to the column, detector, and fraction collector mounted inside the cabinet.

1. Equilibrate the column with low-salt buffer at a flow rate of 7 mL/min to establish a baseline absorbance at 280 nm. Load the soluble filtered cell-lysate fraction described in **Subheading 3.4.** at the same flow rate. Wash with three column volumes of low-salt buffer or until the absorbance of the effluent at 280 nm has reached baseline.
2. Bound protein is eluted with a linear 0–500 mM KCl gradient starting with low-salt buffer and ending with 50% high-salt buffer. Typically the total volume of the gradient is three times the column volume. The column is then stripped using about three column volumes of 70% high-salt buffer (700 mM KCl) and equilibrated with low-salt buffer (*see* **Note 5**).
3. Gradient fractions of about 14–15 mL are collected. Fractions containing cTnC are identified by SDS-PAGE (15% acrylamide; 29.2:0.8 acrylamide to *bis*-acrylamide).
4. **Figure 1** shows a typical protein elution profile of a Macro-Q column, and an SDS gel of gradient fractions containing cTnC3. Cardiac TnC3 usually elutes just before the final large peak of UV absorbing material. This peak does not contain significant protein, and is likely nucleic acid. Occasionally, especially if bacteria are grown in minimal media rather than L-both, expression of cTnC3 will be low. Under these circumstances, prominent peaks caused by cTnC may not be observed since cTnC, which has no *Trp* and only two *Try* residues, absorbs poorly in the UV region.

3.6. Ammonium Sulfate Precipitation

We have found that ammonium sulfate precipitation to be a very useful purification tool. We do not, however, precipitate cTnC3. Rather, the procedure precipitates most proteins in pooled fractions from the Macro-Q column while leaving cTnC3 soluble.

1. The pooled fractions from Macro-Q column that contain cTnC are made 3 M NH_4SO_4 by the slow addition of solid NH_4SO_4 (396.5 gm/L) with stirring.
2. Precipitated protein is removed by centrifugation at 10,000g for 20 min in plastic centrifuge bottles (Sorvall GSA rotor).

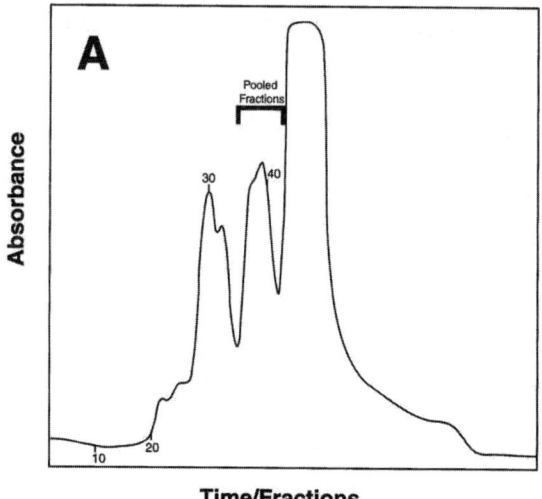

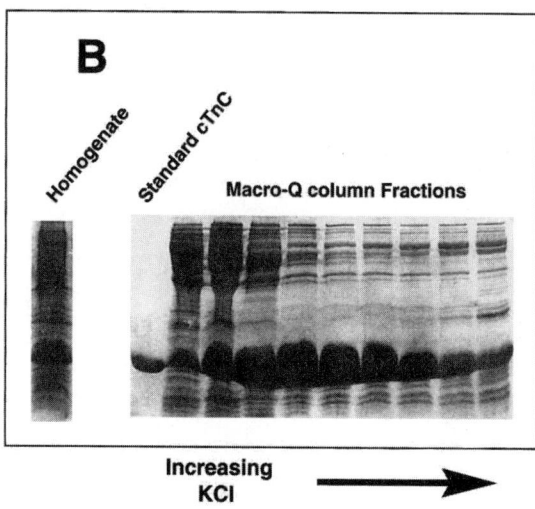

Fig. 1. UV profile of protein eluted from the Macro-Q anion exchange column (**A**), and identification of cTnC3 in eluted fractions (**B**). Cardiac TnC3 in the homogenate and a purified cTnC3 are provided for reference.

3. The soluble fraction is then dialyzed against 8 L (2 × 4 L) of low-salt/Ca^{2+} buffer.
4. The final dialyzed sample is filtered through Gelman 0.45-μm filter.

3.7. DEAE-HPLC Chromatography

The protein sample obtained after the Macro-Q column and ammonium sulfate precipitation is highly enriched in cTnC3. Occasionally, the purity at this

Cardiac Troponin C

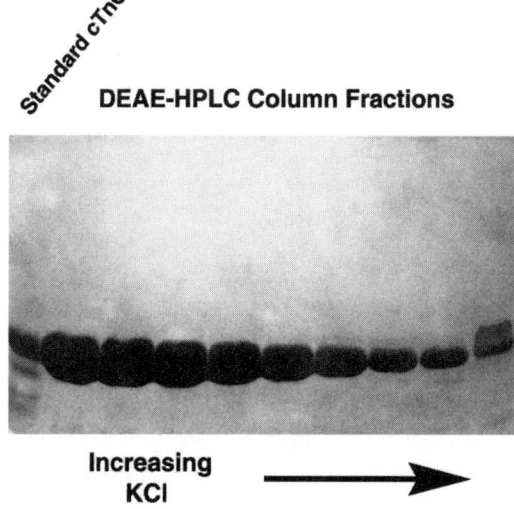

Fig. 2. SDS gel of cTnC3 present in eluted fractions from the DEAE-HPLC column.

stage is sufficient for desired experiments without additional purification. The most common contaminates found at this stage of the purification are larger molecular-weight proteins and UV-absorbing compounds that can not be attributed to protein. Chromatography on a DEAE-HPLC column in the presence of Ca^{2+} effectively removes these contaminants. It also serves to concentrate the cTnC3. The HPLC is configured with low-salt/Ca^{2+} buffer delivered through pump A, and high-salt/Ca^{2+} buffer delivered through pump B.

1. Equilibrate the column with low-salt/Ca^{2+} buffer at flow rate of 5 mL/min. Load the filtered sample from **Subheading 3.6.** We typically load the sample directly through pump A, or through a third pump dedicated for loading large volumes of sample.
2. Wash with three column volumes of low-salt/Ca^{2+} buffer plus 100 mM KCl (90% pump A and 10% pump B) or until baseline absorbance is achieved.
3. Protein is eluted with a linear 100 to 500 mM KCl gradient.
4. Strip the column with about three column volumes 0.7 M KCl and equilibrate with low-salt/Ca^{2+} buffer.
5. Gradient fractions containing cTnC3 are identified by SDS-PAGE and pooled (*see* **Fig. 2**).

3.8. Phenyl-Sepharose Hydrophobic Interaction Chromatography (HIC)

As aforementioned, we have found that HIC is not an efficient step for the routine purification of cTnC3. It is, however, useful if significant proteolysis of cTnC3 occurs. Discrete proteolytic fragments of cTnC3, which are very likely the N- and C-terminal domains, behave similarly to the intact protein on anion exchange columns, and are thus co-purified with cTnC3. We have found that these proteolytic fragments can be separated from cTnC3 by virtue of their differential elution from phenyl-Sepharose in the presence of EDTA. If necessary, pooled fractions from the DEAE column can be processed as described below.

1. Equilibrate the column with HIC high-salt buffer at flow rate of 7 mL/min.
2. Fractions from the DEAE column are made 500 mM in ammonium sulfate, warmed to room temperature, and loaded on the column. Proteins do not salt out at this concentration of ammonium sulfate.
3. Wash the column with about two-column volumes of HIC high-salt buffer and elute-bound proteins with a three-column volume linear gradient of 500–0 mM ammonium sulfate (100% HIC high-salt buffer to 100% low-salt buffer). Generally, the UV elution profile of eluted proteins will show only one peak. The presence and purity of cTnC3 is confirmed by SDS-PAGE.

3.9. Final Sample Preparation and Storage

The final purified protein can be concentrated to a desired volume using an Amicon ultrafiltration unit with 10,000 MWCO filter and dialyzed against a desired buffer. Typically, we dialyze against either 20 mM ammonium bicarbonate, pH ≈ 7.5, for subsequent lyophilization, or against 50 mM MOPS, pH 7.0. Decalcification can be accomplished by passing the protein solution over a small column containing Calcium Sponge polystyrene resin (Molecular Probes).

The UV absorbance spectra of the final purified protein should be recorded as a general assessment of purity and to determine protein concentration. Cardiac TnC has no *Trp*, only three *Tyr*, and nine *Phe* residues. This gives the protein a very diagnostic UV profile shown in **Fig. 3**, which has a peak at 276 nm contributed by *Tyr* and four small peaks or splittings between 250 and 276 nm that are contributed by *Phe*. The UV-absorbance spectra of purified cTnC should reach baseline at 300 nm. Contamination of the protein sample should be suspected if the small *Phe* splittings are not observed or if baseline absorbance is not achieved at 300 nm. The concentration of purified cTnC can be determined using the absorbance at 276 nm and a concentration extinction coefficient of 0.25 mg^{-1}/mL^{-1}, which we have derived by amino acid analysis of highly purified cTnC3.

4. Notes

1. The two endogenous *Cys* residues in cTnC can readily form inter- and intramolecular disulfide bonds, which alter the activity of the protein *(19)*. Various disul-

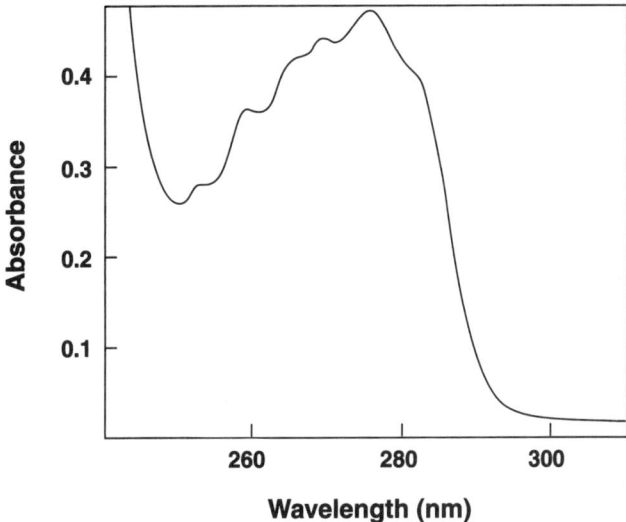

Fig. 3. UV spectra of purified cTnC3.

fide bonded forms of cTnC also have slightly different chromatographic properties, such as split peaks on HIC columns or broadened peaks on ion exchange. Inclusion of dithiothreitol (DTT) or BME will normalize the chromatographic anomalies, but must also be included in all solutions used for biochemical and biophysical characterization of the proteins. We have found that cTnC(A-*Cys*) retains the biological activity of cTnC3 and eliminates the need for reducing reagents. This is particularly advantageous for NMR and crystallographic studies.

2. The bacterial lysate at this stage may be very viscous because of chromosomal DNA, especially after cooling the lysed cells, making it difficult to load on the gel. The viscosity can be decreased by brief sonication for 5 s. EF-hand Ca^{2+}-binding proteins migrate faster on SDS-PAGE in the presence of Ca^{2+}. We typically include 5 mM EDTA in the SDS sample buffer to ensure that cTnC3 appears as a single band on the gel.

3. Occasionally more than 4 L of culture will be prepared for protein isolation. In general, the procedure can be scaled up directly. However, the semipreparative DEAE HPLC column can get saturated with protein leading to a loss of cTnC3. The fall-through from the HPLC column should be checked using SDS-PAGE to ensure that it contains no cTnC3.

4. BL21 (DE3) *pLys* cells are engineered to produce intracellular lysozyme, which binds to basal T7 polymerase and further represses expression of exogenous proteins in the absence of IPTG. *Lys*ozyme released by the freeze/thaw cycles hydrolyzes the bacterial cell wall and promotes efficient lysis during sonication. If another cell line is used for expression, lysozyme should be added prior to sonication.

5. The Macro-Q column should be stripped with 0.1 N NaOH after every 4–5 uses. Eventually, it will lose efficiency over time because of a variety of factors. Acceptable performance can be recovered by removing and repacking the resin. Ultimately, a poorly performing resin will need to be discarded.

References

1. Greaser, M. L. and Gergely, J. (1973) Purification and properties of the components from troponin. *J. Biol. Chem.* **248,** 2125–2133.
2. Potter, J. D. (1982) Preparation of troponin and its subunits. *Methods Enzymol.* **85,** 241–263.
3. Cox, J. A., Comte, M., and Stein, E. A. (1981) Calmodulin-free skeletal-muscle troponin C prepared in the absence of urea. *Biochem. J.* **195,** 205–211.
4. Thulin, E. and Vogel, H. J. (1988) Purification of rabbit skeletal muscle troponin C. *Acta Chem. Scand.* **B42,** 211–215.
5. Tanokura, M., Imaizumi, M., Yamada, K., Shiraishi, F., and Ohtsuki, I. (1992) Preparation and characterization of troponin C from bullfrog skeletal muscle. *J. Biochem. (Tokyo)* **112,** 800–803.
6. Miegel, A., Kobayashi, T., and Maéda, Y. (1992) Isolation, purification and partial characterization of tropomyosin and troponin subunits from the lobster tail muscle. *J. Muscle Res. Cell Motil.* **13,** 608–618.
7. Brekke, C. J. and Greaser, M. L. (1976) Separation and characterization of the troponin components from bovine cardiac muscle. *J. Biol. Chem.* **251,** 866–871.
8. Burtnick, L. D., McCubbin, W. D., and Kay, C. M. (1975) Molecular and biological studies on cardiac muscle calcium-binding protein (TN-C). *Canada J. Biochem.* **53,** 15–20.
9. Putkey, J. A., Sweeney, H. L., and Campbell, S. T. (1989) Site-directed mutation of the trigger calcium-binding sites in cardiac troponin C. *J. Biol. Chem.* **264,** 12,370–12,378.
10. Negele, J. C., Dotson, D. G., Liu, W., Sweeney, H. L., and Putkey, J. A. (1992) Mutation of the high affinity calcium binding sites in cardiac troponin C. *J. Biol. Chem.* **267,** 825–831.
11. Lin, X., Dotson, D. G., and Putkey, J. A. (1996) Covalent binding of peptides to the N-terminal hydrophobic region of cardiac troponin C has limited effects on function. *J. Biol. Chem.* **271,** 244–249.
12. Sia, S., Li, M. X., Spyracopoulos, L., Gagne, S. M., Liu, W., Putkey, J. A., and Sykes, B. D. (1997) NMR structure of cardiac troponin C reveals an unexpected closed regulatory domain. *J. Biol. Chem.* **272,** 18,216–18,221.
13. Putkey, J. A., Slaughter, G. R., and Means, A. R. (1985) Bacterial expression and characterization of proteins derived from the chicken calmodulin cDNA and a calmodulin processed gene. *J. Biol. Chem.* **260,** 4704–4712.
14. Putkey, J. A., Dotson, D. G., and Mouawad, P. (1992) Formation of an intramolecular disulfide bond renders cardiac troponin C calcium independent. *FASEB J.* **6,** A281.

9

Skeletal Muscle Troponin C

Expression and Purification of the Recombinant Intact Protein and Its Isolated N- and C-Domain Fragments

Joyce R. Pearlstone and Lawrence B. Smillie

1. Introduction

The use of skeletal muscle troponin C (TnC) in structural and functional studies ranging from characterization of its Ca^{2+}-binding properties to interaction with troponin I (TnI) and troponin T (TnT), the other two members of the tripartite troponin regulatory complex, has necessitated the purification of large quantities of protein. The procedures for purification of the naturally occurring troponin components from muscle tissue, including preparation of a muscle ether powder, isolation of the crude troponin complex, and further fractionation using column chromatography to obtain the individual subunits, have been described in detail by Potter (1), and therefore, will not be discussed in this chapter. Owing to the problems with heterogeneity arising from copurification of different isoforms and/or variation in the extent of phosphorylation of the final isolated protein from muscle tissue, methodologies have since been developed using molecular biology strategies to express and purify homogeneous preparations of TnC, TnI, and TnT. Other advantages of these recombinant techniques include the ability of producing site-specific mutants and isolated fragments for studying specific regions without interference from the rest of the molecule, as well as ^{15}N- and ^{13}C-labeled derivatives for NMR studies.

Over the past decade, we have been developing procedures that result in consistently high levels of protein expression, while simultaneously streamlining the number of steps used for the purification of intact TnC and its isolated

N (residues 1–90) and C (residue 88–162) domains in high yields. The protocols that have proven to give the best results for purification of these three TnC products will be the focus of this chapter, including transformation and growth of the expression vector, preparation of acetone powder from cell pellet (based on a modification of Potter's ether powder protocol *(1)* for muscle tissue), subsequent extraction of the crude TnC protein, and final purification using affinity chromatographic, ion exchange, and gel exclusion column methods.

2. Materials
2.1. Transformation Procedure

1. A stock of Ca^{2+}-competent *Escherichia coli* BL21(DE3)p*Lys*S cells is prepared every 1–2 yr, and 250-μL aliquots stored in 1.5-mL sterile conical tubes at –80°C. Just prior to use, one tube per transformation is removed and thawed on ice.
2. An environmental shaker equipped with a rack for tubes, set at 37°C and approx 225 rpm.
3. SOC is SOB medium containing 20 m*M* glucose *(2)* prepared as follows: dissolve 2 g bacto-tryptone, 0.5 g bacto-yeast extract and 0.05 g NaCl in 95 mL deionized H_2O. Add 1 mL 250 m*M* KCl and adjust pH to 7.0 with 5N NaOH. Make final volume to 100 mL with deionized H_2O and autoclave (*see* **Note 1**) on liquid cycle. Add 0.5 mL sterile 2 m*M* $MgCl_2$ plus 2 mL sterile 1 *M* glucose.
4. LB plate *(2)* supplemented with ampicillin and chloramphenicol, prepared as follows: dissolve 10 g bacto-tryptone, 5 g bacto-yeast extract, and 10 g NaCl in 950 mL deionized H_2O. (Adding 1 mL 1% vitamin B_1 is optional, but beneficial to cell growth). Adjust pH to 7.0 with 5 *N* NaOH and make volume to 1 L with deionized H_2O. Add 15 g bacto-agar (which dissolves during autoclaving) and autoclave (*see* **Note 1**). Cool the LB medium to 50°C before adding the thermolabile antibiotics ampicillin (1 mL of filter-sterilized stock 100 mg/mL solution in H_2O, final concentration = 100 μg/mL) and chloramphenicol (0.5 mL of unsterilized stock 50 mg/mL solution in ethanol, final concentration = 25 μg/mL). Pour approx 30–35 mL of supplemented medium per plate, flame the surface to remove any air bubbles, and leave plates covered until the agar hardens. Incubate plates at 37°C overnight. Invert and store at 4°C.
5. Incubator oven set at 37°C.
6. Set of three variable volume pipets (20 μL, 200 μL, and 1000 μL).

2.2. Test Expression and Preparative 8 L Cell Culture

1. Each 1 L of 2 × TY medium *(2)* is prepared as follows: dissolve 16 g bacto-tryptone, 10 g bacto-yeast extract, and 5 g NaCl in 900 mL deionized H_2O. Adjust pH to 7.0 with 5 *N* NaOH and volume to 1 L. Autoclave (*see* **Note 1**). Supplement with ampicillin and chloramphenicol as aforementioned.
2. Spectrophotometer set at 600 nm wavelength to monitor cell growth, and cuvet.
3. IPTG (isopropyl β-D-galactopyranoside), 0.4 *M* stock solution in deionized H_2O, filter-sterilized and stored in 1-mL (for test expression) and 8-mL (for 8-L cell culture) aliquots at –20°C.

4. Centrifuges: (a) Eppendorf bench centrifuge set at maximum speed, and 1.5-mL conical tubes; (b) Damon/IEC DPR/1600 centrifuge or equivalent with capacity for 6X 1 L bottles; (c) Sorvall RC5C centrifuge with GSA rotor, and 4×250 mL polypropylene bottles.
5. The apparatus used for sodium dodecyl sulfate-polyacrylamide gel electrophoresis (SDS-PAGE) is a Bio-Rad miniprotean II dual-slab cell with power supply. A two-gel Laemmli system is used, consisting of an upper 6% acrylamide stacking gel and a lower 15% acrylamide separating gel. Buffers are stored in the dark at RT. Chemicals are electrophoresis grade purity.
    a. Sample buffer: 50 mM Tris-HCl, pH 6.8, 1% SDS, 1% mercaptoethanol, 6 M urea, 0.05% bromophenol blue.
    b. Kaleidascope prestained standards (Bio-Rad, Mr range 7500–205,000).
    c. Stain: 0.25% Coomassie brilliant blue R250 (Sigma) in 10% acetic acid, 50% methanol.
    d. Destain: 10% methanol, 10% acetic acid.
    e. Stacking gel: 0.6 mL 40% acrylamide, 3.4 mL stacking buffer (0.14 M Tris-HCl, pH 6.8, 0.12% SDS, 0.01% NaN$_3$), 50 µL 10% ammonium persulfate, 5 µL 100% TEMED (N,N,N',N'-tetramethylethylene-diamine).
    f. Separating gel: 3.0 mL 40% acrylamide, 4.0 mL 2X separating buffer (0.75 M Tris-HCl, pH 8.8, 0.2% SDS, 0.01% NaN$_3$), 1.0 mL H$_2$O, 100 µL 10% ammonium persulfate, 10 µL TEMED.
    g. Running buffer: 0.025 M Tris (do not adjust pH with HCl), 0.15 M glycine, 0.1% SDS, 0.01% NaN$_3$. To save time, the running buffer is stored as a 10X concentrated stock solution, and the required amount diluted 1:10 just before use. For TnC/C domain samples, include 0.1 mM EDTA in both sample and running buffers (*see* **Note 9**).
6. Erlenmeyer flasks (8X 2-L and 1×1 L) for cell growth. Polypropylene centrifuge bottles (4×250 mL). Graduated cylinders (50- and 100-mL volumes).

2.3. Acetone Powder of Cell Pellet

1. Brinkman Polytron homogenizer or equivalent with large probe.
2. Cold ethanol and acetone.
3. Buchner funnel (approx 165 cm diameter), Whatman #3 filter paper disk, and in-house vacuum line.
4. Balance for weighing acetone powder.

2.4. Extraction of Acetone Powder

1. The benign buffer used for extraction varies, and is the same as the equilibration buffer for the first column chromatography procedure as follows: (a) for intact TnC on phenyl-Sepharose, 50 mM Tris-HCl, pH 7.5, 5 mM CaCl$_2$, 1 mM MgCl$_2$, 50 mM NaCl, 1 mM dithiothreitol (DTT); (b) for both TnC/C and/N domains on diethylaminoethyl (DEAE) Sephadex A-25, 50 mM Tris-HCl, pH 8.0, 0.1 M NaCl, 2 mM MgCl$_2$, 1 mM DTT, 0.01% NaN$_3$.
2. Octanoic acid to minimize foaming during homogenization.
3. Sorvall rotor SS-34, 20,000 rpm, 30 min, 4°C, and 12×30-mL bottles.

2.5. Purification of TnC/C and/N Domains Using Column Chromatography

1. DEAE Sephadex A-25 (Pharmacia Biotech) in a 2.6 × 28-cm column at RT; flow rate approx 70 mL/h; fraction size approx 7 mL/tube; profile monitored at 229 nm wavelength. Buffer: 50 mM Tris-HCl, pH 8.0, 0.1 M NaCl, 2 mM MgCl$_2$, 1 mM DTT, 0.01% NaN$_3$. Gradient: 1600 mL from 0.2 M to 0.8 M NaCl.
2. Dialysis tank (14-L capacity with approx 2 mM NH$_4$HCO$_3$ in deionized H$_2$O and 0.1 mM β-mercaptoethanol × 3 changes).
3. Lyophilizer and flasks (600- and 1200-mL volumes).

2.6. Purification of Intact TnC Using Column Chromatography

1. Phenyl-Sepharose CL-4B resin (Pharmacia Biotech) in a 2.5 × 50-cm column at RT; flow rate approx 60 mL/h; fraction size approx 10 mL/tube; profile monitored at 229 nm wavelength. Buffers: (a) equilibration buffer (50 mM Tris-HCl, pH 7.5, 5 mM CaCl$_2$, 1 mM MgCl$_2$, 50 mM NaCl, 1 mM DTT); (b) high-salt wash (50 mM Tris-HCl, pH 7.5, 0.1 mM CaCl$_2$, 1 M NaCl, 1 mM DTT); (c) elution buffer (50 mM Tris-HCl, pH 7.5, 1 mM ethylenediaminetetracetic acid [EDTA], 1 mM DTT).
2. Dialysis tank (14-L capacity with approx 2 mM NH$_4$HCO$_3$ in deionized H$_2$O and 0.1 mM β-mercaptoethanol × 3 changes).
3. Lyophilizer and flasks (600- and 1200-mL volumes).
4. Sephadex G-75 resin (Pharmacia Biotech) in three columns (2 × 100-cm each) joined in tandem, equilibrated with 10 mM NH$_4$HCO$_3$; flow rate 30 mL/h; fraction size 5 mL/tube; profile monitored at 229 nm wavelength.

3. Methods

3.1. Transformation of pET3a.TnC into E. coli BL21(DE3)pLysS

1. Thaw a vial (250 µL) of competent BL21(DE3)p*Lys*S cells on ice, then immediately add 1–2 µL plasmid DNA (25–50 mg pET3a.TnC, pET3a.TnC/ND, or pET3a.TnC/CD), and gently pipet up and down to mix.
2. Incubate on ice 30 min.
3. Add 1 mL SOC medium and shake (approx 225 rpm) at 37°C for 1 h.
4. Plate out 50 and 300 µL of transformation mix onto two LB plates supplemented with ampicillin and chloramphenicol (*see* **Note 1**).
5. Incubate plates in 37°C oven until the liquid is absorbed into the agar (approx 2 h), then invert plates and continue incubating overnight (approx 16 h).

3.2. Screening of Colonies for Test Expression (See Notes 2 and 3)

1. Six colonies are inoculated into six sterile tubes (approx 10-mL capacity), each containing 2 mL 2X-TY medium supplemented with ampicillin plus chloramphenicol, and incubated at 37°C with shaking (approx 225 rpm) for approx 3 h until moderately turbid (OD$_{600nm}$ ≅ 0.3).
2. For induction, 1 mL of the growing culture is then aliquoted into a fresh tube containing IPTG (0.4 mM), and both "noninduced" and "induced" cultures allowed to continue growing for 2.5 h.

3.3. Large-Scale Preparation (8 L) of Cell Culture

3. In order to maintain a log-phase "continuing" culture of cells for later use, another set of six tubes containing 2 mL antibiotic-supplemented 2 × TY medium is freshly inoculated with 2 µL from each of the six "noninduced" growing cultures, and growth continued at 37°C with shaking.
4. To harvest the "noninduced" and "induced" cells, the 1-mL cultures are transferred to 1.5-mL conical tubes and centrifuged at maximum speed in an Eppendorf centrifuge for 10 min.
5. For analysis of expression levels using SDS-PAGE (15%), the centrifuged samples are resuspended in 100 µL 2 × TY buffer + 100 µL Laemmli gel sample buffer + 5 µL mercaptoethanol, boiled for 5 min, and recentrifuged for 10 min prior to application of 10 µL on the gel. Standards used are intact TnC, TnC/ND, or TnC/CD, and a prestained Kaleidascope mixture of proteins (Bio-Rad).

3.3. Large-Scale Preparation (8 L) of Cell Culture

1. The "continuing" culture (4 µL) of the colony showing the highest level of protein expression is used to inoculate 300 mL 2 × TY *media* (supplemented with 100 µg/mL ampicillin and 25 µg/mL chloramphenicol) for overnight growth (approx 15 h) at 37°C with shaking (approx 280 rpm) in a 1-L Erlenmeyer flask.
2. The next morning, each of 8 × 1 L of the same media (in 2-L Erlenmeyer flasks) is inoculated with 30 mL of the overnight culture, and growth of the cells monitored at 600 nm.
3. At $OD_{600nm} \cong 0.8$, 1 mL of "noninduced" cells is removed from one flask for a gel sample and left at RT for 3 h. The cultures (8 × 1 L) are then induced by addition of 1 mL 0.4 M IPTG and growth/induction continued for a further 3 h.
4. A sample (1 mL) of "induced" cells is removed from one flask, and along with the previously removed "noninduced" sample, prepared for analysis by SDS-PAGE as described in **steps 4** and **5** of screening for test expression (**Fig. 1A**, lanes 2 and 3).
5. The 8 L of cells are harvested using a Damon/IEC DPR/1600 centrifuge at 4°C (4200 rpm × 20 min) and the supernatant discarded (*see* **Notes 4** and **5**).
6. The cell pellets are resuspended in approx 50 mL fresh 2 × TY media + 100 µL mercaptoethanol and transferred into 4 × 250-mL polypropylene bottles (i.e., each 250-mL bottle contains the resuspension and washings from 2 L cell culture).
7. After centrifugation (Sorvall GSA rotor, 7000 rpm × 10 min, 4°C) to remove the buffer, the pellets are either stored at –20°C overnight, or processed further to the acetone powder stage.

3.4. Acetone Powder of Cell Pellet

1. If cell pellets were frozen at –20°C overnight, thaw at RT approx 1 h (*see* **Note 8**).
2. Weigh a Whatman #3 filter paper disk, fit into a Buchner funnel (approx 165-cm diameter), connect to in-house vacuum, and leave on benchtop until the end of **step 6**.
3. To each pellet (from 2 L cells) in 250-mL polypropylene centrifuge bottle, add 120 mL cold 95% ethanol (*see* **Note 6** and **Diagram 1**). Homogenize at setting #3 (low speed) until pellet is uniformly suspended, then increase setting to #4–5 (high speed) for 30–40 s. Repeat homogenization 2–three times.

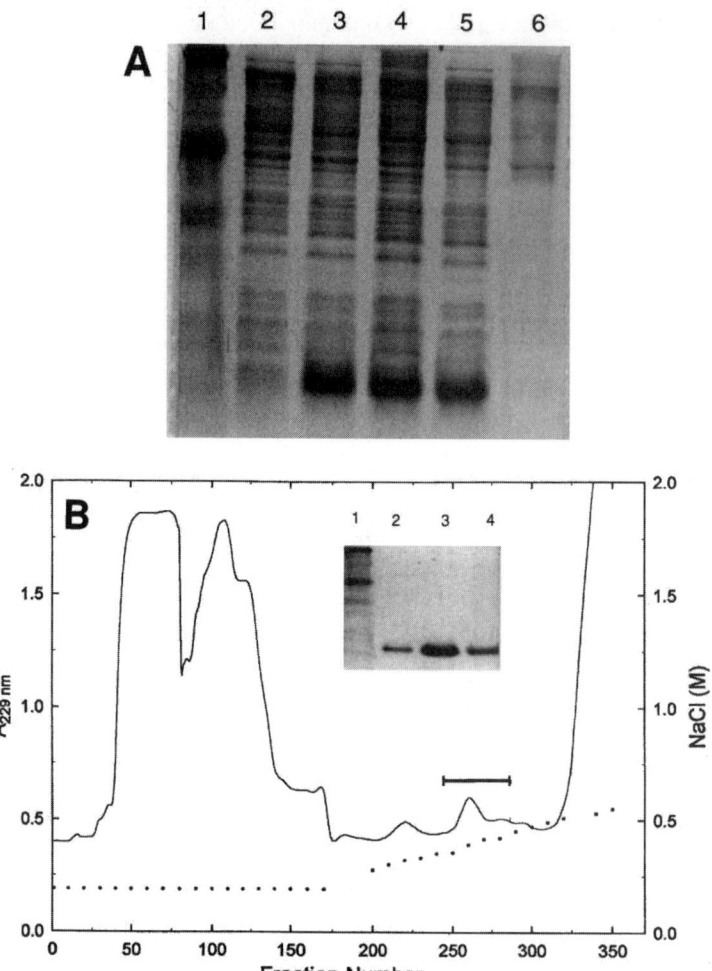

Fig. 1. **(A)** Sodium dodecyl sulfate-polyacrylamide gel electrophoresis (SDS-PAGE) of preparative expression of TnC/C domain fragment, and extraction of acetone powder. Lane 1, 10 µL Kaleidascope prestained standards, 7500–205,000 Mr (Bio-Rad); lanes 2 and 3, noninduced and induced samples, respectively; lane 4, acetone powder before extraction; lanes 5 and 6, supernatant (applied to DEAE Sephadex A-25 column) and pellet, respectively, after extraction of acetone powder. **(B)** DEAE Sephadex A-25 chromatography of TnC/C domain fragment. The extracted supernatant (*see* **[A]**, lane 5; approx 530 mL of 50 mM Tris-HCl, pH 8.0, 2 mM MgCl$_2$, 0.1 M NaCl, 10 mM β-mercaptoethanol, 0.01% NaN$_3$) from approx 4 g acetone powder was applied to a 2.6 × 28-cm column equilibrated with the same buffer, but with [NaCl] increased to 0.2 M. After elution of the flowthrough peak (containing unwanted *E. coli* proteins), a 1600-mL salt gradient from 0.2–0.8 M NaCl was applied to elute the TnC/C domain fragment between approx 0.35–0.45 M NaCl. A final wash with

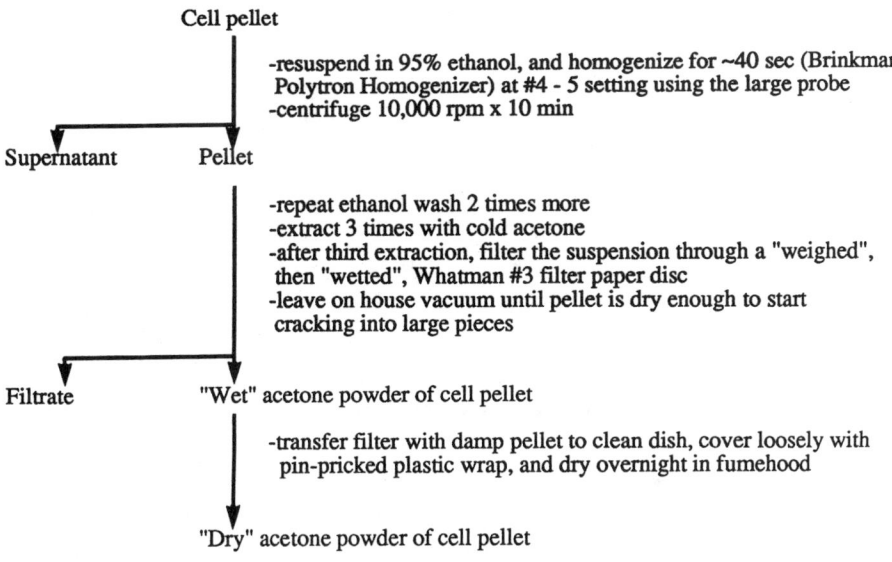

Diagram 1. Cell pellet acetone powder preparation.

4. Centrifuge the four bottles in GSA rotor at $16,300g \times 10$ min, 4°C. If supernatant still has some particulate matter, re-spin. Discard supernatant into an organic waste container.
5. Repeat ethanol extraction twice more (**steps 3** and **4**).
6. Extract the pellet three times with cold acetone, as in **steps 3** and **4**; however, for the third wash, use only 60 mL cold acetone. Instead of removing the acetone wash by centrifugation, filter using the Buchner funnel prepared in **step 2** as follows: (a) wet the filter paper slightly with deionized H_2O and turn on the in-house vacuum to "seal" the paper onto the funnel; (b) filter the acetone suspension by carefully transferring approx 2 mL at a time onto the center of the disk (using a Pasteur pipet) and allowing the powder to be sucked dry before applying the next aliquot; (c) repeat until all the suspension has been applied, leaving the vacuum on for a further 0.5–1 h until the pellet turns from gray to white and "cracks"; (d) place the filter paper with damp acetone powder in a fumehood overnight to dry completely (*see* **Note 7**).

(Fig. 1. *continued*) 0.8 M NaCl buffer was used to elute off any remaining material prior to equilibration with starting buffer for reuse of the column. Tubes 245–287 containing pure TnC/C domain were pooled as indicated. Flow rate, approx 70 mL/h; fraction size, approx 7 mL/tube; profile monitored at 229-nm wavelength. Inset: SDS-PAGE of fractions off the column; lane 1, 10 µL Kaleidascope prestained standards, 7500–205,000 Mr (Bio-Rad); lanes 2–4, tubes 250, 260, and 270, respectively.

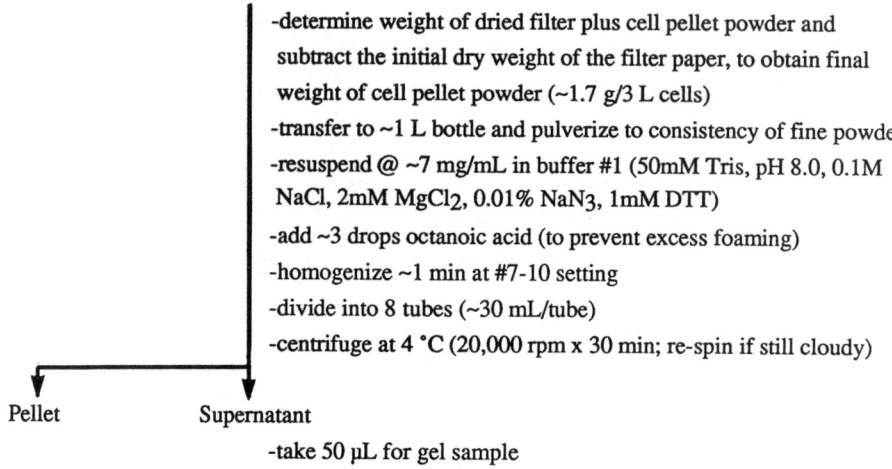

Diagram 2. Extraction of acetone powder for purification of TnC/C domain.

7. The next morning, transfer the powder into a tared tube and determine the dry weight of the acetone powder. The powder can be stored at –20°C, or extracted immediately. Yields vary from approx 2.5–4.5 g/8 L cell culture, depending on the cell density of the culture prior to induction.

3.5. Extraction of Acetone Powder of Cell Pellet

1. The concentration of powder used for extraction is approx 7 mg/mL in the particular benign buffer used for the initial column chromatography step (*see* **Subheading 2.** and **Diagram 2**).
2. Transfer the acetone powder into a cylindrical glass bottle which will be used for homogenization, and pulverize with a pestle to break up the the chunks.
3. Add the required volume of buffer (*see* **Subheading 2.**) and a stirring bar, and allow the mixture to stir at RT for approx 1 h to soften the powder.
4. Remove the stirring bar, and add a few drops of octanoic acid to prevent excess foaming during homogenization. Addition of this small amount of octanoic acid does not affect the pH significantly.
5. Homogenize with the large probe at high speed (~#7–10 setting) approx 1 min. Repeat homogenization 2–three times. Remove 25 µL for a gel sample (*see* **Fig. 1A**, lane 4).
6. Aliquot into 30 mL bottles and centrifuge at 48,000g in SS-34 rotor (20,000 rpm × 30 min, 4°C). Decant supernatant into flask and if still cloudy, re-spin. To determine the yield from extraction, remove 25 µL of the supernatant for a gel

sample. Resuspend the pellet in the same volume of buffer used in **step 3**, homogenize to disperse the particulate matter, and immediately remove 25 µL sample for analysis by SDS-PAGE. To the three gel samples, add an equal volume of Laemmli sample buffer and prepare as before, loading 10 µL onto the gel (*see* **Fig. 1A**, lanes 5 and 6).
7. The clear supernatant is ready for purification by column chromatography.

3.6. Purification of TnC/C Domain Fragment Using Column Chromatography

TnC/ND fragment is purified using the same column.

3.6.1. DEAE Sephadex A-25 Column (see Note 10)

1. The extracted supernatant from approx 4 g acetone powder/530 mL buffer (50 mM Tris-HCl, pH 8.0, 2 mM MgCl$_2$, 0.1 M NaCl, 10 mM β-mercaptoethanol, 0.01% NaN$_3$) is applied to a DEAE Sephadex A-25 (Pharmacia Biotech) column equilibrated in the same buffer, but with [NaCl] increased to 0.2 M (*see* **Fig. 1B**).
2. The column is washed with the 0.2 M NaCl buffer until the off-scale flowthrough peak (containing unwanted *E. coli* proteins) comes on-scale again and starts to approach baseline.
3. A 1600 mL salt gradient from 0.2 M to 0.8 M NaCl is then applied to elute the TnC/CD fragment between approx 0.36–0.43 M NaCl.
4. A final wash with 0.8 M NaCl buffer is applied to elute off any remaining material from the column.
5. Gel samples (5 µL) are taken across the column profile and analyzed by SDS-PAGE as described previously (*see* **Fig. 1B**).
6. Fractions containing pure TnC/C domain fragment (tubes 245–287) were pooled as shown in **Fig. 1B**, dialyzed in membrane tubing (3500 Mr cutoff) against approx 2 mM NH$_4$HCO$_3$ in deionized H$_2$O plus 0.1 mM β-mercaptoethanol (3 changes × 14 L), and lyophilized.
7. In order to minimize the residual salt after dialysis of such a large volume, the initial freeze-dried product from the previous step is redissolved in a smaller volume of H$_2$O (approx 50 mL), and the dialysis repeated in the same manner.
8. In order to eliminate the presence of a mixed disulphide of cysteine/β-mercaptoethanol, the protein is fully reduced by treatment with 5 mM DTT (*see* **Note 11**) to obtain one homogeneous form.
9. Yield = 570 mg/8 L cells. Duplicate amino acid analyses of a weighed sample indicated that the final product is 94% by weight protein (the remainder of the weight most likely caused by H$_2$O of hydration and residual salt). Analysis via mass spectrophotometer indicated a single component having Mr = 8745.

3.7. Purification of Intact TnC Using Column Chromatography

3.7.1. Phenyl-Sepharose Cl-4B Column

1. The extracted supernatant from approx 5 g acetone powder/700 mL buffer (50 mM Tris-HCl, pH 7.5, 5 mM CaCl$_2$, 1 mM MgCl$_2$, 50 mM NaCl, 1 mM

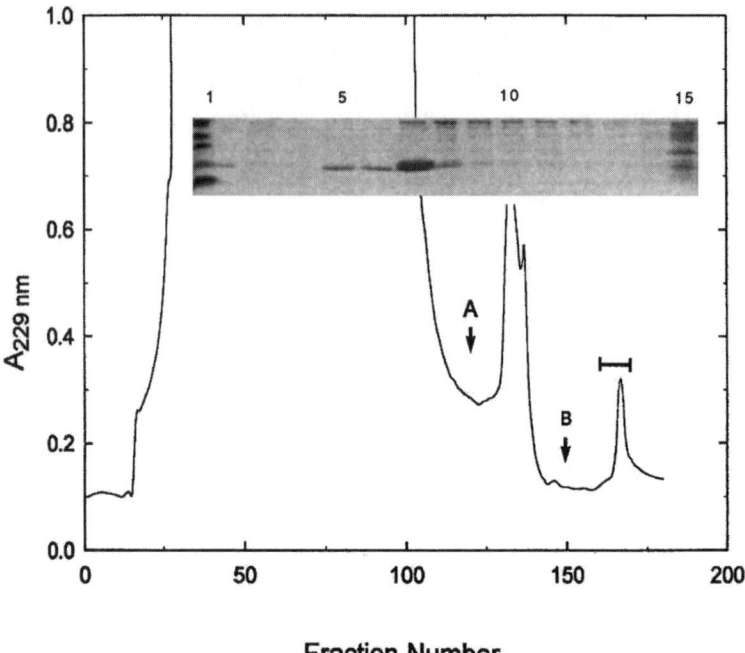

Fig. 2. Phenyl-Sepharose Cl-4B chromatography of intact TnC. The extracted supernatant (approx 700 mL of 50 mM Tris-HCl, pH 7.5, 5 mM $CaCl_2$, 1 mM $MgCl_2$, 50 mM NaCl, 1 mM DTT) from approx 5 g acetone powder was applied to a 2.5 × 50-cm column equilibrated in the same buffer. The column was washed with equilibration buffer (until the flowthrough peak was eluted), then successively with a high-salt wash (50 mM Tris-HCl, pH 7.5, 0.1 mM $CaCl_2$, 1 M NaCl, 1 mM DTT) at (**A**), followed by the elution buffer (50 mM Tris-HCl, pH 7.5, 1 mM EDTA, 1 mM DTT) at (**B**). Tubes 160–168 containing approx 80% pure TnC were pooled as indicated, and treated as described in the text for the next column purification step. Flow rate, approx 60 mL/h; fraction size, 10 mL/tube; profile monitored at 229-nm wavelength. Inset: SDS-PAGE of fractions off the column; lane 1, 10 µL Kaleidascope prestained standards; lane 2, intact TnC standard; lanes 3–14, respectively, tubes 131, 135, 161, 163, 165, 167, 169, 171, 173, 175, 177, 179; lane 15, extracted supernatant from acetone powder loaded onto the column.

DTT) is applied to a phenyl-Sepharose CL-4B (Pharmacia Biotech) column equilibrated in the same buffer (*see* **Fig. 2**), as described by Xu et al. *(3)*.
2. The column is washed successively with equilibration buffer (until the flowthrough peak is eluted), and then with a high-salt wash (50 mM Tris-HCl, pH 7.5, 0.1 mM $CaCl_2$, 1 M NaCl, 1 mM DTT).
3. The intact TnC is finally eluted with 50 mM Tris-HCl, pH 7.5, 1 mM EDTA, 1 mM DTT (*see* **Note 12**).

4. Gel samples (25 µL) are taken across the column profile and analyzed by SDS-PAGE as described previously (*see* **Fig. 2**, inset).
5. Fractions 160–168 containing TnC (approx 80% pure) are pooled, dialyzed against approx 2 mM NH$_4$HCO$_3$ in deionized H$_2$O and 0.1 mM β-mercaptoethanol (3 changes × 14 L), and lyophilized. Yield after this step = 275 mg of approx 80% pure TnC (*see* **Note 13**).
6. The TnC is redissolved in a smaller volume (approx 50 mL) of deionized H$_2$O, redialyzed, and lyophilized to remove the partially volatile residual NH$_4$HCO$_3$. Yield after second dialysis = 255 mg of approx 80% pure TnC (*see* **Note 14**).

3.7.2. Sephadex G-75 Column

1. The 255 mg of lyophilized TnC from the phenyl-Sepharose column is dissolved in approx 12 mL 10 mM NH$_4$HCO$_3$ and applied to a Sephadex G-75 gel exclusion column equilibrated in the same buffer (*see* **Fig. 3**).
2. Gel samples (20 µL) are taken across the column profile and analyzed by SDS-PAGE (**Fig. 3**, inset).
3. Only those fractions containing pureTnC (tubes 37–50) are pooled and lyophilized. The samples are redissolved in a smaller volume (approx 30 mL) of deionized H$_2$O and relyophilized.
4. As done previously with TnC/C domain, the purified intact TnC is treated with 5 mM DTT to obtain one homogeneous reduced form (*see* **Note 11**).
5. Total yield of pure intact TnC from 8 L cells = 235 mg.

4. Notes

1. Sterilization: (a) all glassware and plasticware used for transformation and cell growth procedures were sterilized by steam-autoclaving for 30 min at 15 lb/sq in (approx 250°F) on the "wrapped" cycle; (b) all media and thermostable chemical solutions for bacterial growth were autoclaved for 45 min at 15 lb/sq in on the "liquid" cycle; (c) ampicillin and glucose solutions were sterilized using 0.22 µm Millex-GV and Stericup filter units, respectively (Millipore); (d) chloramphenicol was not treated for sterilization because it was dissolved in ethanol.
2. To allow for adequate aeration of cells during growth, each 1–2 mL of culture for test expression was grown in a 10-mL capacity sterile glass tube loosely fitted with a metal top. For the 8 L preparative growth, 8X 2-L sterile Erlenmeyer flasks were used, each flask containing 1 L of cell culture and stoppered loosely with a gauze plug covered with aluminum foil.
3. Although the preliminary screening of colonies is optional, we have found that not all colonies express protein equally, especially in the case of the other two troponin subunits, TnI and TnT. To ensure the highest yield in the subsequent 8 L preparative culture, therefore, this approx 7 h procedure has been routinely included in each new protein preparation.
4. After harvesting of the 8X 1-L cell cultures, do not freeze the cell pellets in the 1–L bottles, because the pellets become viscous when next thawed, making quantitative transfer difficult. Instead, resuspend the cell pellets and transfer

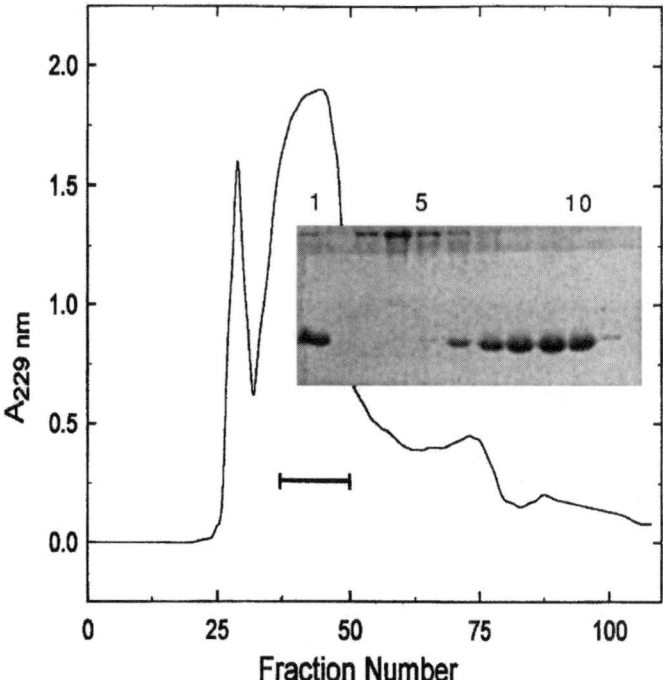

Fig. 3. Sephadex G-75 gel exclusion chromatography of the approx 80% pure TnC fraction. The lyophilized TnC from the phenyl-Sepharose column (255 mg) was dissolved in 12 mL 10 mM NH$_4$HCO$_3$ and applied to three columns connected in tandem (2 × 280-cm final), equilibrated in the same buffer. Fractions containing pure TnC (tubes 37–50) were pooled as indicated, and lyophilized. Flow rate, 30 mL/h; fraction size, 5 mL/tube; profile monitored at 229-nm wavelength. Inset: SDS-PAGE of fractions off the column; lane 1, intact TnC standard; lanes 2–11, tubes 23, 26, 29, 32, 35, 38, 41, 44, 47, and 50, respectively.

immediately to the 250-mL bottles, and recentrifuge before leaving cell pellets at –20°C overnight.
5. During harvesting of the 8-L cell culture and preparation of the acetone powder, centrifuge bottles are kept on ice throughout the procedure to reduce risk of proteolytic degradation of proteins.
6. When making the acetone powder of the cell pellets, use polypropylene centrifuge tubes and not polycarbonate tubes, which are not resistant to ethanol/acetone.
7. Before leaving the damp acetone powder to dry overnight in the fumehood, break up the larger pieces with a spatula to allow for more efficient drying.
8. To minimize proteolysis during preparation of acetone powder, keep the four bottles in an ice bath and work with one bottle at a time.

9. In gel samples of TnC/C domain, the high-affinity Ca^{2+}-binding sites take on different degrees Ca^{2+} that is inherent in the buffers. In order to remove the residual Ca^{2+} and collapse the bands into one *apo* species of the C domain, both the sample and running buffers are supplemented with 0.1 mM EDTA. To saturate the gel in 0.1 mM EDTA, approx 3 µL of tracking dye (bromophenol blue) is placed in one of the sample wells and the gel prerun at 50 mAmps current until the dye has traveled the length of the gel.
10. The DEAE Sephadex A-25 column profile is monitored at 229-nm wavelength, with the monitor set at 0.5 total absorbance deflection, in order to obtain a suitably noticeable peak for the desired TnC/C and TnC/N domain fragments. From SDS-PAGE analysis of many previous runs, the off-scale flowthrough peak is known to contain high Mr *E. coli* proteins. Samples from the off-scale final peak eluted at >0.5 M NaCl are blank on SDS-PAGE, indicating that it does not consist of protein material, and is most likely nucleic acid from the cells.
11. Mass spectrophotometric analysis of the TnC/C domain after the second round of dialysis and lyophilization indicated the presence of two components, one having the correct Mr of the protein, and a second consisting of a mixed disulphide of β-mercaptoethanol with cysteine. In order to obtain one homogeneous product under reducing conditions, the protein (5 mg/mL) was dissolved in 50 mM Tris, 5 mM DTT, 1 mM EGTA, pH 7.5, and stirred overnight at 4°C. The solution was then dialyzed against H_2O (adjusted to pH 6.5 with 0.26 M NH_4HCO_3) with three changes and lyophilized. This final step was incorporated into the procedure of intact TnC as well, so that the single cysteine present would be in the reduced form.
12. After elution of intact TnC from the phenyl-Sepharose column, a final wash is carried out with 6 M GuHCl to remove any residual high Mr contaminants still bound to the resin, prior to reusing the column.
13. After dialysis of TnC against 2 mM NH_4HCO_3, the initial lyophilization does not remove the partially volatile NH_4HCO_3 completely, leading to an erroneously higher yield by weight, and affecting the pH when TnC is subsequently dissolved for use in the next step. In order to remove the residual NH_4HCO_3, the TnC is routinely redissolved in a smaller volume of deionized H_2O and lyophilized a second time. Amino acid analysis after two such steps results in approx 80% by weight pure TnC, the rest of the weight accounted for most likely by H_2O of hydration on the TnC molecule.
14. After the phenyl-Sepharose column, the TnC is approx 80% pure. In order to remove the one high Mr contaminant, TnC is subsequently passed through a series of three Sephadex G-75 gel-filtration columns joined in tandem.

Acknowledgments

We would like to acknowledge the financial support of the Medical Research Council of Canada and the Alberta Heart and Stroke Foundation. We are grateful to Helena Edwards and Helena Lam for providing excellent technical assistance in the protein preparations.

References

1. Potter, J. D. (1982) Preparation of troponin and its subunits. *Methods Enzymol.* **85,** 241–263.
2. Sambrook, J., Fritsch, E. F., and Maniatis, T. (1989) Bacterial media, antibiotics, and bacterial strains, in *Molecular Cloning: A Laboratory Manual*, vol. 3, Cold Spring Harbor Laboratory Press, Cold Spring Harbor, NY, pp A.1–A.6.
3. Xu, G.-Q. and Hitchcock-DeGregori, S. E. (1988) Synthesis of a Troponin C cDNA and expression of Wild-type and mutant proteins in *Escherichia coli. J. Biol. Chem.* **263,** 13,962–13,970.

10

Purification of Recombinant Calbindin D_{9k}

Eva Thulin

1. Introduction

The calcium-binding protein calbindin D_{9k} is a relatively small member (8.5 kDa) of the EF-hand helix-loop-helix family of calcium-binding proteins. Unlike many other calcium-regulatory proteins in this family, binding of the two calcium ions to calbindin does not cause a large conformational change, which exposes a hydrophobic surface. Therefore, calbindin D_{9K} probably acts as a calcium buffer, rather than a calcium-trigger protein. Because there is no calcium-dependent conformational change, hydrophobic interactions as a purification step for recombinant calbindin is not feasible. The purification scheme described here for recombinant calbindin is based on the same procedure as used for purification of calbindin D_{9k} from intestine *(1,2)*. The method starts with a heat shock step that takes advantage of calbindin's unique stability. The second step is anion-exchange chromatography in the presence of ethylenediaminetetracetic acid (EDTA). The third step, gelfiltration, is used to remove small proteins, peptides, and other *Escherichia coli* components that do not elute from the ion exchange column with any distinct pattern. The fourth step is anion-exchange chromatography in the presence of calcium. Because calbindin changes its net charge upon binding calcium, it will elute at a different salt concentration in calcium compared to EDTA. Therefore, proteins that copurify in the EDTA step will be removed in the calcium step. The last anion-exchange step will also remove any deamidated calbindin that may arise during the purification *(3)*. This protocol is optimized for purification of 100–500 mg *E. coli*-expressed bovine calbindin D_{9k} Minor A form *(4)* (*see* **Note 1**). The protocol will also work for mutated calbindins with only small adjustments. For example, if there is a change in charge, the salt gradients should be adjusted upward by 0.05 *M* for every additional negative charge and downward for a positive charge change. Calbindin D_{28k} can also be purified using the same series of purification steps. For a more detailed protocol, *see* **ref. 5**.

2. Materials

2.1. Instruments

1. Cooled centrifuge with capacity for 30,000*g*.
2. Sonicator with 500-W processor and 12.7-mm standard horn.
3. Peristaltic pump with working interval 10–200 mL/h.
4. Fraction collector.
5. Equipment to create a linear gradient.
6. UV monitor.
7. Electrophoresis equipment for:
   a. Protein agarose gels *(6)*.
   b. Polyacrylamide sodium dodecyl sulfate (SDS) gels.
   c. Isoelectric focusing.
8. Lyophilizer.

2.2. Columns

All columns are made of Plexiglas in our workshop, but similar columns are available commercially. They are packed in the laboratory to the following specifications:

1. Anion exchange in EDTA: 3.4 × 25 cm diethylaminoethyl (DEAE) cellulose (Whatman DE 23) (*see* **Note 2**).
2. Gelfiltration 3.4 × 180 cm Sephadex G 50 superfine (Amersham Pharamcia Biotech).
3. Anion exchange in calcium: 2.2 × 15-cm DEAE Sephacell (Amersham Pharamcia Biotech).
4. Gelfiltration 3.4 × 20 cm Sephadex-G 25 superfine (Amersham Pharamcia Biotech).

2.3. Buffers and Solutions

All chemicals were of "pro analysi" or equivalent quality. Solutions were prepared in double-distilled water (ddH$_2$O).

1. Imidazole buffer:1.36 g/L imidazole (20 m*M*), 1.17 g/L NaCl (20 m*M*). Add 800 mL ddH$_2$O and adjust pH to 7.0 with HCl then add up to 1000 mL with ddH$_2$O.
2. 0.5 *M* EDTA pH 7.0: 18.6 g EDTA is suspended in 60 mL ddH$_2$O (it will not dissolve). Add 5 *M* NaOH (ca: 10 mL) until all EDTA is dissolved and the pH is stable at 7.0. Then add ddH$_2$O up to 100 mL.
3. Imidazole buffer with 1 m*M* EDTA: 1000 mL imidazole buffer, 2 mL 0.5 *M* EDTA pH 7.0.
4. Imidazole buffer with 1 m*M* EDTA and 0.3 *M* NaCl: 1000 mL imidazole buffer with 1 m*M* EDTA and 17.53 g NaCl.

5. 0.075 M Barbital buffer with 2 mM EDTA pH 8.6: 2.065 g Barbitone (diethylbarbituric acid), add 500 mL ddH_2O and heat until all is dissolved, then add 13.05 g Barbitone sodium and 0.744 g EDTA. Adjust the volume to 1000 mL with ddH_2O.
6. 0.075 M Barbital buffer with 2 mM calcium lactate pH 8.6: 2.065 g Barbitone then add 500 mL ddH_2O and heat until all is dissolved. Then add 13.05 g Barbitone sodium and 0.575 g calcium lactate. Adjust the volume to 1000 mL with ddH_2O.
7. Fixing solution: 750 mL saturated picric acid in ddH_2O (picric acid is explosive when dry), 150 mL 100% acetic acid.
8. Coomassie brilliant blue stain: 4 g Coomassie blue, 450 mL ethanol, 450 mL ddH_2O, 100 mL 100% acetic acid. Heat the solution to 60°C and then cool to room temperature (RT) before filtering the solution through double Whatman 3M paper filters.
9. Destaining solution: 450 mL ethanol, 450 mL ddH_2O, 100 mL 100% acetic acid.
10. 50 mM ammonium acetate pH 6.0: 800 mL ddH_2O, 3 mL 100% acetic acid. Adjust pH to 6.0 with NH_3 and fill up to 1000 mL with ddH_2O.
11. 1 M $CaCl_2$: 14.7 g $CaCl_2 \cdot 2H_2O$. Fill up to 100 mL with ddH_2O.
12. PIPES buffer: 3.08 g/L PIPES (10 mM), 1 mL 1 M $CaCl_2$ (1 mM). Suspend in 800 mL ddH_2O (it will not dissolve), add 5 M NaOH, (ca: 3 mL) until all PIPES is dissolved and the pH is stable at 6.5, then add ddH_2O up to 1000 mL.
13. PIPES buffer with 0.2 M NaCl: 1000 mL PIPES buffer, and 11.7 g NaCl.
14. 0.1 M EGTA pH 8.0: 3.8 g ethylene glycol-*bis* N,N,N',N'-tetraacetic acid (EGTA) suspended in 50 mL ddH_2O (it will not dissolve), add 5 M NaOH (ca: 4.5 mL) until all EGTA is dissolved and the pH is stable at 8.0, then add ddH_2O up to 100 mL.
15. Chelex 100 treated saturated NaCl: Make a saturated NaCl solution (100g NaCl in 250 mL ddH_2O, you should still have some NaCl crystals). Keep the solution in a plastic container. Put a dialyzing tube (boiled four times in ddH_2O before use) filled with Chelex 100 into the NaCl solution (Chelex 100, 200–400 mesh sodium form from Bio Rad 142-2842). The Chelex 100 is basic, so it needs to be washed with ddH_2O until it reaches a neutral pH. A Büchner-funnel is useful for this. Then transfer a thick slurry into a dialysis tube. 1–2 mL Chelex 100 is enough for 250 mL NaCl solution. Place the solution on a shaker, or turn it over a couple of times every day for 1 wk. The NaCl solution is then calcium free.

3. Methods

All steps are performed at 4°C unless otherwise indicated.

3.1. Heat Precipitation of Bacterial Proteins

1. The bacterial pellet from 4-L culture (ca: 30 mL pellet) was resuspended in 200 mL imidazole buffer on ice.
2. The suspension was poured into 200 mL boiling imidazole buffer under vigorous stirring and heated until the temperature of the mixture reached 95°C.

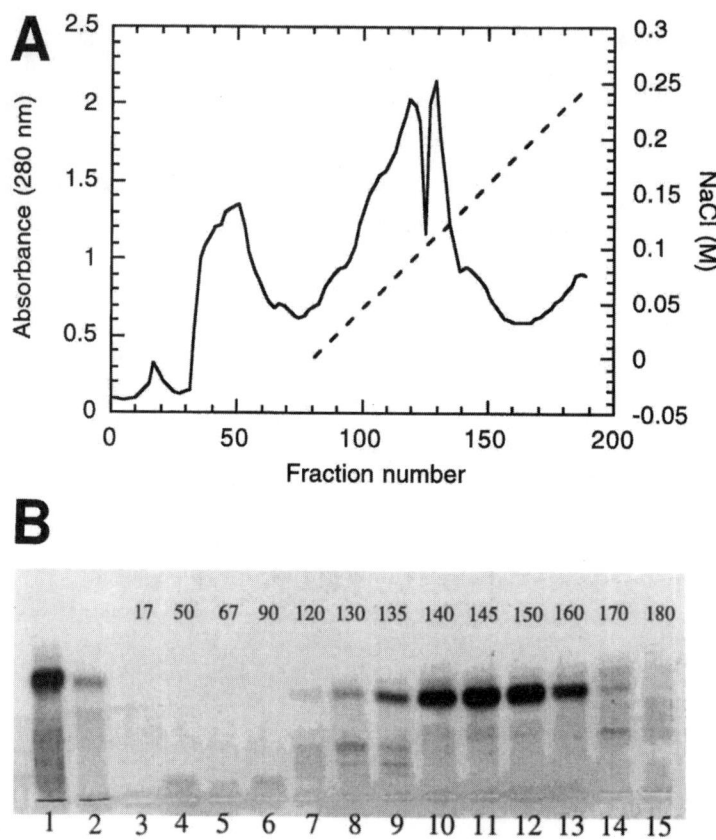

Fig. 1. (**A**) Absorbance at 280 nm of fractions from the DEAE-cellulose column. (**B**) Agarose gel electrophoresis in 0.075 M Barbital buffer with 2 mM EDTA, pH 8.6 (*6*) lane 1: supernatant after first heat treatment, lane 2: supernatant after sonication and heat treatment, and lanes 3–15: fractions from DEAE-cellulose column.

3. It was then immediately cooled on ice-water slurry with stirring to increase the speed of cooling (*see* **Note 3**).
4. Precipitated proteins were removed by centrifugation at 25,000g for 10 min. The supernatant (lane 1 in **Fig. 1B** and lane 1 in **Fig. 5**) was saved for further purification.
5. The pellet was resuspended into 100 mL imidazole buffer on ice and portions of 30–40 mL, kept on ice-water slurry were, sonicated for 5 min with 50% duty cycle and output control 10 using a Sonifier B-30 (Branson Sonic Power Co.) with a 12.7-mm distrupter horn.
6. The sonicated material was poured into 100 mL boiling imidazole buffer under stirring until the temperature of the mixture reached 95°C. The mixture was quickly cooled and precipitated proteins were removed by centrifugation at

Purification of Calbindin D_{9k}

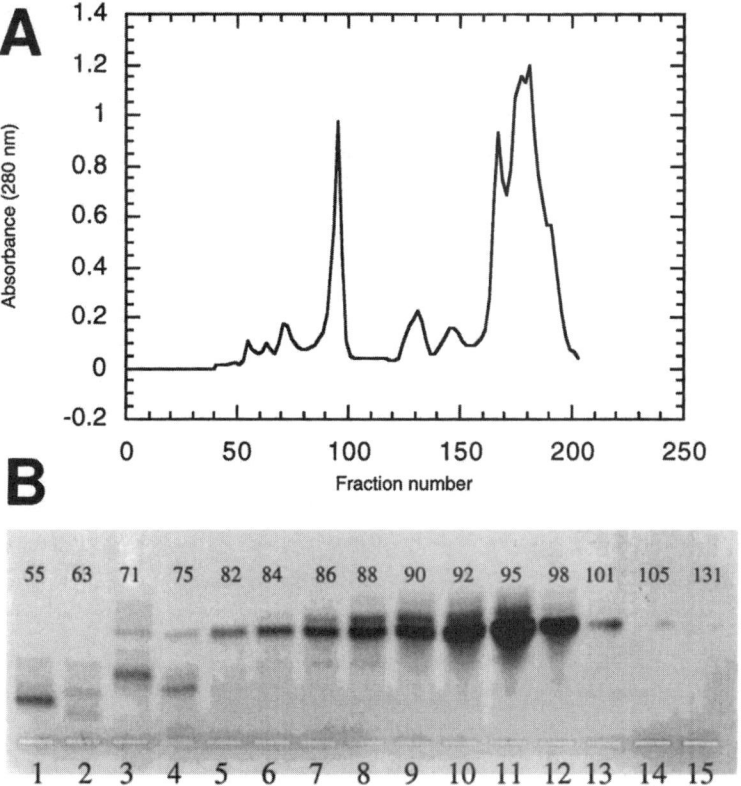

Fig. 2. **(A)** Absorbance at 280 nm of fractions from gel filtration Sephadex G50 superfine. **(B)** Agarose gel electrophoresis in 0.075 M barbital buffer with 2 mM EDTA, pH 8.6 *(6)* lanes 1–15: fractions from the Sephadex G50 column.

30,000g for 10 min. The pellet was discarded and the supernatant (lane 2 in **Fig. 1B** and lane 2 in **Fig. 5**) was combined with the first supernatant (*see* **Note 4**).

3.2. Anion-Exchange Chromatography in the Presence of EDTA

1. The combined supernatants (600 mL) were supplemented with 2 mM EDTA by adding 2.4 mL 0.5 M EDTA pH 7.0 and then pumped (80 mL/h) onto a 3.4 × 25-cm DEAE-cellulose column equilabrated in imidazole buffer with 1 mM EDTA. The column was then eluted with a linear salt gradient of 0–0.3 M NaCl in a total volume of 1400 mL (700 mL imidazole buffer with 1 mM EDTA, 700 mL imidazole buffer with 1 mM EDTA and 0.3 M NaCl).
2. Fractions of 10 mL were collected and analyzed by UV absorbance at 280 nm (*see* **Fig. 1A**) and agarose gel electrophoresis *(6)* (lanes 3–15 in **Fig. 1B**).
3. The fractions containing calbindin (130–165) were identified as those having a lower mobility in the presence of calcium than in the presence of EDTA on the

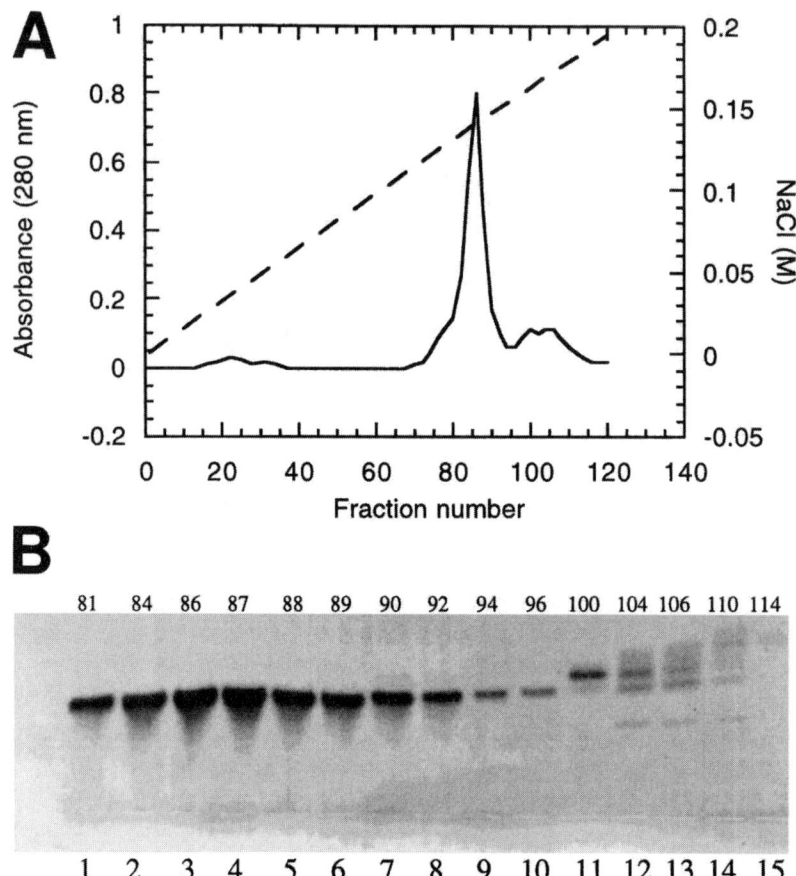

Fig. 3. (**A**) Absorbance at 280 nm of fractions from the DEAE Sephacel column. (**B**) Agarose gel electrophoresis in 0.075 M barbital buffer with 2 mM EDTA pH 8.6 *(6)* lanes 1–15: fractions from DEAE Sephacel column

agarose gel electrophoresis. These fractions were pooled and lyophilized (370 mL lane 3 in **Fig. 5**).

3.3. Gel Filtration

1. The lyophilized fraction was dissolved in 20 mL ddH$_2$O and loaded onto a 3.4 × 180-cm Sephadex G50 superfine column equilibrated in 50 mM ammonium acetate pH 6.0. The pump speed was 35 mL/h, and the fraction size was 7 mL.
2. The fractions were analyzed by UV absorbance at 280 nm (*see* **Fig. 2A**) and agarose gel electrophoresis (lane 1–15 in **Fig. 2B**). Fractions containing calbindin, 82–103, (200 mL lane 4 in **Fig. 5**) were pooled and supplemented with 3 mM in CaCl$_2$ by adding 0.6 mL 1 M CaCl$_2$ and the pH was adjusted to 6.5.

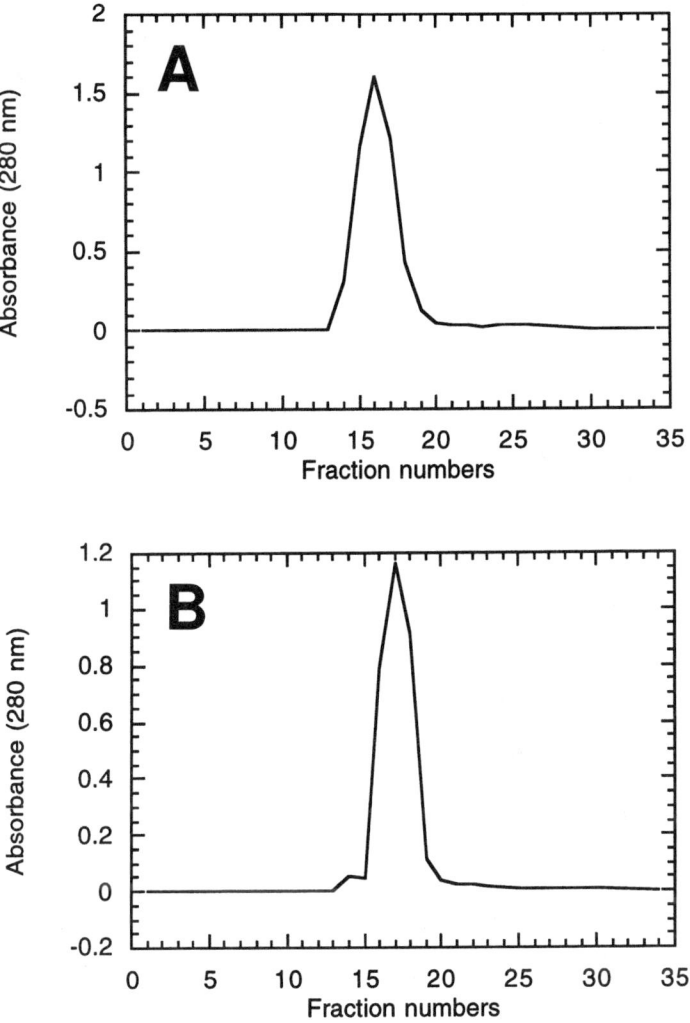

Fig. 4. Absorbance at 280 nm of fractions from desalting on a Sephadex G25 superfine column. **(A)** Ca^{2+} calbindin **(B)** *Apo* calbindin.

3.4. Anion Exchange Chromatography in the Presence of Ca^{2+}

The pooled fractions from the gel filtration were pumped (60 mL/h) on a 2.2 × 15-cm DEAE Sephacel column packed in PIPES buffer. The column was eluted by a 1000-mL linear gradient from 0–0.2 M NaCl (500 mL PIPES buffer, 500 mL PIPES buffer with 0.2 M NaCl). Fractions of 7 mL were collected and analyzed by UV absorbance at 280 nm (*see* **Fig. 3A**) and agarose gel electrophoresis (lanes 1–15 in **Fig. 3B**). Calbindin fractions 74–96 (225 mL line 5 in

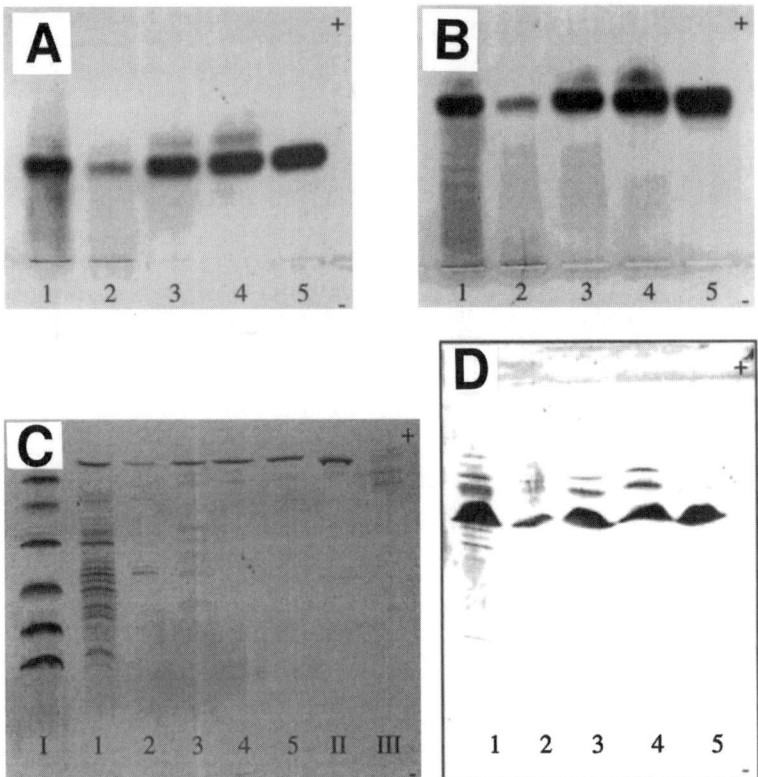

Fig. 5. Electrophoresis from the different purification steps. (**A**) Agarose gel electrophoresis in 0.075 M Barbital buffer with 2 mM calcium lactate pH 8.6. (**B**) Agarose gel electrophoresis in 0.075 M Barbital buffer with 2 mM EDTA, pH 8.6. (**C**) SDS polyacrylamide gradient gel 8–18% from Pharmacia Biotech. (**D**) Isoelectric focusing gel pH 4–6.5 from Pharmacia Biotech. Lane 1, supernatant after first heat treatment; lane 2, supernatant after sonication and heat treatment; lane 3, pooled fractions after the DEAE-cellulose column; lane 4, pooled fractions after the Sephadex G50 column; lane 5, pooled fractions after the DEAE Sephacel column. I Amersham Pharmacia Biotech molecular standard (M) 94, 67, 43, 30, 20.1,14.4 kDa. II Calmodulin TR$_2$C 8 kDa. III Amersham Pharmacia Biotech molecular standard (L) 17.2, 14.6, 8.2, 6.4, 2.6 kDa.

Fig. 5) were pooled and divided into two equal fractions of 112 mL and lyophilized.

3.5. Desalting and Removal of Calcium

1. Ca^{2+}-calbindin: One of the two lyophilized calbindin fractions was dissolved in 3 mL 10 mM CaCl$_2$ pH 7.0 and applied to a 3.4 × 20-cm Sephadex G25 superfine column equilibrated and eluted with ddH$_2$O. The pump speed was 30 mL/h. Fractions of 4 mL were collected and analyzed by UV absorbance at 280 nm

(**Fig. 4A**). Fractions 14–20 (27 mL) were pooled and lyophilized. The resulting yield was 93 mg Ca^{2+} loaded calbindin D_{9k}.
2. *Apo*-calbindin: The other half of the lyophilized calbindin fraction was dissolved in 5 mL 0.1 M EGTA pH 8.0 (*see* **Note 5**) and applied to a 3.4 × 20-cm Sephadex G25 superfine column equilibrated and eluted with H_2O. Before applying the calbindin solution, 15 mL of a saturated solution of NaCl treated with Chelex 100 was applied to the column (*see* **Note 6**). The pump speed was 30 mL/h. Fractions of 4 mL were collected and analyzed by UV absorbance at 280 nm (**Fig. 4B**). Fractions 14–19 (24 mL) were pooled and lyophilized. The resulting yield was 86 mg *apo* calbindin. The residual Ca^{2+} was quantified using Ca^{2+} titrations in the presence of *quin*2, as described under dye-binding calcium assays (*see* Chapter 2, Volume 2).

4. Notes

1. To purify small amounts of calbindin, an FPLC system (Pharmacia Biotech) could be used with a Mono Q-ion exchange column and a Superdex G75 or Superdex peptide column for gel-filtration.
2. It may seem strange to use an old-fashioned low-capacity ion-exchange matrix, but there is a good reason for doing so in this case. Calbindin is one of the most negatively charged proteins in the protein solution to be fractionated. You do not need high capacity to bind the calbindin, and it is preferable that lower charged proteins are constricted. To achieve this with a high-capacity matrix, the column has to be very small and, thereby, the flow speed will be lower. Preferably, the time needed in the first purification steps should be as short as possible to avoid degradation of calbindin.
3. It is important that the protein suspension immediately reaches a temperature above 50°C and reaches 95°C within 2–3 min. Calbindin itself is a very temperature-stable protein, but all proteases, which will be present in the suspension are very active in the temperature range 20–55°C.
4. If you are satisfied with ca: 70% yield you can exclude the sonication and the second heat-treatment step. You will also get less impurities, so the first ion-exchange column may not be necessary.
5. The amount of EGTA required should be enough to bind the Ca^{2+} from the PIPES buffer and four times the estimated calbindin concentration. The pH is chosen to favor Ca^{2+}-binding to EGTA over calbindin.
6. The salt is used to prevent EGTA binding to calbindin, which will occur at low ionic strength (*7*). The protein is much larger than EGTA and will leave EGTA behind while passing through the NaCl-solution. The protein will elute from the column in H_2O, free of both Ca^{2+} and EGTA. The volumes of the sample, NaCl, and column are adapted to each other.

References

1. Hitchman, A. J. W., Kerr M.-K., and Harrison J. E. (1973) The purification of pig vitamin D-induced intestinal calcium binding proteins. *Arch. Biochem. Biophys.* **155,** 221–222.

2. O'Neil, J. D. J., Dorrington, K. J., Kells, D. I. C., and Hofmann, T. (1982) Fluorescence and circular-dichroism properties of pig intestinal calcium-binding protein (Mr = 9000), a protein with a single tyrosine residue. *Biochem. J.* **207,** 389–396.
3. Chazin, W. J., Kördel, J., Thulin, E., Hofmann, T., Drakenberg, T., and Forsén, S. (1989) Identification of an isoaspartyl linkage formed upon deamidation of bovine calbindin D9k and structural characterization by 2D'H NMR. *Biochemistry* **28,** 8646–8653.
4. Brodin, P., Grundström, T., Hofmann, T., Drakenberg, T., Thulin E., and Forsén, S. (1986) Expression of bovine intestinal calcium binding protein from a synthetic gene in *Escherichia coli* and characterization of the product. *Biochemistry* **25,** 5371–5377.
5. Thulin, E. and Linse, S. (1999) Expression and purification of human calbindin D_{28k}. *Prot. Express. Purificat.* **15,** 265–270.
6. Johansson, B. G. (1972) Agarose gel electrophoresis. *Scand. J. Clin. Lab. Invest.* **29(124),** 7–19.
7. Chiancone, E., Thulin, E., Boffi, A., Forsén, S., Brunori, M. (1986) Evidence for the interaction between the calcium indicator 1,2-*bis*(o-aminophenoxy)ethane-*N,N,N',N'*-tetraacedic acid and calcium-binding proteins. *J. Biol. Chem.* **261,** 16,306–16,308.

11

S100 Proteins

From Purification to Functions

Jean Christophe Deloulme, Gaëlh Ouengue Mbele, and Jacques Baudier

1. Introduction

The first description of S100 proteins was made by B. W. Moore who initially characterized a group of abundant low-molecular-weight (10–12) acidic (pH 4.0–5.0) proteins that were enriched in the nervous system *(1)*. Their name derives from their unusual solubility in 100% ammonium sulfate at nearly neutral pH (7.0–7.5). The S100 fraction, isolated from brain, was subjected to intensive investigation, and it was then demonstrated that there were, in fact, multiple protein species. Two distinct S100 proteins were first characterized, the S100A1 (S100αα) and S100B (S100ββ) by Isobe and co-workers *(2)*. Over the past few years, this S100 protein family has expanded with the description of at least 17 other S100 proteins *(3)*. All the characterized members of the S100 protein family have significant physicochemical properties in common. They show different degrees of homology, ranging from 25–65% identity at the amino acid level. In solution, S100 proteins associate as homodimers, but they are also prone to form heterodimers by combination with other S100 species *(4)*. The S100 proteins contain two EF-hand calcium-binding sites. When these sites are filled with Ca^{2+}, a number of physicochemical and structural properties of the proteins are altered. Calcium binding induces conformational changes that result in the exposure of hydrophobic patches to solvent, allowing hydrophobic interactions with target proteins, hydrophobic dyes, and hydrophobic matrix *(5)*.

Several human disorders such as cancer, neurodegenerative diseases, cardiomyopathies, and autoimmune disorders are associated with an altered expression of S100 proteins *(3)*. For these reasons, there is considerable interest in understanding the exact functions of the S100 proteins. To reach that goal, efforts have been made to produce recombinant proteins for in vitro studies. Expression vectors adapted to S100B production in eukaryotic cells for *in situ* investigation have also been developed. We will describe here an expression vector construct for production of recombinant S100 proteins in bacteria and a rapid purification method for recombinant S100A1, S100A6 (calcyclin), S100A11 (S100C, calgizzarin), and S100B. We will also describe an inducible eukaryotic expression vector for S100B and methods to analyze S100B expression in transfected cells. These methods should be adaptable to other S100 proteins.

1.2. Expression and Purification of Human Recombinant S100 Proteins in Bacteria

The biological functions of S100 proteins are still unknown. Purification of large quantities of proteins is essential for structural studies and to develop activity studies and analytical tools. For example, purified S100 proteins may be required for antibody production. Large-scale purification of S100 proteins may also be used to synthesize matrix for affinity purification of antibodies and affinity chromatography directed toward identification and purification of target proteins. Several methods for production and purification of recombinant of S100 protein have been developed in different laboratories *(6–8)*. We have adapted these methods to develop a single protocol that is applicable to the simultaneous purification of human S100A1, S100A6, S100A11, and S100B.

2. Materials
2.1. pET Expression Vector Construct

1. Thermal cycler.
2. *Taq* polymerase and 10X *Taq* reaction buffer: 500 mM KCl, 100 mM Tris-HCl, pH 9.0, 1% Triton X-100, 25 mM MgCl$_2$.
3. 2.5 mM deoxynucleotide 5'-triphosphate (dNTP) stock solution.
4. pET28 plasmid (a, b, or c) (Invitrogen).
5. Restriction enzymes *Nco*I, *Bam*HI, and *Bsp*HI, and 10X *Bam*HI reaction buffer (Gibco-BRL).
6. UV transilluminator (312-nm wavelength).
7. Agarose gel apparatus.
8. 10X TAE buffer: 400 mM Tris-acetate, pH 8.3, 10 mM ethylenediaminetetracetic acid (EDTA).
9. DNA purification kit from agarose gel (e.g., Qiagen).
10. pGEM-T easy vector system from Promega (*see* **Note 1**).
11. Plasmid-DNA purification kit (*see* **Note 2**).
12. DH5α competent *Escherichia coli*.

2.2. Bacterial Growth, Expression, and Purification of Human Native S100 Proteins

1. BL-21 competent *E. coli*.
2. Isopropyl-β-D-thiogalactopyranoside (IPTG).
3. *Ly*sis buffer: 30 m*M* Tris-HCl, pH 8.0, 150 m*M* NaCl, 1 m*M* ethylene glycol-*bis* N,N,N',N'-tetraacetic acid (EGTA), 5 m*M* dithiothreitol (DTT), 0.5 mg/mL lysozyme.
4. Protease inhibitors: leupeptin (25 mg/mL in water), aprotinin (25 mg/mL in water), phenylmethylsulfonyl fluoride (PMSF) at 250 m*M* in 100% ethanol.
5. Phenyl-Sepharose CL-4B (Sigma, P7892).
6. Column buffer: 25 m*M* imidazol, pH 7.7, 200 m*M* NaCl, and 5 m*M* DTT.
7. Sodium dodecyl sulfate-polyacrylamide gel electrophoresis (SDS-PAGE) vertical apparatus.
8. Gel solutions: 10X cathode running buffer (1 *M* Tricine, 1 *M* Tris-HCl, 1% SDS); 10X anode running buffer (2 *M* Tris-HCl, pH 8.9) (*see* **Note 3**).
9. 4X Laemmli buffer: 160 m*M* Tris-HCl, pH 6.8, 20% glycerol, 4% SDS, 200 m*M* DTT, 0.0024% bromophenol blue.
10. Bradford solution: 0.01% (w/v) Coomassie blue G250, 4.7% (v/v) pure ethanol, 8.5% (v/v) H_3PO_4.

3. Methods

3.1. pET Expression Vector Constructs for S100 Protein Production

To produce human recombinant S100 proteins in bacteria, we have chosen the pET expression vector system that contains the strong, regulated T7 prokaryotic promoter. The coding sequence of human S100A1, S100A6, S100A11 and S100B were first polymerase chain reaction (PCR)-amplified using specific oligonucleotides with the addition of new subcloning sites (*see* **Table 1**). Subsequently amplified DNA fragments were subcloned in pET28 expression vector.

3.1.1. PCR

1. Mix, in a 0.5 mL Eppendorf microtube: 1 µg of template (*see* **Table 1**), 500 n*M* 5' primer, 500 n*M* 3' primer, 5 µL 10X *Taq* reaction buffer (*see* **Note 4**), 0.2 m*M* dNTPs, and distilled water to 45 µL.
2. Overlay the reaction mixture with 50 µL mineral oil. Heat the sample for 2 min at 94°C and start (hot start) the PCR reaction by adding the *Taq* polymerase (2.5 U in 5 µL of 1X *Taq* reaction buffer).
3. PCR programmation used is as follows:
   a. Step 1: hot start 2 min at 94°C;
   b. Step 2: denaturation 1 min at 94°C;
   c. Step 3: annealing 45 s at 50°C;
   d. Step 4: extention 45 s at 72°C;
   e. Step 5: 14 cycles **steps 2–4**;
   f. Step 6: final extention 5 min at 72°C.

Table 1
Templates and Primers[a]

	5' primers	3' primers	Templates
S100A1	GAAT<u>CCATG</u>GGC TCTGAGCTGG	TCGGATCCGGTCGAC TCAACTGTTCTCCC	pHSA52S100A1
S100A6	CCTCAG<u>CCATG</u>G CATGCCCCC	CCATGGATCCTATTTAT TTTCAGCCCTTGAGG	pCDN3S100A6
S100A11	CGCTCAGCTCCA <u>CCATG</u>GC	CCAAGGGATCCTCAGG TCC	pCDNAS100A11
S100B	TTCC<u>TCATG</u>ATGTC TGAGCTGGAGA AGGCCATGGTG	TTGGATCCAGTCACTCA TGTTCAAAGAACTCG	pCDNneoS100B

[a]The 5' and 3' primers used to amplify the S100 cDNA for cloning in pET28 are summarized in this table. The 5' primers for S100A1, S100A6, and S100A11 amplification introduce a *Nco*I cleavage site and the 5' primer for S100B amplification introduces a *Bsp*HI cleavage site. The 3' primers introduced a *Bam*HI restriction site in the amplified sequences. Restriction sites are underlined and ATG start codons are shown in bold. Templates used in PCR protocol are indicated in the last column. The pHSA52S100A1 containing human S100A1 cDNA was generously provided by R. Kuwano. pCDNA neoS100B, pCDNAMycS100A11, and pCDNAmycS100A6 were from our laboratory.

3.1.2. Purification of PCR Products and Cloning in pGEMT Easy Vector

1. Add 5 µL of 10X loading dye in the reaction mixture (*see* **Note 5**).
2. Vortex and briefly spin the samples.
3. Load the blue aqueous phases containing the PCR fragments and run on a 2% agarose gel (*see* **Note 6**).
4. Identify the PCR products by UV transilluminator and excise the bands with a clean sharp scalpel (*see* **Note 7**).
5. Purify each PCR product using a DNA purification kit from agarose gel (*see* **Note 2**).
6. Clone purified products using pGEM-T easy vector system from Promega (*see* **Note 1**).
7. To test individual colonies for the presence of desired insert, perform a small-scale (miniprep) isolation and purification of plasmid DNA using alkaline lysis method (*see* **Note 8**) or a commercial kit (*see* **Note 2**).

3.1.3. Subcloning in pET28 Expression Vector

1. Digest 2 µg of pGEMT-S100A1, S100A6, and S100A11 constructs with *Nco*I (5U) and *Bam*HI (5U) restriction enzymes or pGEMT-S100B construct with BspHI (5U) (*see* **Note 9**) and *Bam*HI (5U) restriction enzymes in 1X *Bam*HI reaction buffer.

S100B Proteins

2. Digest 1 µg of pET28 vector with *Nco*I (5U) and *Bam*HI (5U) restriction enzymes in 1X *Bam*HI reaction buffer (*see* **Note 11**).
3. Run on a 0.8% and 2% agarose gel for digested pET28 sample and digested pGEMT-S100 samples, respectively.
4. Identify the digested S100 DNA bands (*see* **Note 7**) and pET28 DNA band (*see* **Note 12**) by UV transilluminator and excise them with a sharp scalpel.
5. Purify each PCR product using a DNA purification kit from agarose gel.
6. Ligate *Nco*I/*Bam*HI digested S100A1, S100A6, S100A11 inserts and BspHI/*Bam*HI digested S100B insert into *Nco*I/*Bam*HI digested pET28 vector (*see* **Notes 8** and **10**).
7. Transform DH5α competent *E. coli* with the ligation reaction products (*see* **Notes 8** and **Note 13**).
8. To test individual colonies for the presence of the desired insert, perform a small-scale (miniprep) isolation and purification of plasmid DNA using alkaline lysis method or a commercial kit.
9. The S100A1, S100A6, S100A11, and S100B DNA inserted in pET28 plasmid should be fully sequenced.

3.2. Expression, Extraction, and Purification of Human Native S100 Proteins from E. coli

The extraction procedure of recombinant S100 proteins takes advantage of their unusual solubility in ammonium sulfate at nearly neutral pH. In combination with calcium-dependent affinity chromatography on phenyl-Sepharose, selective solubilization of S100 proteins from *E. coli* extract in ammonium sulfate will yield a pure S100 preparation.

3.2.1. Expression and Extraction of Recombinant S100 Proteins from E. coli

1. Transform Bl-21 competent *E. coli* with pET28-S100 constructs and plate them on LB medium containing kanamycin (*see* **Note 14**).
2. Inoculate 5 mL of LB medium containing 30 µg/mL of kanamycin with 1 colony of each construct. Incubate overnight with orbital shaking (250 to 300 rpm) at 37°C.
3. Dilute the preculture (1:100) in 500 mL of LB medium containing kanamycin (30 µg/mL), and incubate with orbital shaking at 37°C.
4. At the mid-log phase of growth (OD of 0.5 at 600 nm), add IPTG to a final concentration of 1 mM and incubate for an additional 2 h at 37°C with shaking at 250 rpm. The maximal recombinant S100 expression is reached after 2 h of IPTG stimulation (*see* **Fig. 1**).
5. Transfer cultures in 250-mL bottles and sediment the bacteria at 5000g for 15 min at 4°C.
6. Resuspend the pellet in 20 mL of ice-cold lysis buffer (composition *see* **Subheading 2.2.**) containing 2 µg/mL aprotinin, 2 µg/mL leupeptin, 1 mM PMSF (*see* **Note 15**), and sonicate 3 × 10-s bursts.

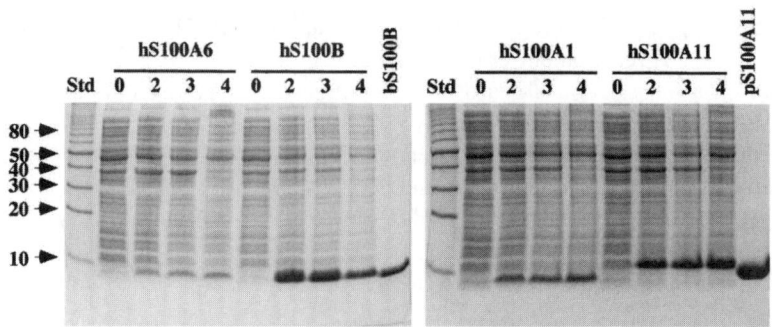

Fig. 1. Expression of human S100 proteins in *E. coli* BL-21 strain induced by IPTG. *E. coli* BL-21 transformed with pET28-S100A1, -S100A6, -S100A11 or -S100B expression vectors were stimulated with IPTG (1 mM). After 2, 3, and 4 h of treatment (as indicated on top), 500 µL of each culture were pelleted in microtube and resuspended in 100 µL 1X Laemmli buffer. Samples were briefly sonicated and proteins (15 µL) were run on 11% SDS Tris-tricine PAGE and stained with Coomassie blue. S100 protein expressions are maximal after 2 h of IPTG treatment. As aforementioned (7), the expression level of recombinant S100A6 is much lower than those of S100A1, S100A11, and S100B. Molecular-mass markers are in kDa on the left. Bovine S100B (bS100B) and porcine S100A11 (pS100A11) were run as a reference.

7. Ultracentrifuge the lysate at 100,000g for 20 min at 4°C, and save the supernatant containing soluble proteins.

3.2.2. Partial Purification of Recombinant S100 Proteins Using Ammonium Sulfate Precipitation

3.2.2.1. AMMONIUM SULFATE EXTRACTION OF S100A1 AND S100B PROTEINS

1. Measure the volume of the soluble *E. coli* extracts and add $(NH_4)SO_4$ to 80% of saturation (561 g/L at 4°C).
2. Stir for 30 min at 4°C and centrifuge at 15,000g for 30 min.
3. Retain the supernatants, which contain S100A1 or S100B, and adjust the pH to 4.0 by adding 25% v/v H_2SO_4 solution. At this acidic pH, S100A1 and S100B precipitate.
4. Stir for 40 min at 4°C.
5. Centrifuge at 15,000g for 30 min at 4°C and discard the supernatant.
6. Resuspend the pellet in the column buffer (composition *see* **Subheading 2.2.**) plus 5 mM $CaCl_2$.
7. Load on the phenyl-Sepharose column (*see* **Subheading 3.2.3.** below).

3.2.2.2. AMMONIUM SULFATE EXTRACTION OF S100A11 PROTEIN

Add $(NH_4)_2SO_4$ to 60% of saturation (390 g/L at 4°C) and follow the same steps as described in **Subheading 3.2.2.1.**

S100B Proteins

3.2.2.3. Ammonium Sulfate Extraction of S100A6 Protein

1. Measure the soluble extract volume and add $(NH_4)_2SO_4$ to 50% of saturation (313 g/L at 4°C).
2. Stir for 30 min at 4°C and centrifuge at 15,000g for 30 min.
3. Retain the supernatant, which contains S100A6 and add $CaCl_2$ to a final concentration of 5 mM.
4. Load on the phenyl-Sepharose column.

3.2.3. Phenyl-Sepharose Chromatography

The method outlined below takes advantage of the hydrophobic property of the Ca^{2+}-bound S100 proteins. The S100 proteins bind to phenyl-Sepharose column in calcium buffer and are subsequently eluted with EGTA buffer.

1. Pour the phenyl-Sepharose (5 mL) in a 10-mL plastic syringe plugged with a glass fiber filter.
2. Equilibrate the column with 100 mL of column buffer plus 5 mM $CaCl_2$ (*see* **Subheading 2.2.**).
3. Load the protein extract on the top of the column and wash the column with at least 500 mL of column buffer plus 5 mM $CaCl_2$.
4. Perform a second wash with 30 mL of column buffer plus 1 mM $CaCl_2$.
5. Elute recombinant S100 protein with the column buffer plus 2 mM EGTA.
6. Collect 1-mL fractions and analyze protein content in 20 µL elution fraction using Bradford reagent (200 µL) in 96 microwell plate.

Figure 2 shows a typical elution profile of the S100B from the phenyl-Sepharose column analyzed by SDS/Tris-tricine PAGE (*see* **Note 3**).

3.3. Purified S100 Proteins: Conclusions

The described method allows large scale purification of the individual S100 protein species with a high degree of purity (*see* **Fig. 3**). The purification procedure yields about 5–10, 25–30, 20–25, and 25–30 mg of S100A6, S100B, S100A1, and S100A11 protein per liter of culture, respectively. The proteins can be stored in solution at 4°C for several weeks in the presence of reducing agent (5–10 mM DTT). For long-term storage at –20°C, we recommend to lyophilize the proteins.

3.4. Inducible Expression of S100B in Eukaryotic Cells and Characterization

Transient or stable expression of proteins in eukaryotic cells appears as a powerful tool to analyze protein functions. Many S100 protein members have been characterized by isolation of mRNA species or proteins whose expression increases depending on the state of cellular growth, transformation, or differentiation, suggesting a direct implication of the S100 proteins in cell-cycle

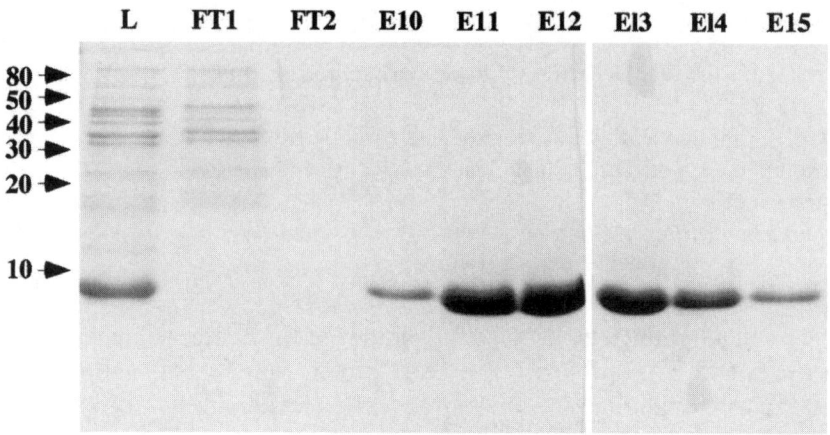

Fig. 2. Purification of human recombinant S100B on phenyl-Sepharose column analyzed by 11% SDS/Tris-tricine PAGE and Coomassie blue staining. Human recombinant S100B was partially purified from *E. coli* BL-21 by ammonium sulfate solubilization and acidic precipitation. The resuspended pellet (*lane L*) was loaded onto a phenyl-Sepharose column in the presence of calcium. The column is then extensively washed with calcium buffer. Analysis of the flowthrough fractions revealed that S100B remains bound to the column (*lanes FT1* and *FT2*). The S100B is then eluted with EGTA containing buffer (*lanes E10–E15*). Molecular mass markers are in kDa on the *left*.

regulation. In some cases, forced S100B expression is detrimental for normal cell growth *(9)*. Because of possible negative-growth regulation activity exerted by S100 proteins, we thought that inducible S100 expression vectors might overcome problems encountered in establishing cell lines carrying S100 expression plasmid. We have chosen the ecdysone-inducible expression system (Invitrogen) because it is the most tightly regulated system available for expression in mammalian cultured cells. The ecdysone system is composed of two vectors: pVgRXR and a pIND-based inducible vector. The pVgRXR vector constitutively expresses the ecdysone heterodimeric receptor and contains the zeocin resistance gene for selection of stable cell lines. The pIND-based inducible expression vector contains the ecdysone receptor-responsive element, which ultimately drives expression of the gene of interest (in our case, the *S100B* gene). It contains multiple cloning sites for the gene of interest, and a hygromycin- or neomycin-resistance gene to generate double-stable cell lines (*see* **Note 16**). We will first describe the construction of an ecdysone-inducible expression vector for S100B and then a refined method to analyze S100B production in transfected cells by immunoblot.

S100B Proteins

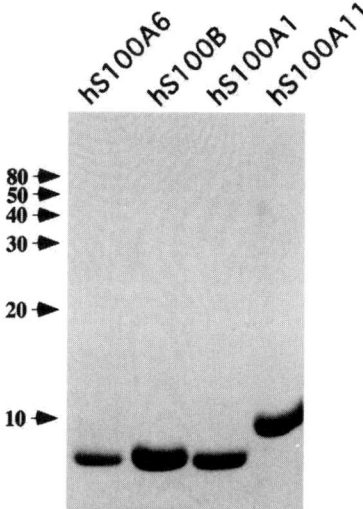

Fig. 3. SDS-PAGE analysis of purified human recombinant S100A1, S100A6, S100A11, and S100B. Purified S100 protein fractions eluted from phenyl-Sepharose columns were pooled and analyzed by 11% SDS/Tris-tricine PAGE and Coomassie blue staining. Molecular-mass markers are in kDa on the *left*.

3.4.1. Materials for Expression in Eukaryotic Cells

3.4.1.1. EQUIPMENT

1. Tissue culture facilities.
2. SDS-PAGE apparatus and power pack.
3. Western blotting apparatus.

3.4.1.2. PLASMIDS AND TRANSFECTION REAGENTS

1. pcDNA I *neo*/hS100b plasmid is as described in **ref. 9**.
2. Ecdysone-inducible expression vectors are purchased from Invitrogen.
3. Inducing agent (*see* **Note 17**).
4. Transfection reagent: Fugene 6 transfection reagent (Boehringer Mannheim) (*see* **Note 18**).

3.4.1.3. WESTERN BLOT ANALYSIS

1. Rabbit anti-S100B antibodies are from Dako (Ref: Z0311).
2. Horseradish peroxidase-conjugated secondary antibodies.
3. Nitrocellulose membrane: Protran BA 85 from Schleicher and Schuell.
4. Transfer buffer: 95 mM glycine, 12.5 mM Tris, 10% ethanol and 1 mM $CaCl_2$.
5. Tris buffer saline (TBS): 20 mM Tris-HCl, pH 7.5, 150 mM NaCl.

6. Blocking buffer: TBS containing 1 mM CaCl$_2$, 5% skimmed milk, and 0.05% Tween-20.
7. Incubation buffer: TBS containing 1 mM CaCl$_2$, 2% skimmed milk, and 0.05% Tween-20.

3.4.2. Methods for Expression in Eukaryotic Cells

3.4.2.1. SUBCLONING OF S100B GENE IN pIND PLASMID

The 5' untranslated sequence of subcloned cDNA might be, in certain cases, crucial for S100B expression. In order to optimize transcription of the S100B cDNA in Pind/Hygro, two pIND/hS100B plasmids were constructed with different 5' untranslated sequences. A 500-bp hS100b cDNA fragment, encoding the complete 92 amino acid sequence of S100B flanked by the 5' and 3' untranslated sequences, was excised from pcDNA *neo*/hS100b at the *Hin*dIII site and subcloned in *Hin*dIII site of pIND. That plasmid will be called hereafter pIND/5' S100B.

In the same manner, a 300-bp hS100b cDNA fragment, encoding only the 92 amino acid sequence of S100B flanked by the 3' untranslated sequence (but not the 5' one), was excised from pcDNA *neo*/hS100b at the *Sac*I and *Bam*HI sites and subcloned in *Sac*I and *Bam*HI sites of pIND. That plasmid will be called hereafter pIND/S100B.

3.4.2.2. PREPARATION OF DNA FOR TRANSFECTION

DNA used for transfection can be prepared by classical methods, such as CsCl gradient purification (*see* **Note 8**), or by column kit methods (*see* **Note 2**). The DNA concentration should be accurately determined by measuring its optical density at 260 nm. An OD reading of 1 at 260 nm corresponds to 50 µg/mL.

3.4.2.3. TRANSFECTION OF *COS*-7 CELLS WITH ECDYSONE-INDUCIBLE EXPRESSION PLASMIDS

The *Cos*-7 cells are maintained in DMEM high glucose (4.5 g/L) containing 10% fetal bovine serum (FBS). The day before the transfection experiment, *Cos*-7 cells are plated (2.3×10^5/dish) in 60-mm Petri dishes. Cells should be transfected at 50–70% of confluency. The Cos-7 cells are transfected with pVgRXR and pIND/5'S100B or pIND/S100B using the Fugene 6 transfection reagent as described later.

1. Take 6 µL of Fugene 6 and mix with 94 µL serum-free medium in a small sterile tube. Do not allow undiluted Fugene 6 reagent to come in contact with the tube, to prevent adsorption to the plastic tube and solubilization of plasticizers from the surface, both of which adversely affecting transfection efficiency. Incubate the diluted Fugene 6 at room temperature for 5 min.

S100B Proteins

2. In a second sterile tube, mix 1 μg of PVgRXR plasmid with 1 μg pIND5'S100B or pINDS100B DNA. Total DNA volume should not exceed 10 μL. DNA amount (μg) and Fugene volume (μL) can be modified proportionally, but you should keep a ratio of 1/3 (w/v).
3. Dropwise, add diluted Fugene 6 transfection reagent to the tube containing DNA. Gently tap the tube to mix contents and then incubate for 15 min at RT.
4. While the Fugene-DNA complexes are forming, add new culture medium to the cells.
5. To perform the transfection, dropwise the DNA-Fugene mixture onto the cells, distributing it around the flask. Swirl the flask to ensure even dispersal. Return the cells to the incubator for 33 h.

3.4.2.4. Induction of S100B Production in Cos-7 Cells Transfected with Ecdysone-Inducible Expression Plasmids

Thirty-three h after the transfection procedure, the culture medium is replaced with fresh culture medium containing 2 μM ponasterone A. Cells are incubated for an additional 24 h prior to lysis in 1X Laemmli buffer omitting DTT (composition *see* **Subheading 2.2.**). Protein contents are determined with bicinchoninic acid protein assay kit (Pierce), with bovine serum albumin (BSA) as standard. Cell extracts can be stored at $-20°C$ for several weeks.

3.4.2.5. Analysis of S100B Production in Cos-7 Cells by Immunoblotting

Several problems can be encountered in analysis of low-molecular-weight proteins by immunoblotting. Particularly, in the case of S100 proteins their low molecular weight and their acidic characters interfere with membrane binding. We have optimized immunoblot procedure for the detection of S100B by including 1 mM $CaCl_2$ in transfer and incubation buffers (*see* **Subheading 3.4.1.3.**). We believe that in the presence of Ca^{2+}, the hydrophobic character of the Ca^{2+}-bound S100B increases its binding to nitrocellulose membrane.

3.4.2.5.1. Immunoblot Analysis of Transfected Cos-7 Cells.

Samples containing 50 μg total protein are subjected to 11% SDS/Tris-tricine polyacrylamide gel according to the method developed by Schagger and Von Jagow adapted to low-molecular-weight proteins *(10)*.

After electrophoresis, gel is soaked in transfer buffer for 15 min and proteins are transferred electrophoretically to nitrocellulose membrane in transfer buffer at 90 V for 1 h. The nitrocellulose membrane is then incubated in blocking buffer at room temperature for 1 h and subsequently incubated with rabbit antibovine S100B antibody (1:4000 dilution) in incubation buffer. After 1 h, the blot is washed six times for 5 min each with incubation buffer and then incubated with HRP-conjugated goat anti-rabbit IgG (1:10000 dilution) in

7incubation buffer for 1 h at room temperature. After six washes with incubation buffer, the blot is developed using enhanced chemo-luminescence reagent (Pierce) and exposed to Kodak biomax film.

3.4.2.5.2. Results.

Immunoblot analysis of ponasterone-induced S100B in transfected *Cos*-7 cells is shown in **Fig. 4**. Lane 1 contains non transfected *Cos*-7 cells extract. A faint S100B-like immunoreactivity is present in *Cos*-7 cells. In lanes 2 and 4, are shown extracts of *Cos*-7 cells transfected with pIND/5'S100B or pIND/S100B plasmid, respectively, but not stimulated with ponasterone A. There was no significant increase in basal S100B expression after plasmids transfection. In lanes 3 and 5 are shown transfected *Cos*-7 cells with pIND/5'S100B or pIND/S100B plasmids, respectively, and stimulated with 2 μM ponasterone A. A drastic increase in S100B immunoreactivity is observed in ponasterone-stimulated cells indicating that the pIND-derived inducible expression vectors were both optimized for specific S100B expression.

The production of an inducible expression vector for S100B protein in mammalian cells opens up new perspectives in the understanding of the biological functions of the protein. One can now envision to generate double-stable cell lines carrying the pVgRXR and pIND/5'S100B or pIND/S100B plasmids. These cells should be powerful tools to delineate the functions of S100B in the regulation of cytoskeletal proteins organization, transcription factor activities, cell cycle, and apoptosis. S100B inducible expression vectors should also be promising in elucidating and comparing the cellular functions of each S100 protein species.

4. Notes

1. Other kits are available from different companies (e.g., TA Cloning Kit from Invitrogen). These cloning kits take advantage of the nontemplate-dependant activity of thermostable polymerase used in PCR that add a single deoxyadenosine to the 3' end of duplex molecule. So, DNA polymerases such as *Vent* or *Pfu*, which possess 3' to 5' exonuclease activity, should not be used.
2. Different kits are commercially available. Usually and with practice, their procedure can be completed quickly, yielding pure plasmids suitable for enzymatic manipulation including PCR, DNA sequencing, and for cell transfection.
3. 11% SDS/Tris-tricine PAGE were used according to the method described by Shagger and Von Jagow *(10)* with some modifications. Mix the following component in a 50-mL Bëcher: 10 mL of Tris-HCl 3 M pH 8.5, 11 mL 30% acrylamide/0.8% *bis*-acrylamide solution, and 10 mL of water then add 200 μL of 10% ammonium persulfate and 22 μL of TEMED solution. Pour the gels.
4. 10X *Taq* reaction buffer contains $MgCl_2$ to give a final concentration of 1.5 mM, which is sufficient for most PCRs.
5. 10X loading buffer: 0.4% bromophenol blue, 0.4% xylene cyanol, 30% glycerol
6. The agarose gel is dissolved in 1X TAE containing 40 μg/L of ethidium bromide.

S100B Proteins

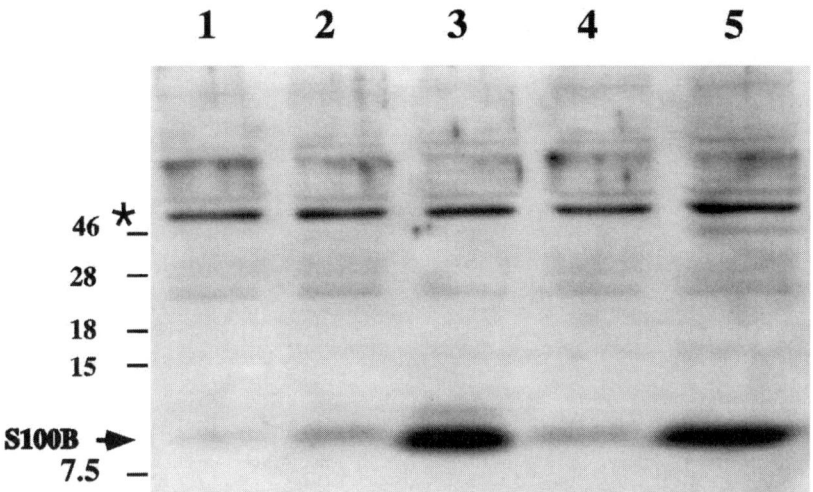

Fig. 4. Immunoblot analysis of ponasterone-induced hS100B in transfected *cos*-7 cells. Cos-7 cells were cotransfected with PvgRxR and Pind/hS100b-*Hin*dIII (*lanes 2 and 3*) or Pind/hS100b-*Sac*I/*Bam*HI (*lanes 4* and *5*) for 33 h are then incubated in the absence (*lanes 2* and *4*) or in the presence of 2 μM of ponasterone A (*lanes 3* and *5*) for 24 h. *Lane 1* shows nontransfected *cos*-7 cell. Arrow points to S100B. The asterisk indicates a crossreacting protein that serves as internal loading control. Molecular-mass markers are in kDa on the *left*.

7. The size of S100A1, S100A6, S100A11, and S100B PCR products is 306, 300, 343, and 298 base pairs, respectively.
8. This method is described in detail in reference: Sambrook, J., Fritsch, E. F., and Maniatis, T. (1989) *Molecular Cloning: A Laboratory Manual*, 2nd ed. Cold Spring Harbor Laboratory, Cold Spring Harbor, NY.
9. The restriction enzyme *Bsp*HI is an isoschizomer of *Rca*I.
10. The restriction endonucleases *Nco*1 and *Bsp*HI produce compatible cohesive ends, but they cannot recleave the ligation product.
11. pET28 a, b, or c expression vector may be used for subcloning. The N-terminal poly*His*- and T7 tags in pET28 DNA sequence is eliminated by the *Nco*I/*Bam*HI double digest. The distance between the ribosomal binding site (rbs) and the ATG start codon of the fourth S100 cDNA in the resulting plasmid is 7 bp.
12. The size of *Nco*I/*Bam*HI digested pET28 vector is around 5300 bp.
13. The pET28 expression vectors contain kanamycin resistance gene. Kanamycin working concentration in LB medium is 30 μg/mL.
14. The BL-21 *E. coli* strain is deficient in both *lon* and *omp*T proteases.
15. At this step, you can freeze the resuspended bacteria at −20°C.
16. More information about the Ecdysone Inducible Expression System is available on the Invitrogen Web site: http://www.invitrogen.com.

17. Several inducing agents are available. We chose the ecdysone analog ponasterone A. Stock solution of ponasterone A in ethanol (1 mM) is stored at $-20°C$.
18. We recommend Fugene 6 reagent because it gives the highest transfection efficiency compared to other transfection reagent tested. Also, it causes minimal death to most cell lines.

References

1. Moore, B. W. (1965) A soluble protein characteristic of the nervous system. *Biochem. Biophys. Res. Commun.* **19,** 739–744.
2. Isobe, T. and Okuyama, T. (1981) The amino acid sequence of S-100 protein (PAP-Ib protein) and its relation to the calcium binding proteins. *Eur. J. Biochem.* **116,** 79–86.
3. Schafer, B. W. and Heizmann, C. (1996) The S100 family of EF-Hand calcium-binding proteins: functions and pathology. *Trends Biochem. Sci.* **21,** 134–140.
4. Baudier, J., Glasser, N., and Gérard, D. (1986) Ions binding to S100 protein. Calcium and zinc-binding properties of bovine brain S100$\alpha\alpha$, S100a ($\alpha\beta$), and S100B ($\beta\beta$) protein: Zn^{2+} regulates Ca^{2+} binding on S100B protein. *J. Biol. Chem.* **261,** 8192–8203.
5. Baudier, J. and Gérard, D. (1983) Ions binding to S100 proteins: structural changes induced by calcium and zinc on S100a and S100b proteins. *Biochemistry* **22,** 3360–3369.
6. Van Eldik, L. J., Staecker, J. L., and Winningham-Major, F. (1988) Synthesis and expression of a gene coding for the calcium-modulated S100b and designed for cassette-based, site-directed mutagenesis. *J. Biol. Chem.* **263,** 7830–7837.
7. Pedrocchi, M., Schäfer, B. W., Durussel, I., Cox, J. A., and Heizmann, C. W. (1994) Purification and characterization of recombinant human calcium-binding S100 proteins CAPL and CACY. *Biochemistry* **33,** 6732–6737.
8. Seemann, J., Weber, K., and Gerke, V. (1996) Structural requirements for annexin-I-S100C complex-formation. *Biochem. J.* **319,** 123–129.
9. Scotto, C., Deloulme, J. C., Rousseau, D., Chambaz, E., and Baudier, J. (1998) Calcium and S100B regulation of p53-dependent cell growth arrest and apoptosis. *Mol. Cell Biol.* **18,** 4272–4281.
10. Schagger, H. and Von Jagow, G. (1987) Tricine-sodium dodecyl sulfate-polyacrylamide gel electrophoresis for the separation of proteins in the range from 1 to 100 kDa. *Anal. Biochem.* **166,** 368–379.

12

Cadherins

Jean-René Alattia, Kit I. Tong, Masatoshi Takeichi, and Mitsuhiko Ikura

1. Introduction

It is now firmly established that calcium is a key component of cellular adhesiveness, as shown from live tissues and cultured cells dissociating whenever exposed to a calcium-depleted environment *(1)*. Cadherins are Ca^{2+}-dependent cell-adhesion molecules that bind homotypically to identical cadherin molecules on opposing cells *(2)*. The homotypic adhesion and Ca^{2+}-binding activities reside in the extracellular region of cadherins, whereas the intracellular domain, together with adapter proteins such as α-, β-, and γ-catenins, is responsible for linking cadherin to cytoskeletal actin filaments. The coordinated intercellular and intracellular protein–protein interactions are crucial for the formation of tissue structure and the proper physiological functioning of tissues *(3)*. Recent reviews have compiled the findings of several structural studies showing the intricate partnership between calcium binding and protein–protein interactions in cadherins *(4,5)*. In this chapter, we will describe a variety of techniques that have been employed to study the function and structure of cadherin molecules. Because there is an excellent review of the protocols for the investigation of cadherin cellular functions *(6)*, we will focus on the molecular and structural studies of this family of Ca^{2+}-binding proteins. A brief description of a cell aggregation assays and its inhibition has also been provided.

2. Materials
2.1. Cell Aggregation

1. 24-well plate (Falcon 3047).
2. Agar (1% Agar Noble, Difco) solution prepared with the standard cell culture medium (e.g., Dulbecco's minimum essential medium [DMEM]) with 10% fetal calf serum (FCS).

3. Bovine serum albumin (BSA) solution, 1% (w/v): Dissolve 1.0 g of BSA (Sigma A-4503) and make up to 100 mL with HCMF buffer. Adjust pH to 7.4 with NaOH. Filter sterilize through a 0.45-μm filter. Store 1 wk at 4°C.
4. 100 mM CaCl$_2$.
5. CO$_2$ incubator.
6. DMEM.
7. FCS.
8. Gyratory shaker (New Brunswick Scientific, Model G2).
9. HCMF (HEPES-buffered salt solution without CaCl$_2$ and MgCl$_2$): 0.8% (w/v) NaCl, 0.04% (w/v) KCl, 0.012% (w/v) Na$_2$HPO$_4$.12H$_2$O, 0.1% (w/v) glucose in 10 mM HEPES. Adjust pH to 7.4 with NaOH. Filter with 0.45-μm to remove any dusty materials and store at 4°C.
10. TC: 0.01% (w/v) trypsin, 1 mM CaCl$_2$ in HCMF buffer. Prepare fresh.
11. TE: 0.01% (w/v) trypsin, 1 mM ethylenediaminetetracetic acid (EDTA) or 1 mM ethylene glycol-*bis* (β-aminoethyl ether)-*N,N,N',N'*-tetraacetic acid (EGTA) in HCMF buffer. Prepare fresh.
12. Paraformaldehyde, 8% in HCMF buffer (pH 7.4).
13. 100 mM MgCl$_2$.
14. Trypsin solution: 0.1% (w/v) crystalline trypsin (Sigma T-8253) in HCMF buffer. Store at –20°C for 1 mo in aliquots.
15. Soybean trypsin inhibitor: 0.5% (w/v) in HCMF buffer. Store at –20°C for 1 mo in aliquots.

2.2. Expression of Ecad(1–145)

1. Ampicillin, 100 mg/mL.
2. LB broth: 1% (w/v) tryptone, 0.5% (w/v) yeast extract, 0.5% (w/v) NaCl. Autoclave at 121°C for 20 min.
3. LB agar: 1% (w/v) tryptone, 0.5% (w/v) yeast extract, 0.5% (w/v) NaCl, 1.5% (w/v) Bacto-agar. Autoclave at 121°C for 20 min.
4. 32°C incubator shaker.
5. 42°C incubator shaker.
6. 65°C water bath.
7. Spin concentrator: e.g., Amicon's centriplus 10.
8. Refrigerated high speed centrifuge: e.g., Dupont's Sorvall refrigerated super speed centrifuge or equivalent.

2.3. Purification for Ecad(1–145)

1. Buffer A: 50 mM Tris-HCl, pH 7.9, 5% glycerol, 1 mM EDTA, 1 mM dithiothreitol (DTT), 1 mM phenylmethylsulfonyl fluoride (PMSF).
2. Equilibration buffer: 25 mM NaCl in buffer A.
3. Lysis buffer: 50 mM NaCl, 0.01 mg/mL pepstatin A, 0.01 mg/mL leupeptin, 1% (v/v) aprotinin, 3 mM MgCl$_2$, 10 μg/mL DNaseI, 0.1% (v/v) Nonidet P-40 (NP40) in buffer A.
4. Gel-filtration buffer: 200 mM NaCl in buffer A.

5. Buffer B: 1 M NaCl in buffer A.
6. Cell disrupter/sonicator: e.g., Branson's sonifier 450.
7. DEAE Sepharose Fast Flow, Sephacryl S100 HR, MonoQ HR5/5 (all Pharmacia).
8. FPLC system (Pharmacia).

2.4. Calcium Blot

1. Incubation buffer: 60 mM KCl, 5 mM MgCl$_2$, 10 mM imidazole, adjust pH to 6.8 with HCl.
2. ^{45}Ca: e.g., Calcium-45 (NEN Life Science Products) in 10 mCi/mL. 10 µL of this stock is needed for each blot.

3. Methods
3.1. Cell Aggregation

Cell–cell adhesion mechanisms are subdivided into the Ca^{2+}-dependent system (CDS) and Ca^{2+}-independent system (CIDS); cadherins are the major components of the former. To prepare for assaying cadherin-dependent cell–cell adhesion activity, one can make use of the unique trypsin and Ca^{2+} sensitivity exhibited in cadherins. Cultured cells should first be TC-treated in a 1% BSA precoated container to remove other adhesion molecules such as those grouped under the CIDS. The cell suspension is then spun down to remove the TC buffer. In order to limit the tryptic activity, 100 µL of soybean trypsin inhibitor is added to cells in every culture plate. Clumped cells are then dissociated into healthy single cells by washing repeatedly with Ca^{2+} free HCMF buffer using 1% BSA precoated pipets. The cells are then ready for the aggregation assay. This preparation method is satisfactory when applied to fibroblasts. Other cell types, e.g., epithelial cells, may require TE treatment, which results in complete disaggregation of cells *(6)*.

There are two types of established aggregation assays: short-term and long-term assays. The long-term aggregation method should be applied to single cells obtained by TE treatment *(6)*. Under this method, the TE-treated cells will restore the adhesion molecules that were lost during the treatment, thus enabling them to reaggregate. The cells are assayed during a period ranging from several hours to several days, using a standard culture medium (e.g., DMEM with 10% FCS) in a 1% agar-coated 24-well plate under CO$_2$ with periodic replenishment of fresh medium. The remainder of the procedure is inherently similar to the short-term method described below (*see* also **Note 1**).

1. The suspension made of dissociated cells in 0.5 mL HCMF buffer is added on ice to a clean (speck- and dust-free) 24-well plate precoated with 1% BSA.
2. CaCl$_2$ (1 mM final concentration) is added to activate cadherins on TC-treated cells. DNaseI (10 µg/mL final concentration) and MgCl$_2$ (1 mM final concentration) are added to prevent nonspecific cell aggregation. A negative control with-

out Ca^{2+} should also be prepared. Immediately following, the aggregation assay is performed on a gyratory shaker at 80 rpm and 37°C. The aggregation process should be monitored every 30 min with an inverted microscope.
3. TC-treated cells should manifest cadherin-dependent cell aggregation induced by the presence of Ca^2. TE-treated cells should not normally aggregate under this condition and should only be assayed with the long-term aggregation method.
4. After 30–60 min of incubation, cadherin mediated aggregation can be inhibited by placing the culture on ice. An equal volume of 8% paraformaldehyde or other aldehyde-based fixatives is added to each well to stabilize the aggregated cells. To quantify cell aggregation, cells are stirred gently and sampled with a 1% BSA coated Pasteur pipet. Cell count can be obtained with a hemocytometer or an automated cell counter.

3.2. Inhibition of the Adhesion Function

There are at least three possible ways to block the cadherin activity: removal of Ca^{2+}, use of inhibitory antibodies, and expression of dominant negative cadherin constructs. Of all three, the removal of Ca^{2+} is the most effective method. For most cell types, cadherins are inactivated at a Ca^{2+} concentration lower than 0.1 mM. Incubating cells with chelators such as EGTA or EDTA (e.g., at a 1 mM concentration), greatly facilitates Ca^{2+} removal, and cell dissociation can usually be observed within a few minutes at 37°C. Chelator-treated cells tend to have a round shape. However, other cell types, especially epithelial cells, are not completely separated from each other by the same treatment. These chelator-resistant connections can only be disrupted by TE preparation.

The cadherin activity can also be blocked by treating monolayer cultures with antibodies raised against the extracellular domain of cadherins. Gaps are often seen between the antibody-treated cells as compared to nontreated cells.

There are two groups of cadherin cDNA constructs that can inhibit cadherin mediated adhesion activity. One group (N) encodes only the conserved cytoplasmic domain and the other group (C) encodes only the intact extracellular domain. Products of N-type constructs can interfere non-specifically with a wide variety of classic cadherins by competing with endogenous, intact cadherins for interactions with cytoplasmic components. On the other hand, products of C-type constructs interfere with homophilic interactions at the extracellular domain of endogenous cadherins. Detailed methods and other experimental tips are given by Takeichi and Nakagawa *(6)*.

3.3. Overexpression and Isolation of Recombinant Cadherins

A number of cDNA constructs were made encoding different regions of extracellular domain of mouse E-cadherin, Ecad(1–145) *(7)*, and Ecad(1–225) *(8,9)*. cDNA fragments were introduced into bacterial expression vectors, pAS *(10)*, by standard PCR procedures. The resulting plasmids contain the λP_L promoter, and the P_L-directed transcription is activated by a temperature shift. Therefore, cells harboring the vector could be initially grown to a high density without expression of the cloned gene at 32°C and subsequently induced to rapidly synthesize the product at 42°C *(10)*. Recombinant mouse E-cadherins purified as mentioned below reproducibly achieved higher than 95% purity and were used for in vitro biochemical and biophysical experiments, including NMR solution structure determination *(11,12)*, X-ray crystallography *(8)*, calcium binding *(9)*, multiangle laser light scattering (MALLS) *(9)*, analytical ultracentrifugation *(9,12)*. The techniques of MALLS and ultracentrifugation are thoroughly described in Chapter 11 of Volume 2.

3.3.1. Expression of Ecad(1–145)

1. For overexpression of Ecad(1-145), *Escherichia coli* strain AR58 is transformed with the expression plasmid encoding Ecad(1–145) and grown at 32°C on LB agar plates containing 100 µg/mL ampicillin.
2. A single colony is inoculated into 15 mL of LB medium containing 100 µg/mL of ampicillin, in a 125-mL flask. The culture is incubated in a 32°C water bath shaker for 16–18 h at 230 rpm.
3. A 1.5 L of LB medium, in a 6-L Erlenmyer's flask, is inoculated with 15 mL of the overnight saturated culture from **step 2**. When A_{600} reaches approx 0.8, 0.5 L of warm medium (prewarmed in a 65°C water bath) is added to each 1.5 L of growing cells. The flask is then shifted to a 42°C shaker incubator for another 3–4 h. Cells are then harvested by centrifuging at 6,000g for 20 min (e.g., 6000 rpm in a Dupont's SLA-3000 rotor) and are either kept at –20°C or carried onto the next step.

3.3.2. Purification of Ecad(1–145)

1. Lysis: The cell pellet from 4 L of LB culture is resuspended in 100 mL of lysis buffer. The cell suspension is lysed by sonicating on ice at setting 10 and 10% duty cycle for 3 × 5 min with 5 min cooling periods between cycles. Cells that are not lysed are removed by spinning at 6000g for 10 min. The resulting pellet is resuspened in another 100 mL of the same lysis buffer and the same sonication procedure is repeated. All cell extracts are then pooled and centrifuged at 30,000g for 30 min (e.g., 16,000 rpm in Dupont's SS34 rotor) to remove all cell debris before loading onto purification columns. Approximately 80% of Ecad(1–145) remains in the soluble fraction (*see* also **Note 2**).
2. DEAE Sepharose (Pharmacia): A DEAE column of 40 mL (2.5 × 10 cm) is poured and equilibrated with two column volumes of equilibration buffer. An equal volume of buffer A is added to the soluble fraction prior to loading onto DEAE Sepharose so as to reduce NaCl concentration to 25 mM. After loading, the column is washed extensively with the equilibration buffer at a flow rate of approx 2 mL/min until the UV absorbance is back to the baseline (usually after running 2–3 column volumes of the equilibration buffer). The adsorbed proteins are eluted

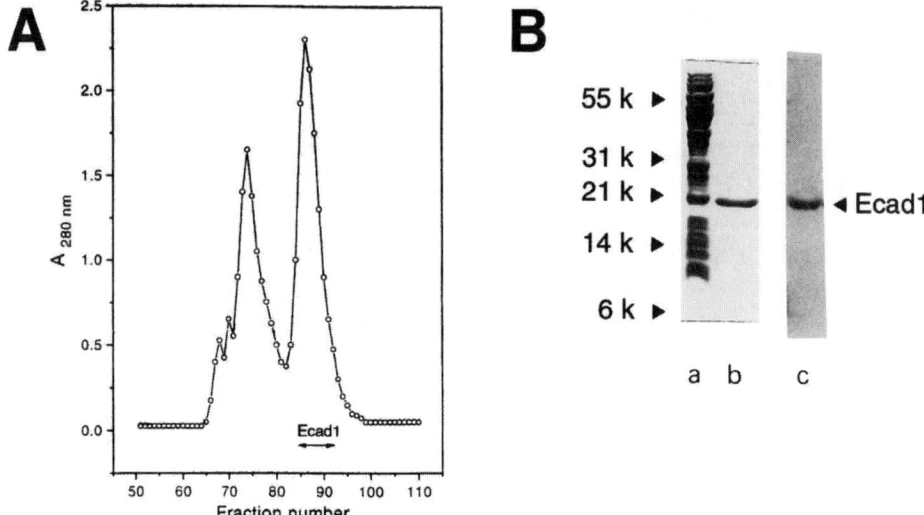

Fig. 1. (**A**) Gel filtration chromatography of Ecad1 (residues 1–145). (**B**) SDS-PAGE of ECAD1. Lanes a–c contain, respectively, the total soluble extract from *E. coli.*, purified Ecad1, and an autoradiogram obtained from a ^{45}Ca blot of Ecad1.

by a linear NaCl gradient (25–125 mM) in a total of 400 mL of buffer A. Ecad(1–145) is eluted at about 85–100 mM NaCl. Fractions should be checked on using 15% glycine sodium dodecyl sulfate-polyacrylamide gel electrophoresis (SDS-PAGE).

3. Sephacryl S-100HR gel-filtration: A Sephacryl S-100HR (Pharmacia) column (2.5 × 88 cm) is equilibrated with approx 400 mL of gel-filtration buffer. Fractions from DEAE Sepharose that contain Ecad(1–145) are pooled and concentrated down to about 4–5 mL using centriplus 10 (Amicon) devices. The protein sample is then loaded onto the column and run at 12 mL/h with the same gel-filtration buffer. Ecad(1–145) is eluted as the latest peak at about 170–200 mL (*see* **Fig. 1**). Peak fractions are checked on 15% glycine SDS-PAGE.

4. Mono Q HR5/5(FPLC/Pharmacia): The peak fractions containing Ecad(1–145) are pooled and the NaCl concentration is reduced to 25 mM by: (a) eightfold dilutions with buffer A; (b) exchange to equilibration buffer by centriplus 10 (amicon) according to supplier's instructions; or (c) dialyzing against equilibration buffer before loading onto a MonoQ column, previously equilibrated with the same equilibration buffer. Buffer A and buffer B are used for pump A and pump B, respectively, in a FPLC system (Pharmacia). The protein sample is loaded onto the MonoQ column at 0.5 mL/min. The column is then washed with 5 mL of buffer B (2.5–3.5%), and the protein is eluted with 50 mL of buffer B (3.5–13.5%) at a flow rate of 0.5 mL/min. The protein peak of Ecad(1–145) is typically observed at 3.5–10.5% of buffer B (i.e., 55–105 mM NaCl). Eluted frac-

tions should be checked on 15% glycine SDS-PAGE and stored at –20°C for later use. The typical yield is approx 3–4 mg/L of culture.

3.4. Calcium Blot

The ability for proteins to bind Ca^{2+} ions can be tested using the method developed by Maruyama et al. *(13)*. It involves running the protein sample on SDS-PAGE, followed by a blot transfer to a PVDF or nitrocellulose membrane. The membrane is then incubated with the $^{45}Ca^{2+}$ radioactive isotope and exposed to a photographic film. This is a useful method to quickly check whether or not a protein (or domain) binds Ca^{2+}. However, this is not the method of choice to characterize Ca^{2+}-binding to proteins quantitatively. For the latter purpose, methods involving Ca^{2+} electrodes, flow dialysis, and equilibrium dialysis should be employed (*see* **Subheading 3.5.**). A protocol for a ^{45}Ca blot is given below (*see* also **Note 3**).

1. The purified protein is resolved on SDS-PAGE gels and electrotransferred onto polyvinylidene difluoride (PVDF) membranes using standard methods *(13)*.
2. Following the transfer, the blot is then soaked in 100 mL of the incubation buffer for 15 min with gentle shaking; a clean container must be used.
3. The incubation should be repeated three times with fresh buffer.
4. To another 100 mL of incubation buffer, 100 µCi of ^{45}Ca (i.e., 10 µL of the 10 mCi/mL stock, NEN) is added and the blot is incubated with gentle shaking for 10 min (the membrane, especially the side in contact with gel, should be submerged into the buffer).
5. Nonbound ^{45}Ca should then be removed by washing once with 200 mL of 50% (v/v) ethanol for 5 min.
6. The membrane is then air-dried and wrapped with a plastic film and exposed to X-ray film (XAR–5 film, Kodak) for 2 d or until a satisfactory signal is obtained (*see* **Fig. 1**).

3.5. Ca^{2+} Binding

As mentioned in **Subheading 1.**, Ca^{2+} ions play a fundamental role in the structure and function of cadherin molecules. Calcium binding mediates dimer formation, which is a prerequisite for cadherin-mediated cell–cell adhesion. In this subsection, we describe an optimal method to measure the Ca^{2+}-binding parameters of cadherin extracellular fragments using an ion-selective electrode.

The measuring system consists of an Orion ionplus Sure-Flow electrode equipped with a Ca^{2+}-selective membrane. The electrode is coupled to an Orion perpHecT meter. The reading provided by the system is a difference in potential that varies with the calcium concentration in the solution. At low calcium concentrations, which is the case in most calcium binding experiments, the electrode's response is nonlinear and it is necessary to establish a calibration curve giving the differences in potentials for several standard solutions. The calcium binding measurements for the protein sample must be conducted immediately following the calibration of the calcium electrode.

3.5.1. Electrode Calibration Curve

The calibration curve should encompass the range of experimental calcium concentrations. Because calcium contaminants may affect the measurements at low levels, Ca^{2+}/EDTA or Ca^{2+}/EGTA buffers are used as standards *(14,15)*. In fact, the calibration curve is obtained by titrating an EDTA or EGTA solution with $CaCl_2$ standard solutions. The actual concentration of free calcium at equilibrium can be calculated theoretically using a specialized software (*see* **Note 4**).

1. A solution is prepared with the following composition: 5 mM EGTA, 100 mM KCl, 4.62 mM $CaCl_2$. The electrode is equilibrated in 5 mL of this solution placed in a 10-mL beaker. A microstirrer bar is used to ensure the homogeneity of the solution throughout the measurement. The stirrer bar should not come into contact with the calcium-selective membrane to avoid any damages. The solution is then discarded and replaced with 5 mL of the fresh solution.
2. Successive fractions of a standard $CaCl_2$ solution are added to the beaker. The reading is allowed to stabilize before being recorded. As the free Ca^{2+} concentration is affected by the pH, which undergoes minor fluctuations during the titration, the solution's pH is simultaneously measured and recorded for each titration point to calculate a more accurate value of $[Ca^{2+}_{free}]$. The Ca^{2+} electrode reading is given in mV and is plotted as a functions of $[Ca^{2+}_{free}]$ (*see* **Fig. 2**).

3.5.2. Calcium Titration of the Protein Solution

1. An 8 µM solution of the protein is prepared using previously decalcified water or buffer. Approximately 60 g of Chelex resin is needed to remove any traces of calcium ions present in the protein solution. The protein solution should contain at least 2.5 mM KCl to ensure a sufficient ionic strength for the proper function of the electrode. The final volume of the sample is 5 mL.
2. As for the calibration curve, successive fractions of a standard $CaCl_2$ solution are added to the sample. The electrode reading in mV is allowed to stabilize before being recorded.
3. The calibration curve is used to determine the free (Ca^{2+}) for each addition. This is performed best by using interpolation routines as provided by the MATLAB package (The Mathworks, Inc.).
4. A plot of r vs $[Ca^{2+}_{free}]$, where r is $[Ca^{2+}_{bound}]$/[protein], is fitted to an appropriate binding equation in order to determine the calcium dissociation constant(s). The most general description of the binding equilibrium is given by the Klotz-Adair model *(16)*. However, if additional assumptions are made about the binding sites, it is possible to interpret the data in terms of a multiple-site equation, or the Hill model *(17,18)*. The number of binding sites is determined by the bound Ca^{2+} concentration at saturation, i.e., from the curve's plateau, and the protein concentration. An aliquot of the protein sample is taken for amino acid analysis to obtain

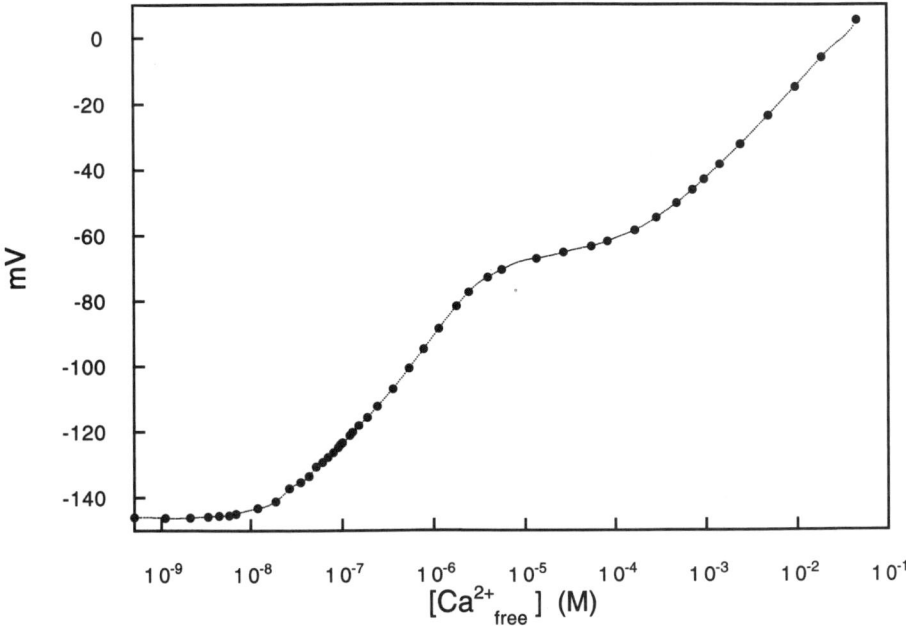

Fig. 2. Calibration curve of the Ca^{2+}-selective electrode using Ca^{2+}/EGTA buffers.

an accurate estimation of the protein concentration. **Figure 3** shows an example of a Ca^{2+}-binding measurement for E-cadherin.

4. Notes
1. Cell aggregation: There are two recently reported in vitro cell adhesion assays *(19)*: laminar flow adhesion assay using glass capillaries, and bead aggregation assay.
2. Expression of Ecad(1–225): The protocol described in **Subheading 3.3.2.** was originally developed for Ecad(1–145), but can be applied to Ecad(1–225). However, the NaCl gradients used with the ion exchange columns are different: 40–170 mM for the DEAE Sepharose column, and 50–165 mM for the FPLC/MonoQ.
3. ^{45}Ca blot: PVDF membranes have to be activated by 100% methanol, prior to the transfer. The membranes have to be completely air-dried before exposure to films. It is always useful to mark the side of the membrane that comes in contact with the gel so as to expose the same side to the photographic film. PVDF can be substituted by nitrocellulose. The half-life of ^{45}Ca is approx 163 d. If the ^{45}Ca stock is not fairly recent, then the amounts used should be adjusted to compensate for the decay. In most cases, the signal caused by cadherin-bound ^{45}Ca is not remarkably intense, especially when detecting truncated recombinants. There-

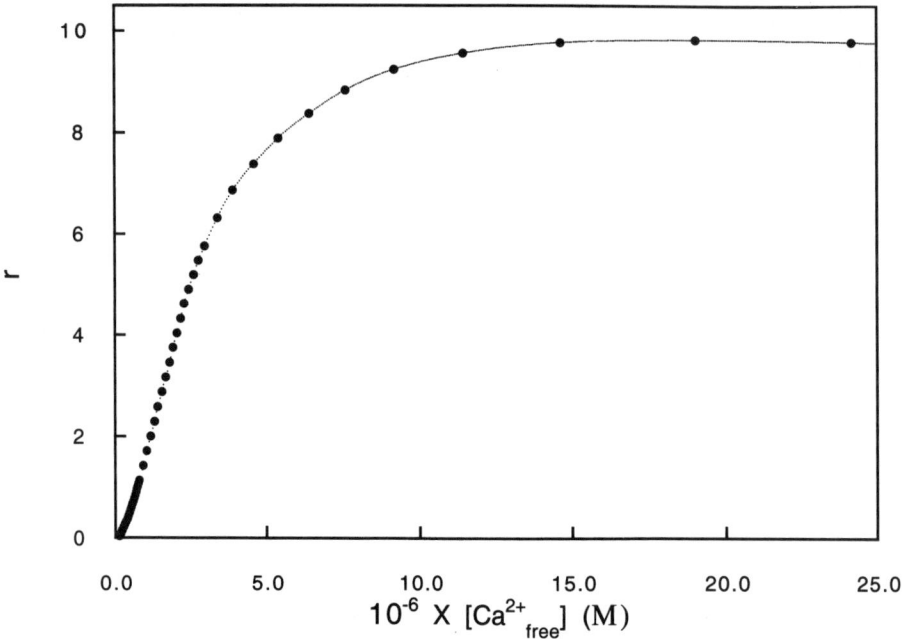

Fig. 3. Calcium-binding data obtained for the extracellular fragment of E-cadherin using the Ca^{2+}-selective electrode. The apparent dissociation constant is approx 10×10^{-6} M.

fore, it is advisable to include noninduced and subsequently induced total cell extracts on the same blot, so as to distinguish the new band corresponding to the expressed molecule. To our experience, recombinant cadherins always behave on SDS-PAGE as having an apparent molecular weight significantly larger than their actual ones.

4. Ca^{2+} electrode: The program "Free Calcium" calculates the free calcium concentration for a given combination of total Ca^{2+} and EDTA (or EGTA) concentrations. The calculation takes into account the effect of other ions like Mg^{2+}, Na^+, and K^+. The program was initially writen by Thierry Jean and ported to Apple Macintosh ZBasic by Werner Klee. The program is based on a method described by Fabiato and Fabiato *(20)*.

Acknowledgments

We are grateful to Professor Cyril Kay for his communication to the early version of this manuscript. This work was supported by grants to Mitsuhiko Ikura from the National Cancer Institute of Canada. Jean-René Alattia acknowledges the Caven post-Doctoral fellowship. Mitsuhiko Ikura is a Howard Hughes Medical Research International Research Scholar and a Scientist of the Medical Research Council of Canada.

References

1. Kartenbeck, J., Schmid, E., Franke, W. W., and Geiger, B. (1982) Different modes of internalization of proteins associated with adhaerens junctions and desmosomes: experimental separation of lateral contacts induces endocytosis of desmosomal plaque material. *EMBO J.* **1,** 725–732.
2. Takeichi, M. (1991) Cadherin cell adhesion receptors as a morphogenetic regulator. *Science* **251,** 1451–1455.
3. Gumbiner, B. M. (1996) *Cell* adhesion: the molecular basis of tissue architecture and morphogenesis. *Cell* **84,** 345–57.
4. Alattia, J.-R., Kurokawa, H., and Ikura, M. (1999) Structural view of cadherin-mediated cell-cell adhesion. *Cell. Mol. Life Sci.* **55,** 359–367.
5. Koch, A. W., Bozic, D., Pertz, O., and Engel, J. (1999) Homophilic adhesion by cadherins. *Curr. Opin. Struct. Biol.* **9,** 275–281.
6. Takeichi, M. and Nakagawa, S. (1999) Cadherin-dependent cell–cell adhesion in: *Current Protocols in Cell Biology* (Bonifacino, J. S., Dasso, M., Lippincott-Schwartz, J., Harford, J. B., and Yamada, K. M., eds.) Wiley, New York.
7. Tong, K. I., Yau, P., Overduin, M., Bagby, S., Porumb, T., Takeichi, M., and Ikura, M. (1994) Purification and spectroscopic characterization of a recombinant amino-terminal polypeptide fragment of mouse epithelial cadherin. *FEBS Lett.* **352,** 318–322.
8. Nagar, B., Overduin, M., Ikura, M., and Rini, J. M. (1996) Structural basis of calcium-induced E-cadherin rigidification and dimerization. *Nature* **380,** 360–364.
9. Alattia, J.-R., Ames, J. B., Porumb, T., Tong, K. I., Heng, Y. M., Ottensmeyer, P., et al. (1997) Lateral self-assembly of E-cadherin directed by cooperative calcium binding. *FEBS Lett.* **417,** 405–408.
10. Shatzman, A. R. and Rosenberg, M. (1986) Efficient expression of heterologous genes in *Escherichia coli*. The pAS vector system and its applications. *Ann. NY Acad. Sci.* **478,** 233–248.
11. Overduin, M., Harvey, T. S., Bagby, S., Tong, K. I., Yau, P., Takeichi, M., and Ikura, M. (1995) Solution structure of the epithelial cadherin domain responsible for selective cell adhesion. *Science* **267,** 386–389.
12. Overduin, M., Tong, K. I., Kay, C. M., and Ikura, M. (1996) 1H, 15N, and 13C resonance assignments and monomeric structure of the amino-terminal extracellular domain of epithelial cadherin. *J. Biomol. NMR* **7,** 173–189.
13. Maruyama, K., Mikawa, T., and Ebashi, S. (1984) Detection of calcium binding proteins by 45Ca autoradiography on nitrocellulose membrane after sodium dodecyl sulfate gel electrophoresis. *J. Biochem. (Tokyo)* **95,** 511–519.
14. Morton, R. W., Chung, J. K., Miller, J. L., Charlton, J. P., and Fager, R. S. (1986) Extended sensitivity for calcium selective electrode. *Anal. Biochem.* **157,** 345–352.
15. Linse, S., Brodin, P., Drakenberg, T., Thulin, E., Sellers, P., Elmdén, K., et al. (1987) Structure-function relationships in EF-hand Ca^{2+}-binding proteins. protein engineering and biophysical studies of calbindin D_{9k}. *Biochemistry* **26,** 6723–6735.
16. Fletcher, J. E., Spector, A. A., and Ashbrook, J. D. (1970) Analysis of macromolecule—ligand binding by determination of stepwise equilibrium constants. *Biochemistry* **10,** 4580–4587.

17. Dahlquist, F. W. (1978) The meaning of Scatchard and Hill plots. *Methods Enzymol.* **48,** 270–299.
18. Porumb, T. (1994) Determination of calcium binding constants by flow dialysis. *Anal. Biochem.* **220,** 227–237.
19. Brieher, W. M., Yap, A. S., and Gumbiner, B. M. (1996) Lateral dimerization is required for the homophilic binding activity of C-cadherin. *J. Cell Biol.* **135,** 487–496.
20. Fabiato, A. and Fabiato, F. J. (1979) Calculator programs for computing the composition of the solutions containing multiple metals and ligands used for experiments in skinned muscle cells. *J. Physiol. (Paris)* **75,** 463–505.

13

α-Lactalbumin and (Calcium-Binding) Lysozyme

Katsutoshi Nitta

1. Introduction

Lysozyme and α-lactalbumin have undoubtedly evolved from a common ancestor because of the similarity of their amino acid sequences *(1)*. Studies of α-lactalbumin and lysozyme published before 1990 have been reviewed by McKenzie and White *(2)*. α-Lactalbumin, which is a major milk component of milk whey, is a calcium-binding metalloprotein *(3)*. It is the so-called B component of lactose synthase *(4)* and acts as a specificity modifier of galactosyltransferase to convert it to lactose synthase *(5)*. The original discovery of its calcium-binding property was made when an effect of ethylenediaminetetracetic acid (EDTA) on the conformational stability of bovine α-lactalbumin was demonstrated. With the addition of 1 mM EDTA, the unfolding temperature decreased by 20–32 degrees. α-Lactalbumin was confirmed by flame spectrophotometry to bind one calcium ion tightly *(3)*. The binding constant for calcium ions to α-lactalbumin has been determined with a variety of methods and the results have been summarized by Kronman *(6)*.

Upon unfolding, α-lactalbumin changes its conformation from the native state through a molten globule state to the unfolded state *(7)*. Decalcification promotes the partial unfolding from the native to the molten globule state *(8)*. According to Kuwajima et al. *(9)*, the binding constant of bovine α-lactalbumin decreases from 10^{6-7} M^{-1} for the native state to 8.8×10^2 M^{-1} for the molten globule state; for the activated state between the native to the molten globule state, the binding constant has been determined to be 3.1×10^5 M^{-1} from unfolding-refolding kinetics.

In 1986, the structure of the calcium- binding site of α-lactalbumin was determined by X-ray structure analysis of the baboon α-lactalbumin *(10)*. The binding site consists of the two backbone carbonyl groups of residues 79 and

**Table 1
Binding Constants of a Calcium Ion to Calcium-Binding Lysozymes and α-Lactalbumin Determined with Quin-2 or Fura-2**

	n	K_{Ca}/M^{-1}
Canine lysozyme[b]	1.16	8.9×10^6
Equine lysozyme[a]	1.00	2.0×10^6
Ursine lysozyme[c]	1.06	5.2×10^7
Echidna lysozyme[b]	1.02	8.6×10^6
Pigeon lysozyme[a]	0.79	1.6×10^7
Bovine α-lactalbumin[a]	0.83	4.0×10^7

[a]Data from ref. *12*.
[b]Data from ref. *22*.
[c]Unpublished result (Miyasaka, Y., Kurokawa, Y., and Nitta, K.).

84 and three side-chain carboxyl groups of *Asp*-82, *Asp*-87, and *Asp*-88. These three aspartyl residues are all conserved throughout the α-lactalbumin subfamily. In the case of hen egg-white lysozyme, which does not bind calcium, the backbone conformation of the α-lactalbumin elbow *(10)*, the calcium-binding loop, is essentially conserved, although the side chains are radically altered. Furthermore, it was pointed out that the above three aspartyl residues are all conserved in equine milk lysozyme and it was suggested that equine milk lysozyme might be able to bind a calcium ion *(10)*.

The calcium-binding property of equine milk lysozyme was confirmed by Bio-Gel P–4 chromatography. With this method, contaminating free calcium and lysozyme were separated and the remaining bound calcium was analyzed with atomic absorption spectrophotometry *(11)*. Equine milk lysozyme binds calcium as tightly as α-lactalbumin. Further, pigeon egg-white lysozyme was also confirmed to bind a calcium ion *(12)*, and with the amino acid sequences of two calcium-binding lysozymes in hand, the evolution of α-lactalbumin and lysozyme could be discussed *(13,14)*. The binding constants for calcium ions to calcium-binding lysozymes are listed in **Table 1**. Amino acid sequences of α-lactalbumins and lysozymes are summarized in **refs.** *2* and *15*, respectively.

The binding site of calcium-binding lysozyme was later confirmed to have the same structure as that of α-lactalbumin with X-ray crystallography *(16)*. With the aid of Fourier-transformed infrared spectroscopy studies it was shown that three aspartyl residues of both bovine α-lactalbumin and equine lysozyme interact with the calcium ion in a pseudobridging mode *(17,18)*, which is distinct from that of the EF-hand where the calcium ion binds in a bidentate mode *(19)*.

Calcium-binding lysozymes have further been found in milk of donkey *(20)*, dog *(14)*, and echidna *(21)*. Milk lysozyme from brown bear (*Ursus arctos esoensis*) has preliminary been shown to be a calcium-binding lysozyme as well (Miyasaka, Y., Kurokawa, Y., and Nitta, K., unpublished results). Furthermore, from partial amino acid sequences, feline and hoofed seal milk lysozymes, and egg-white lysozymes from red-winged black birds are also classified as calcium-binding lysozyme *(15)*. Hoatzin lysozymes are clearly phylogenetically related to the other calcium-binding avian lysozymes, but they may not actually bind calcium in light of the absence of the aspartyl residue at position 85 *(15)*. Lysozymes from these calcium-binding lysozymes seem to be orthologous to conventional, noncalcium-binding lysozymes such as hen egg-white lysozyme and human lysozyme *(13)*. Namely, the lineages of calcium-binding lysozymes and conventional noncalcium-binding lysozymes were separated before the radiation of amniotes by gene duplication *(13)*. In fact, dogs have both calcium-binding and conventional lysozymes *(14)*. There is no evidence to indicate that the bound calcium ion is directly involved in the function of α-lactalbumin nor calcium-binding lysozyme. It seems to contribute merely to the stability of the protein. There are two lineages of calcium-binding lysozyme; one is eutherian milk lysozyme from perissodactyls (horse and donkey) and a carnivore (dog) and the other is prototherian milk lysozyme from echidna and avian calcium-binding lysozyme from pigeon *(14,15,22)*. The former shows a stable folding intermediate like α-lactalbumin, but the latter does not show it like conventional lysozyme *(22)*. Calcium-binding lysozymes of mammals identified to date are all from milk, including echidna. Many studies of the folding behavior of α-lactalbumin and calcium-binding lysozymes have been performed *(23–29)*. It is one of the characteristics of calcium-binding lysozyme that the molten globule state is much more stable and more native-like than that of α-lactalbumin. It is noteworthy that the native state of calcium-binding lysozyme is also more stable than that of α-lactalbumin and that it crystallizes without bound calcium *(30,31)*. For the methods to study the folding behavior of these proteins, consult another volume of this series (Methods in Molecular Biology, vol. 40, *Protein Stability and Folding*, Shirley, B.A., ed., 1995, Humana Press, NJ).

By introducing aspartyl residues to human lysozyme, which is not a calcium-binding lysozyme, Inaka et al. converted it into mutant lysozyme with a calcium-binding site *(32)*. This mutant lysozyme shows a high affinity for the calcium ion, as can be anticipated from the backbone conformation of the α-lactalbumin elbow *(33)*. Furthermore, several attempts have been made to partially introduce the binding site to human lysozyme, and to modify the calcium-binding ligands of bovine α-lactalbumin, and the results are summarized in **Table 2**. Among two-*Asp* mutants of human lysozyme, Q86D/D91Q/A92D

Table 2
Calcium-Binding Constants of Mutant Lysozymes of Which the Binding Site is Partially Introduced and Mutant Bovine α-Lactalbumins of Which the Calcium-Binding Ligands are Modified

Amino acid site		Binding constant (M^{-1})
Mutant of human lysozyme	86 91 92	
D86/92	D D D	5×10^{6a}, 8.2×10^{4c}, 1.9×10^{5e}
A83K/Q86D/A92D[e]	D D D	3.9×10^5
M4 lysozyme[b]	D D D	1×10^7, 0.9×10^7
A92D	Q D D	8×10^{3b}, ND[c,e]
Q86D[c,e]	D D A	ND
Q86D/D91Q/A92D[c]	D Q D	2.25×10^2
Mutant of bovine α-lactalbumin[d]	82 87 88	
Wild Type	D D D	4.8×10^6
D82A	A D D	2.5×10^6
D87A	D A D	ND
D88A	D D A	ND

ND: not detected, [a](*45*), [b](*46*), [c](*37*) (pH 4.5, 73°C), [d](*47*), [e](*48*) (pH 5.5, 30°C).

binds a calcium ion weakly and A92D binds even more weakly. However, in the case of α-lactalbumin, the mutant D82A (which corresponds to Q86D/D91Q/A92D of human lysozyme) binds as strongly as the wild type does, and the other two mutants do not bind a calcium ion, as shown in **Table 2**. On the other hand, Kim and Baum (*34*) have discussed, based on NMR studies of guinea pig α-lactalbumin, that D82 and D87 have the largest contribution to the stability of the calcium-saturated native state. These discrepancies have not yet been explained.

2. Materials
2.1. Materials for HPLC Determination

1. Biofine PO-4K gel-filtration column material.
2. 0.01 M HEPES buffer pH 7.5, containing 0.1 M KCl.
3. Chelex 100 (Bio-Rad).
4. 8 M urea stock solution.
5. Spectrofluorimeter.
6. Highly purified α-lactalbumin and lysozyme samples.

2.2. Materials for Quin-2 Competition Assay

1. Quin-2 calcium-binding dye.
2. Calcium stock solution 0.8 mM.
3. Spectrofluorimeter.

2.3. Materials for FT-IR Binding Studies

1. Bio-Gel P-4 column.
2. 0.01 M HCl.
3. D_2O.
4. 20 mM Tris-HCl, 0.2 M NaCl, pH 7.3.
5. Fourier transform infrared spectrometer.

3. Methods

3.1. Determination of Bound Calcium Ions by HPLC

In order to investigate a calcium-binding protein, one must first determine the number of calcium ions bound to that protein by means of, for example, atomic absorption spectrometry. However, using this method, contamination from labware is frequently encountered. One must take great care to eliminate contaminating nonbound calcium ions, and additionally with this method, the sample — once burned — is not recoverable.

To overcome such disadvantages, a microanalytical method has been developed using high-performance gel-filtration chromatography — together with the use of a calcium-specific fluorescent reagent (35). With this method, one can determine the number of bound calcium ions with less than 200 pmol of the protein. The binding strengths of calcium-binding proteins are high and, in many instances, exceed those of typical chelating reagents. Most calcium-binding proteins bind a calcium ion only when they are in their native conformation. Unfolded proteins bind the cation with far lower binding strength, if it binds at all. The reduction in the conformational stability of the protein results from the reduction of the binding strength. In this method, aqueous urea is used for destabilization of the protein, together with Quin-2, which binds calcium ions the most tightly among the fluorescent calcium ion indicators (36). It should be noted that the sample protein need not be denatured. It is sufficient if the apparent binding constant is decreased to be sufficiently lower than that of Quin-2 ($10^{7.1}$ M^{-1}) for this determination. With this method, one can determine bound calcium ions with sufficient accuracy by using less than 200 pmol of the protein. Moreover, this method is doubly specific for calcium; not only does Quin-2 show a high stability constant for calcium against magnesium, but also the emission spectrum of the calcium complex of Quin-2 is totally different from that of the magnesium complex (35).

A schematic diagram of the apparatus used is shown in **Fig. 1**. A Biofine PO–4K gel filtration column (Japan Spectroscopic, 300 mm × 7.5 mm ID, exclusion limit 4000 Dalton) was equilibrated with 0.01 M HEPES buffer (pH 7.5, 0.1 M KCl), which had been decalcified with a Chelex–100 ion-exchange column. The same buffer was introduced to the column at a flow rate of 1 mL/min by a high-performance liquid chromatographic (HPLC) pump (1) (Model

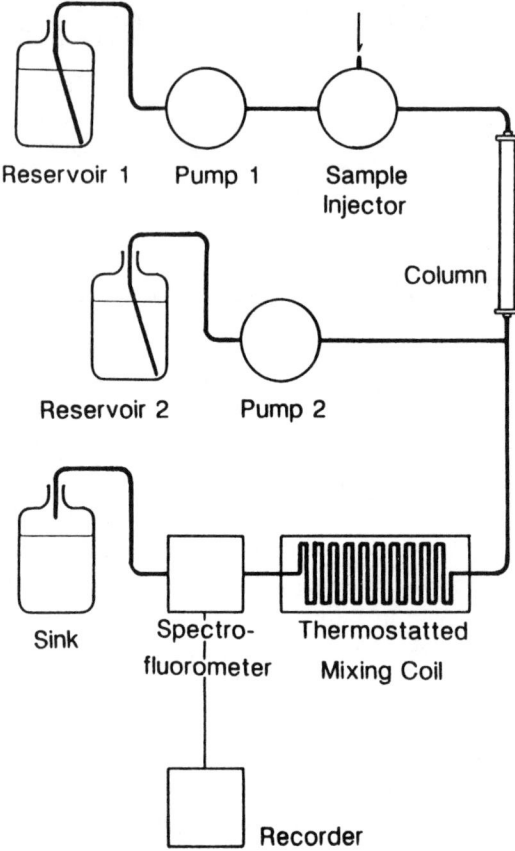

Fig. 1. Schematic diagram of the apparatus.

885PU; Japan Spectroscopic). A volume of 10 µL of sample or standard solution was applied from a sample injector with a microsyringe. With this column, macromolecular proteins and low-molecular-weight ions were separated. From reservoir 2, 8 M aqueous urea solution containing 20 µM Quin-2, which had also been decalcified, was transmitted with another pump *(2)* at a flow rate of 1 mL/min and combined with the eluate from the column. After passing through a 150-cm mixing coil, the solution was introduced to the flow cell of a spectrofluorimeter (Model FP 210; Japan Spectroscopic), with excitation at 335 nm and emission at 492 nm. The resulting fluorescence intensity was recorded with a Chromatocoder 12 (System Instruments, Tokyo, Japan) and integrated with time coordinates by an internal computer. Bound and contaminating calcium ions were detected at approx 5 and 13 min after injection, respectively (*see* **Fig. 2**).

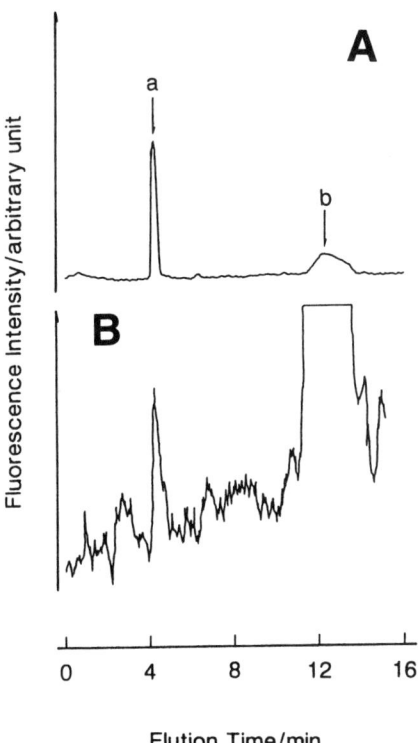

Fig. 2. Elution profiles of calcium bound to α-lactalbumin. Amount of α-lactalbumin loaded: **(A)** 10 µL of 100 µM; **(B)** 10 µL of 1 µM. Peak 1, elution peak of calcium bound to α-lactalbumin; peak 2, nonbound, contaminating calcium.

3.2. Determination of Binding Constant Using Quin-2 as a Competitor of the Binding

There are several methods to determine binding strength *(6)* including differential scanning calorimetry *(37,38)* and NMR *(39,40)*. Here we describe a microanalytical method using Quin-2 as a competitor of calcium-binding. In the case of α-lactalbumin, the protein changes conformation after decalcification at neutral pH and the ellipticity at 270 nm changes considerably. Consequently, the ellipticity can be a monitor of the amounts of calcium-binding to α-lactalbumin. EDTA is used to control the activity of the calcium ion, and the binding constant can be determined with a few mL of 30–40 µM α-lactalbumin solution *(8)*. Calcium-43 NMR competition experiments have also been used *(40)*, but these require much higher calcium and protein concentrations.

On the other hand, calcium-binding lysozyme does not change its conformation by decalcification in neutral pH *(12)*. In this case, the binding constant

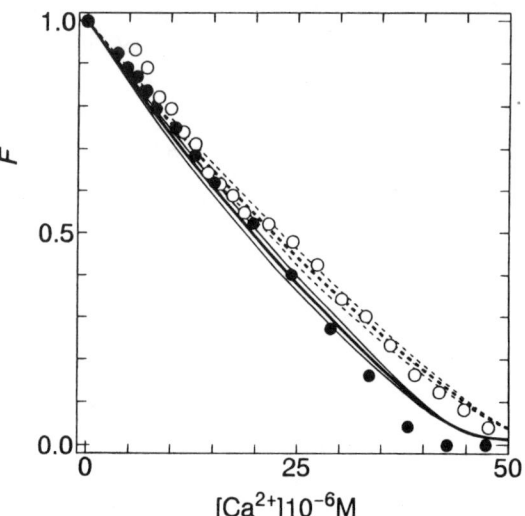

Fig. 3. Titration of Quin-2 in the presence of echidna milk lysozyme (*open circles*) and canine milk lysozyme (*filled circles*). Free fraction (F) of Quin-2 is plotted against the total concentration of calcium ions. The concentration of Quin-2 were 15.5 µM for echidna lysozyme and 13.5 µM for canine milk lysozyme. The concentration of lysozymes were 34.5 and 26.6 µM for echidna and canine milk lysozymes, respectively. Thick lines are the best fitted curves calculated with the parameters listed in **Table 1** by **Eq. 1**. Accompanying thin lines are calculated using the binding constants 1×10^6 M^{-1} less/more than the best-fitted values.

can be determined with the competition titration of decalcified lysozyme and Quin-2. The complex of Quin-2 and calcium can be detected with spectrophotometry.

Apo-lysozyme was dissolved to approx 30 µM in 0.01 M HEPES buffer (pH 7.5) with 0.1 M KCl and ca. 15 µM Quin-2. The solution was titrated with 0.8 mM calcium dichloride, and the absorbance was measured at 354 nm. The result is shown in **Fig. 3**.

The number of the binding sites n and binding constant of calcium ion to lysozymes K_{ca} were determined by the following equation *(12)*:

$$[Ca]_t = (1-F)/(FK_d) + (1-F)[Quin2]_t + \{nK_{Ca}(1-F)[L]_t\}/\{FK_d + (1-F)K_{Ca}\} \quad (1)$$

where F is the free fraction of Quin-2, which can be calculated from the change of the absorption at 354 nm, $[Ca]_t$, $[Quin2]_t$, and $[L]_t$ are total concentration of calcium ions, Quin-2, and lysozyme, respectively. Solid lines represent the best-fitted curves calculated by a trial-and-error method. As the value F is an implicit function of the total concentration of calcium ions, $[Ca]_t$, the usual least-squares method cannot be used in this case.

3.3. Determination of Calcium-Binding Mode with FT-IR

The COO⁻ groups can coordinate to metal ions in a number of ways, i.e., unidentate, bidentate, bridging, and pseudobridging modes. These coordinating modes are reflected in infrared absorption spectra. In the second derivative, and Fourier-self-deconvolved IR spectra for the Ca^{2+}-bound pike parvalbumin, a strong band found at 1553 cm⁻¹ reflects the bidentate coordinations of the COO⁻ groups of *Glu* residues in the Ca^{2+}-binding sites *(19)*. In the case of α-lactalbumin, the calcium is coordinated by seven oxygen atoms that form a slightly distorted pentagonal bipyramid. These oxygens are supplied by two main-chain carbonyl groups and three side chain carboxyl groups, and two water molecules *(10,41–43)*. Although there are certain similarities between the typical EF-hand motif of pike parvalbumin and the α-lactalbumin elbow, distinct differences have been found from the X-ray analysis *(10)*. The COO⁻ groups of D82, 87 and 88 of α-lactalbumin in the calcium-binding site are accessible to water molecules, so it is reasonable to consider that these side-chain COO⁻ groups coordinate to Ca^{2+} in the pseudobridging mode, where a divalent metal cation is bound to only one of the two oxygens in the COO⁻ group and a water molecule is hydrogen bonded to the other oxygen. ^{113}Cd NMR *(44)* studies, as well as ^{43}Ca NMR *(40)*, have suggested that the geometry of the EF-hand motif is different from that of the α-lactalbumin elbow.

It is described below how to use FT-IR spectra of bovine α-lactalbumin for the purpose of establishing the correlation between the COO-stretching frequencies of *Asp* side chains and the type of coordination, namely the pseudobridging mode.

The calcium-free form of α-lactalbumin, *apo*-α-lactalbumin, was obtained by chromatography on a Bio-Gel P-4 column equilibrated with 0.01 *M* HCl, followed by lyophilization. Complete deuteration of the NH groups of the main chain is required to examine the region of the COO⁻ antisymmetric stretching vibration (1610–1550 cm⁻¹), where the amide II band (the NH bending mixed with the CN stretching) gives rise to a strong absorption. The freeze-dried bovine *apo*-α-lactalbumin was dissolved in D_2O (99.8%) and incubated at 38°C for at least 60 min, long enough to ensure the complete exchange of any amide protons for deuterons. The solution was lyophilized after cooling at room temperature. *Apo*-protein solution was obtained by dissolving the powder of deuterated Ca^{2+}-free bovine α-lactalbumin in a D_2O solution containing 0.2 *M* NaCl and 20 m*M* Tris-HCl buffer (pH 7.3). No NH proton signals were observed in the ^1H NMR spectra of the deuterated proteins dissolved in D_2O

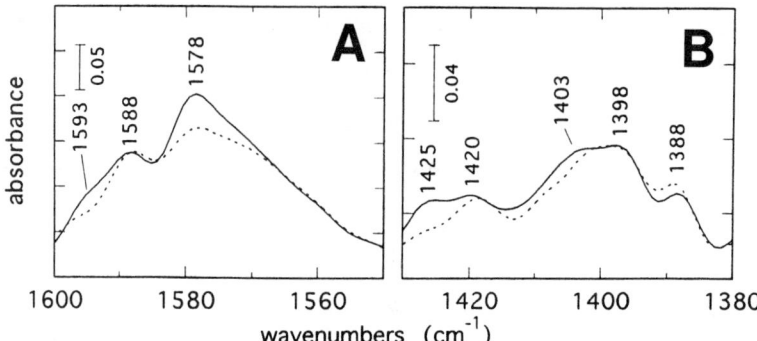

Fig. 4. Fourier-selfdeconvolved spectra of the regions of the COO-antisymmetric and symmetric stretching vibrations for bovine α-lactalbumin in D_2O (**A**) and H_2O (**B**) solutions. Solid and dotted lines represent the Ca^{2+}-bound, holo-, and Ca^{2+}-free, *apo*-form, respectively. Deconvolving parameters: 2, 16 cm^{-1}; L, 0.162 cm.

solution. Therefore, the bands around 1580 cm^{-1} are not contributed by the absorption of unexchanged amide II. Sample solution for the Ca^{2+}-bound, holo-form was obtained by adding $CaCl_2$ stock solution to the apo-protein solution described above. All reagents used in this experiment had been D_2O exchanged before use. The protein concentration employed was 5.5 mM.

Fourier-transform infrared (FT-IR) measurements were carried out at room temperature on a JEOL JIR-7000 FT-IR spectrophotometer equipped with an MCT detector (IR-DET 100) at 2 cm^{-1} resolution. To improve the signal to noise ratio, a total of 1000 sample scans and 1000 reference scans were taken for each spectrum, by using a shuttle devise. Dry air was constantly pumped into the spectrophotometer to eliminate water vapor that absorbs in the spectral region of interest. About 13 μL of the protein sample was placed between two CaF_2 plates separated by a 15-μm thick Teflon spacer. FT-IR absorption spectra of the buffer solution were collected for each spectrum in the same way and carefully subtracted from the sample spectrum. Fourier-self-deconvolved spectra were calculated using the software supplied by JEOL. A (triangle)^2 apodizing function was used with deconvolving parameters 2σ and L given in the figure captions.

The COO$^-$ antisymmetric and symmetric stretching bands of the deconvolved spectra for bovine α-lactalbumin are given in **Fig. 4**. The Fourier-selfdeconvolved spectrum of bovine *apo*-α-lactalbumin mainly shows two component bands at 1588 and 1578 cm^{-1} in the COO$^-$ antisymmetric stretching region whose frequencies are identical with those of holo-form. However, the intensity of the band at 1578 cm^{-1} of the holo-form seems to be stronger than that of the *apo*-form. The other apparent difference is the presence of the band at 1593 cm^{-1} in the spectrum of the holo-form, which is not present in the spectrum of the *apo*-form (*see* **Fig. 4A**).

Marked differences are clearly seen in the region of the COO⁻ symmetric stretching vibration as shown in **Fig. 4B**. The most notable features in the deconvolved spectra of bovine α-lactalbumin are as follows: Ca^{2+}-binding results in the loss of the peak area at about 1388 cm^{-1} and gains of intensities around 1425 and 1403 cm^{-1} bands compared with bovine *apo*-α-lactalbumin. These observations indicate that parts of the absorption around 1388 cm^{-1} in the spectrum of *apo*-form shifts to higher wavenumbers with splitting into two components at 1403 and 1425 cm^{-1} by binding of Ca^{2+} to the COO⁻ groups. The same results have been obtained with second derivative spectroscopy and difference spectroscopy *(17)*. These shifts of the frequencies of COO⁻ stretching were concluded to be caused by the coordination in the pseudobridging mode *(17)*, and the same shifts as those of bovine α-lactalbumin have been observed for equine milk lysozyme *(18)*.

References

1. Brew, K., Castellino, F. J., Vanaman, T. C. and Hill, R. L. (1970) The complete amino acid sequence of bovine α-lactalbumin. *J. Biol. Chem.* **245,** 4570–4582.
2. McKenzie, H. A. and White, F. H., Jr. (1991) Lysozyme and α-lactalbumin: structure, function, and interrelationships. *Adv. Protein Chem.* **41,** 173–315.
3. Hiraoka, Y., Segawa, T., Kuwajima, K., Sugai, S., and Murai, N. (1980) α-Lactalbumin: a calcium metalloprotein. *Biochem. Biophys. Res. Commun.* **95,** 1098–1104.
4. Brodbeck, U., Denton, W. L., Tanahashi, N., and Ebner, K. E. (1967) The isolation and identification of the B protein of lactose synthetase as α-lactalbumin. *J. Biol. Chem.* **242,** 1391–1397.
5. Brew, K., Vanaman, T. C., and Hill, R. L. (1970) The role of α-lactalbumin and the A protein in lactose synthase: a unique mechanism for the control of a biological reaction. *Proc. Natl. Acad. Sci. USA* **59,** 491–497.
6. Kronman, M. J. (1989) Metal-ion binding and the molecular conformational properties of α-lactalbumin. *Crit. Rev. Biochem. Mol. Biol.* **24,** 565–667.
7. Kuwajima, K., Nitta, K., Yoneyama, M., and Sugai, S. (1976) Three-state denaturation of α-lactalbumin by guanidine hydrochloride. *J. Mol. Biol.* **106,** 359–373.
8. Segawa, T. and Sugai, S. (1983) Interactions of divalent metal ions with bovine, human and goat α-lactalbumin. *J. Biochem.* **93,** 1321–1328.
9. Kuwajima, K., Mitani, M., and Sugai, S. (1989) Characterization of the critical state in protein folding. Effects of guanidine hydrochloride and specific Ca^{2+} binding on the folding kinetics of α-lactalbumin. *J. Mol. Biol.* **206,** 547–561.
10. Stuart, D. I., Acharya, K. R., Walker, N. P. C., Smith, S. G., Lewis, M., and Philips, D. C. (1986) α-Lactalbumin possesses a novel calcium binding loop. *Nature* **324,** 84–87.
11. Nitta, K. Tsuge, H., Sugai, S., and Shimazaki, K. (1987) The calcium-binding property of equine lysozyme. *FEBS Lett.* **223,** 405–408.

12. Nitta, K., Tsuge, H., Shimazaki, K., and Sugai S. (1988) Calcium-binding lysozymes. *Biol. Chem.* **369,** 671–675.
13. Nitta, K. and Sugai, S. (1989) The evolution of lysozyme and α-lactalbumin. *Eur. J. Biochem.* **182,** 111–118.
14. Grobler, J. A., Rao, K. R., Pervaiz, S., and Brew, K. (1994) Sequences of two highly divergent canine type c lysozymes: implications of the evolutionary origins of the lysozyme/α-lactalbumin superfamily. *Arch. Biochem. Biophys.* **313,** 360–366.
15. Prager, E. M. and Jolles, P. (1996) Animal lysozymes c and g: an overview, in *Lysozymes: Model Enzymes in Biochemistry and Biology* (Jolles, P., ed.), Bazel-Boston-Berlin, Birkhauzer Verlag, pp. 9–31.
16. Guss, J. M., Messer, M., Costello, M., Hardy, K., and Kumar, V. (1997) Structure of the calcium-binding echidna milk lysozyme at 1.9 Å resolution. *Acta Crystallogr.* D53, 355–363.
17. Mizuguchi, M., Nara, M., Kawano, K., and and Nitta, K. (1997) FT-IR study of the Ca^{2+}-binding to bovine α-lactalbumin. Relationships between the types of coordination and characteristics of the bands due to the Asp COO⁻ groups in the Ca^{2+}-binding site. *FEBS Lett.* **417,** 153–156.
18. Mizuguchi, M., Yue K., Kawano, K., Hiraoki, T., and Nitta, T. (1997) FT-IR Studies on the coordination of the side-chain COO⁻ groups to Ca^{2+} in equine lysozyme. *Eur. J. Biochem.* **250,** 72–76.
19. Nara, M., Tasumi, M., Tanokura, M., Hiraoki, T., Yazawa, M., and Tsutsumi, A. (1994) Marker bands for identifying the types of coordination of the side-chain COO⁻ groups to metal ions in pike parvalbumin (pI = 4. 10). *FEBS Lett.* **349,** 84–88.
20. Godovac-Zimmermann, J., Conti, A., and Napolitano, L. (1988) The primary structure of donkey (equus asinus) lysozyme contains the Ca(II) binding site of α-lactalbumin. *Biol. Chem.* **369,** 1109–1115.
21. Teahan, C. G., McKenzie, H. A., Shaw, D. C., and Griffiths, M. (1991) The isolation and amino acid sequences of echidna (tachyglossus aculeatus) milk lysozyme I and II. *Biochem. Int.* **24,** 85–95.
22. Kikuchi, M., Kawano, K., and Nitta, K. (1998) Calcium-binding and structural stability of echidna and canine lysozymes. *Protein Sci.* **7,** 2150–2155.
23. Nitta, K., Tsuge, H., and Iwamoto, H. (1993) Comparative study of the stability of the folding intermediates of the calcium-binding lysozymes. *Int. J. Protein Res.* **41,** 118–123.
24. Mizuguchi, M., Arai, M., Yue Ke, Nitta, K., and Kuwajima K. (1998) Equilibrium and kinetics of the folding of equine lysozyme studied by circular dichroism spectroscopy. *J. Mol. Biol.* **283,** 265–277.
25. Van Dael, H., Haezebrouck, P., Morozova, L., Ariko-Muendel, C., and Dobson, C. M. (1993) Partially folded states of equine lysozyme. Structural characterization and significance for protein folding. *Biochemistry* **32,** 11,886–11,894.
26. Griko, Y. V., Freire, E., Privalov, G., Van Dael, H., and Privalov, P. L. (1995) The unfolding thermodynanics of c-type lysozymes: a calorimetric study of the heat denaturation of equine lysozyme. *J. Mol. Biol.* **252,** 447–459.

27. Morozova, L., Haynie, D. T., Ariko-Muendel, C., Van Dael, H., and Dobson, C. M. (1995) Structural basis of the stability of a lysozyme molten globule. *Nat. Struct. Biol.* **2,** 871–875.
28. Morozova, L., Ariko-Muendel, C., Haynie, D. T., Emelyanenko, V. I., Van Dael, H., and Dobson, C. M. (1997) Structural characterisation and comparison of the native and A-states of equine lysozyme. *J. Mol. Biol.* **268,** 903–921.
29. Van Dael, H. (1998) Chimeras of human lysozyme and α-lactalbumin: an interesting tool for studying partially folded states during protein folding. *Cell. Mol. Life Sci.* **54,** 1217–1230.
30. Yao, M., Tanaka, I., Hikichi, K., and Nitta, K. (1992) Crystallization and preliminary X-ray structure analysis of pigeon egg-white lysozyme. *J. Biochem.* **111,** 1–3.
31. Tsuge, H., Ago, H., Noma, M., Nitta, K., Sugai, S., and Miyano, M. (1992) Crystallographic studies of a calcium binding lysozyme from equine milk at 2.5 Å resolution. *J. Biochem.* **111,** 141–143.
32. Inaka, K., Kuroki, R., Kikuchi, M., and Matsushima, M. (1991) Crystal structures of the apo- and holomutant human lysozymes with an introduced Ca^{2+} binding site. *J. Biol. Chem.* **266,** 20,666–20,671.
33. Kuroki, R., Nitta, K., and Yutani, K. (1992) Thermodynamic changes in the binding of Ca^{2+} to a mutant human lysozyme (D86/92). *J. Biol. Chem.* **267,** 24,297–24,301.
34. Kim, S. and Baum, J. (1998) Electrostatic interactions in the acid denaturation of α-lactalbumin determined by NMR. *Protein Sci.* **7,** 1930–1938.
35. Nitta, K. and Watanabe, A. (1991) Determination of calcium ions tightly bound to proteins. *J. Chromatogr.* **585,** 173–176.
36. Tsien, R. Y. (1980) New calcium indicators and buffers with high selectivity against magnesium and protons: design, synthesis, and properties of prototype structures. *Biochemistry* **19,** 2396–2404.
37. Koshiba, T., Tsumoto, K., Masaki, K., Kawano, K., Nitta, K., and Kumagai, I. (1998) Thermal study of mutant human lysozymes with partially introduced Ca^{2+} binding sites by using efficient refolding system from inclusion bodies. *Protein Eng.* **11,** 683–690.
38. Koshiba, T., Hayashi, T., Ishido, M., Kumagai, I., Ikura, T., Kawano, K., et al. (1999) Expression of a synthetic gene encoding canine milk lysozyme in *Escherichia coli* and characterization of the expressed protein. *Protein Eng.* **12,** 429–435.
39. Tsuge, H., Koseki, K., Miyano, M., Shimazaki, K., Chuman, T., Matsumoto, T., et al. (1991) A structural study of calcium-binding equine lysozyme by two-dimensional 1H-NMR. *Biochim. Biophys. Acta* **1078,** 77–84.
40. Aramini, J. M., Drakenberg, T., Hiraoki, T., Nitta, K., Yue, K., and Vogel, H. J. (1992) Calcium-43 NMR studies of calcium binding lysozymes and α-lactalbumin. *Biochemistry* **31,** 6761–6768.
41. Acharya, K. R., Stuart, D. I., Walker, N. P. C., Lewis, M., and Philips, D. C. (1989) Refined structure of baboon α-lactalbumin at 1.7 Å resolution comparison with C-type lysozyme. *J. Mol. Biol.* **208,** 99–127.
42. Acharya, K. R., Ren, J., Stuart, D. I., Philips, D. C., and Fenna, R. E. (1991) Crystal structure of human α-lactalbumin at 1.7 Å resolution. *J. Mol. Biol.* **221,** 571–581.

43. Pike, A. C. W., Brew, K., and Acharya, K. R. (1996) Crystal structures of guinea-pig, goat and bovine α-lactalbumin highlight the enhanced conformational flexibility of regions that are significant for its action in lactose synthase. *Structure* **4,** 691–703.
44. Aramini, J. M., Hiraoki, T., Yue, K., Nitta, K., and Vogel, H. J. (1995) Cadmium-113 NMR studies of bovine and human α-lactalbumin and equine lysozyme. *J. Biochem.* **117,** 623–628.
45. Kuroki, R., Kawakita, S., Nakamura, H., and Yutani, K. (1989) Entropic stabilization of a mutant human lysozyme by calcium binding. *Proc. Natl. Acad. Sci. USA* **89,** 6803–6807.
46. Haezebrouck, P., Baestlier, A. D., Joniau, M., Van Dael, H., Rozenberg, S., and Hanssens, I. (1993) Stability effects associated with the introduction of a partial and a complete Ca^{2+}-binding site into human lysozyme. *Protein Eng.* **6,** 643–649.
47. Anderson, P. J., Brooks, C. L., and Berliner, L. J. (1997) Functional identification of calcium binding residues in bovine α-lactalbumin. *Biochemistry* **36,** 11,648–11,654.
48. Kuroki, R. and Yutani, K. (1998) Structural and thermodynamic responses of mutations at a Ca^{2+} binding site engineered into human lysozyme. *J. Biol. Chem.* **273,** 34,310–34,315.

14

Recombinant Annexin II Tetramer

Hyoung-Min Kang, Nolan R. Filipenko, Geetha Kassam, and David M. Waisman

1. Introduction

The annexins are a family of approx 13 proteins that bind to acidic phospholipids in a Ca^{2+}-dependent manner *(1)*. The amino acid sequence of the annexins indicates four repeats (eight repeats in the case of annexin VI) of about 70 amino acids, which are highly homologous. The crystal structures of several annexins have been determined, providing the basic structure of an annexin *(2–4)*. Each repeated sequence comprises one compact domain that consists of five α-helices wound into a right-handed superhelix. The domains are arranged in a planar, cyclic array with a convex and concave side. The convex side faces the biological membrane and contains the Ca^{2+}- and phospholipid-binding sites. The concave side contains an N-terminus of about 40 amino acid residues and the C-terminal domain.

Annexin II (p36) is unique among the annexins in that the N-terminus of the protein contains a high-affinity binding site for a dimeric protein (monomeric M_r 11,000, p11), which is a member of the S100 family of Ca^{2+}-binding proteins *(5)*. The heterotetrameric complex formed by these proteins ($p36_2p11_2$), referred to as annexin II tetramer or AIIt, is the predominant species in most cells (*see* **Fig. 1**). Studies of the total annexin II content in endothelial, epithelial, and MDCK cells have established that at least 90–95% of the total annexin II content of these cells is in the heterotetrameric form (AIIt) *(6,7)*. AIIt was initially shown to be present at the cytosolic surface of the plasma membrane of many cells *(8,9)*. In the case of secretory cells, such as adrenal medulla or anterior pituitary cells *(8,9)*, AIIt has been shown to form crosslinks between secretory granules and plasma membrane. This has led to the suggestion that AIIt is involved in exocytosis or endocytosis *(10)*.

From: *Methods in Molecular Biology*, vol. 172:
Calcium-Binding Protein Protocols, Vol. 1: Reviews and Case Studies
Edited by: H. J. Vogel © Humana Press Inc., Totowa, NJ

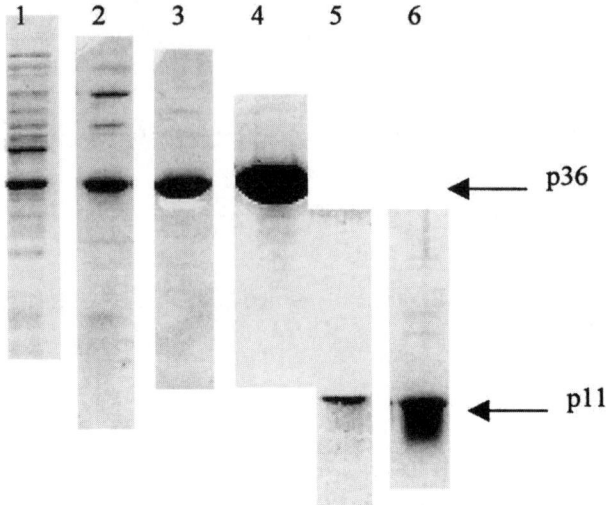

Fig. 1. SDS-PAGE analysis of each step in the purification of recombinant annexin II (*lane 1*, *E. coli* lysate; *lane 2*, hydroxyapatite fraction; *lane 3*, heparin-Sepharose fraction; *lane 4*, Sephacryl S-100 gel-filtration fraction) and recombinant p11 (*lane 5*, Fast S fraction; *lane 6*, Sephacryl S-100 gel-filtration fraction).

Annexin II and AIIt are fundamentally distinct proteins in terms of in vitro activities and subcellular distribution *(11,12)*. Both of the proteins have been shown to exist at the extracellular surface of the plasma membrane of several cell lines including endothelial cells *(13–17)*. Extracellular annexin II has been suggested to be important in several biological processes, such as fibrinolysis *(18)*, cell–cell adhesion, tenascin-mediated cell signaling *(19)*, and virus infection *(20)*. However, because in the majority of experiments annexin II and p11 antibodies were not used together, it was unclear if the annexin II immunoreactivity corresponded to annexin II alone or to the p36 subunit of AIIt. Recently, annexin II *(16,21)* or annexin II tetramer *(17)* has been suggested to be a plasminogen receptor that stimulates the activation of plasminogen.

Previous attempts at producing AIIt from recombinant p11 and annexin II have been unsuccessful *(22)* because their subcloning strategy allowed the expression of a recombinant annexin II protein that contained six additional amino acid residues (MRGSFK) plus the cDNA initiation codon-derived *Met* at the N-terminus. The recombinant annexin II with this extended N-terminus could not bind the p11 subunit and therefore did not form AIIt *(23)*. In this chapter, we describe a recombinant bacterial expression system that allows the expression and purification of recombinant p11 and annexin II on the order of milligrams (at least approx 15 mg/L culture for each p11 and annexin II), and the formation of recombinant AIIt. Overall comparison of the tissue AIIt and recombinant AIIt showed that both of the proteins were structurally and functionally similar to each other *(12)*. The availability of large amounts of recombinant AIIt will facilitate further characterization of the structure–function relationships of the protein.

2. Materials
2.1. Materials for Bacterial Expression Vector Construction

1. pAED4.91 or pET bacterial expression vectors.
2. BL21(DE3) *Escherichia coli* competent cells.
3. Human annexin II and p11 cDNA.
4. PCR primers (*see* **Subheading 3.1.1.** for primer sequences).
5. DNA thermal cycler.
6. PCR buffer: 10 mM Tris-HCl, pH 8.3, 50 mM KCl, 2.5 mM MgCl$_2$.
7. Agarose gel electrophoresis apparatus.
8. DNA ligation kit (Takara).
9. dNTPs (dATP, dCTP, dTTP, dGTP).
10. Amp *Taq* (Perkin-Elmer) polymerase.
11. DNA sequencing kit (Perkin-Elmer/Applied Biosystems Inc.).
12. Isopropyl thio-β-D-thiogalactopyranoside (IPTG) (dilute 1 g with ddH$_2$O to a final volume of 8 mL to give a 0.5 M stock solution).
13. Modified LB (Luria-Bertani) bacterial growth media: 10 g/L bacterial tryptone, 5 g/L yeast extract, 5 g/L NaCl (*see* **Note 1**).
14. *Pyrococcus furiosis* (*Pfu*) DNA polymerase (Stratagene).
15. *Nde*I restriction endonuclease (Pharmacia).
16. Polyacrylamide gel-electrophoresis (PAGE) apparatus.
17. Western blot apparatus.
18. Anti-annexin II and anti-p11 monoclonal antibody (Calbiochem).

2.2. Materials for Protein Purification

1. Modified LB bacterial growth media.
2. IPTG.
3. French press.
4. Ceramic hydroxyapatite columnn (Bio-Rad) equilibrated in 10 mM potassium phosphate (KPi) buffer, pH 7.5 and 1 mM dithiothreitol (DTT).
5. Heparin-Sepharose affinity column (Pharmacia) equilibrated in 50 mM HEPES buffer, pH 7.5, and 1 mM DTT.
6. Sephacryl S-100 gel-filtration column (Pharmacia) equilibrated in 40 mM HEPES, pH 7.5, 150 mM NaCl, 0.1 mM ethylene glycol-*bis N,N,N',N'*-tetraacetic acid (EGTA), 1 mM DTT (gel-filtration buffer).
7. PAGE apparatus.
8. Fast Q column (Pharmacia) equilibrated in 20 mM imidazole buffer, pH 7.5, and 1 mM DTT.

9. Fast S column (Pharmacia) equilibrated in 50 mM MES, pH 6.0, and 1 mM DTT.
10. p11 lysis buffer: 200 mM Tris-HCl, pH 7.5, 200 mM NaCl, 10 mM MgCl$_2$, 1 mM NaN$_3$, 2 mM DTT.
11. p11 dialysis buffer: 20 mM imidazole, pH 7.5, 50 mM NaCl, 2 mM DTT, 0.5 mM ethylenediaminetetracetic acid (EDTA), 0.5 mM EGTA, 1 mM NaN$_3$, 0.2 mM phenyl methyl sulfonyl fluoride (PMSF), 1 mM diisopropyl fluorophosphate (DIFP), and 10 µg/L leupeptin.
12. Annexin II lysis buffer: 20 mM Tris-HCl, pH 7.5, 150 mM NaCl, 5 mM EGTA, 1 mM DTT, 0.2 mM PMSF, 1 mM DIFP, and 10 µg/L leupeptin.
13. Annexin II dialysis buffer: 20 mM Tris-HCl, pH 7.5, 25 mM NaCl, 1 mM DTT.
14. Formation of recombinant annexin II tetramer.

2.3. Materials for Formation of Recombinant Annexin II Tetramer

1. Sephacryl S-100 gel-filtration column equilibrated in gel-filtration buffer.
2. PAGE apparatus.

3. Methods

3.1. Construction of Human Annexin II and p11 Bacterial Expression Vector

1. Perform PCR in a thermal cycler using the following primers: 5'-ATGCCAT CTCAAATGGAACACGCCATG-3' (5' primer) and 5'-CTACTTCTTTCCCT TCTGCTTCATGTGTACTACAAAATAG-3' (3' primer) for p11 and 5'-ATGTC TACTGTTCACGAAATCC-3' (5' primer) and 5'-TCAGTCATCTCCACCACA CAG-3' (3' primer) for annexin II. These sequences are derived from the nucleotide sequence of human p11 and annexin II reported by *(24,25)*, respectively. The 50 µL PCR mixture contains 0.2 mM of each dNTP, 1 µM of each primer, 2.5 U of Amp*Taq* polymerase (Perkin-Elmer), 5 ng of cDNA templates, and enough PCR buffer to bring to volume. Denature the sample at 94°C for 2 min followed by 35 cycles of 1 min of denaturation at 94°C, 1 min of annealing at 55°C, and 2 min of extension at 72°C. Complete the PCR by having one cycle of 7 min of extension at 72°C. [The cDNA template vectors, YEp51-p11 and pcDX-p36, were kindly donated by Dr. Carl Creutz, University of Virginia and Dr. Tony Hunter, Salk Institute, respectively.]
2. Analyze the PCR product by agarose gel electrophoresis. The expected size for p11 and annexin II is 294 basepairs and 1020 basepairs, respectively.
3. Blunt-end ligate the PCR product to *Nde*I (Pharmacia)-digested (*see* **Notes 2** and **3**) and *Pfu* DNA polymerase (Stratagene)-filled-in pAED4.91 (kindly donated by Dr. Donald Doering, Massachusetts Institute of Technology) or pET (Stratagene) using a DNA ligation kit (Takara).
4. Transform the BL21(DE3) *E. coli* competent cells with the ligation mixture.
5. Grow a 2 mL culture of the *E. coli* colonies at 37°C in modified LB containing 50 µg/mL ampicillin until the A$_{600\ nm}$ reaches 0.5.
6. Induce for expression by adding 0.8 mM IPTG. After 4 h of induction, harvest the cells in an Eppendorf tube by centrifugation at 12,000*g* for 2 min using a microcentrifuge.

7. Confirm the expression of recombinant p11 and annexin II by sodium dodecyl sulfate-polyacrylamide gel electrophoresis (SDS-PAGE), Western blot using anti-p11 and antiannexin II monoclonal antibody (Calbiochem), and DNA sequencing of the plasmid DNA using T7 and reverse pET primer (*see* **Note 4**).

3.2. Bacterial Cell Culture for Protein Overexpression

1. Grow a 20-mL culture of BL21(DE3) *E. coli*, carrying the expression constructs pAED4.91-annexin II or pAED4.91-p11 overnight at 37°C in modified LB media.
2. Add the culture to 1 L of modified LB media with 50 µg/mL of ampicillin and grow it at 37°C until the $A_{600\,nm}$ reaches 0.5.
3. Induce for expression by adding 0.8 m*M* IPTG. After 4 h of induction, harvest the cells by centrifugation at 6000*g* for 20 min.

3.3. Purification of Recombinant p11

1. *Lyse* the bacterial pellet containing recombinant p11 in p11 lysis buffer using a French press (1000 psi) (*see* **Note 5**). After centrifugation at 100,000*g* for 1 h, dialyze the supernatant overnight against p11 dialysis buffer.
2. Apply the dialyzed supernatant to a Fast Q column (Pharmacia) and wash with 10-fold volume of column buffer.
3. Adjust the Fast Q flowthrough and wash to pH 6.0 with solid MES, (2-[N-Morpholino ethanesulfonic acid]) apply to a Fast S column (Pharmacia) and wash with 10-fold volume of column buffer.
4. Elute the bound proteins with a linear salt gradient of 0–0.5 *M* NaCl in column buffer. The recombinant p11 is eluted at 250 m*M* NaCl (*see* **Note 6**).
5. Concentrate the fractions containing the recombinant p11 to have a volume of less than 4 mL and apply to a Sephacryl S-100 gel-filtration column equilibrated with gel-filtration buffer. The recombinant p11 is eluted from the Sephacryl S-100 column at an elution volume of approx 160 mL, corresponding to a molecular weight of about 27 kDa for a globular protein (*see* **Note 7**).
6. Collect the purified recombinant p11 and store it at –70°C at a concentration of at least 1 mg/mL in gel-filtration buffer (*see* **Note 8**) (*see* **Fig. 1**).

3.4. Purification of Recombinant Annexin II

1. Lyse the bacterial pellet containing pAED4.91-annexin II in annexin II LB using a French press apparatus at 1000 psi (again, *see* **Note 5**).
2. After centrifugation at 100,000*g* for 1 h, dialyze the supernatant overnight against 8 L of annexin II dialysis buffer.
3. Apply the dialyzed sample to a ceramic hydroxyapatite column (Bio-Rad) and wash with 10-fold volume of column buffer (*see* **Note 9**).
4. Elute the bound proteins with a linear phosphate gradient of 0–0.5 *M* potassium phosphate, pH 7.5, which results in the elution of recombinant annexin II at 300 m*M* potassium phosphate.
5. Dialyze the fractions containing recombinant annexin II overnight against 8 L of annexin II dialysis buffer.
6. Apply the dialyzed fractions to a heparin-Sepharose affinity column (Pharmacia) and wash with 10-fold volume of column buffer (*see* **Note 10**).

7. Elute the bound proteins using a linear gradient of 0–0.5 M NaCl in column buffer, resulting in the elution of recombinant annexin II at 150 mM NaCl.
8. Concentrate the fractions containing recombinant annexin II to a volume of less than 4 mL and apply to a Sephacryl S-100 gel-filtration column (Pharmacia). The recombinant annexin II is eluted from the Sephacryl S-100 column at an elution volume of approx 145 mL, corresponding to a molecular weight of about 38 kDa for a globular protein. Collect the purified recombinant annexin II and store it at –70°C at a concentration of at least 1 mg/mL in gel-filtration buffer (*see* **Fig. 1**).

3.5. Formation of Recombinant Annexin II Tetramer

1. Concentrate approx 10 mg of the purified recombinant annexin II and approx 4 mg of purified recombinant p11 to approx 2 mL of volume each.
2. Mix them together with a 1:1 molar ratio (i.e., combine the two 2-mL fractions) and incubate on ice overnight (*see* **Note 11**).
3. Centrifuge the solution at 14,000g for 15 min using a microcentrifuge at 4°C.
4. Apply the supernatant to a Sephacryl S-100 gel-filtration column. The elution profile gives a large peak of recombinant annexin II tetramer (rAIIt) at the elution volume of approx 110 mL, followed by small peaks of recombinant annexin II and p11 (*see* **Fig. 2**).
5. Upon concentrating analysis of fractions using SDS-PAGE, collect the fractions that have recombinant annexin II and recombinant p11 in the same lane. After concentration to at least 1 mg/mL, store the protein at –70°C in gel-filtration buffer. Typically, approx 90% of the recombinant annexin II combines with approx 90% of the recombinant p11 and forms the rAIIt.

4. Notes

1. The original constituents of LB media were 10 g/L of bacterial tryptone, 10 g/L of NaCl, and 5 g/L of yeast extract. However, we have found that BL21(DE3) *E. coli* grows better in a modified LB media in which the NaCl concentration has been reduced by half (5 g/L).
2. There are several commercially available expression vectors that allow the expression of a protein of interest as a fusion protein for the convenience of purification. Even after removal of a purification tag of a fusion protein by a specific proteolytic enzyme, the protein usually contains a few additional amino acid residues that are not part of genuine coding sequence. Depending on the protein, which is the case with annexin II, this may cause serious problems with functionality.
3. *Nde*I is not an efficient restriction enzyme. Sometimes overnight incubation at 37°C with mineral oil on top of the incubation mixture is required. Based on our own experience, Pharmacia *Nde*I is highly recommended. After digestion of the vector DNA with *Nde*I, it is recommended to run an agarose gel electrophoresis and extract the linearized vector DNA from the gel.
4. To screen the transformed *E. coli* colonies, perform the SDS-PAGE first and pick the colonies that give the p11 and or annexin II (p36) band. Then, analyze the band

Recombinant AIIt

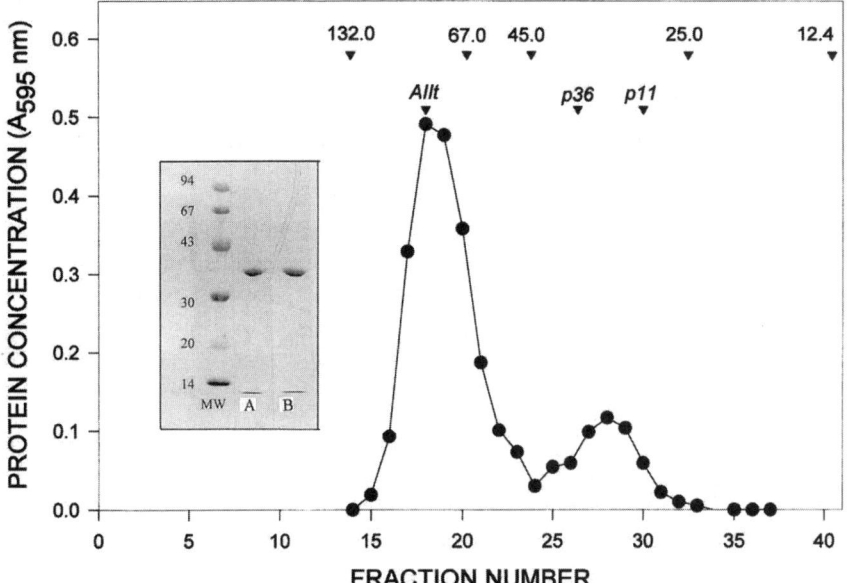

Fig. 2. Formation of recombinant annexin II tetramer. Recombinant annexin II (0.260 µmol) and recombinant p11 (0.258 µmol) were mixed and incubated overnight on ice in 40 mM Tris-HCl, pH 7.5, 1 mM DTT, 0.15 M NaCl, and 0.1 mM EGTA. The mixture was applied to a Sephacryl S-100 gel-filtration column equilibrated with the same buffer and fractions were analyzed for protein concentration. The position in the elution profile of several globular proteins of known molecular weight is provided. For comparison, the position in the elution profile of bovine lung AIIt (*AIIt*), recombinant annexin II (*p36*), and recombinant p11 (*p11*) is also indicated. Inset: SDS-PAGE analysis of pooled recombinant AIIt (fractions 18–21), (**A**) and bovine lung AIIt, (**B**).

by Western blot using anti-p11 or antiannexin II monoclonal antibody. DNA sequencing of the plasmid DNA isolated from colonies that have shown either positive p11 or annexin II (p36) band in SDS-PAGE and Western blot analysis completes the screening of the colonies.

5. When resuspending the *E. coli* paste in LB, it is crucial to ensure that the resultant solution has the proper viscosity. We have found that the more dilute the solution is, the better the French press lyses the *E. coli*. The optimal dilution is a minimum 20-fold dilution in relation to the volume of the *E. coli* paste. For example, a 5 mL *E. coli* paste should be diluted in at least 100 mL of lysis buffer. If the solution is not dilute enough, the *E. coli* will not be lysed completely, and also the p11 and annexin II will be relatively insoluble and located in the pellet after the 100,000*g* centrifugation step following cell lysis.

6. When eluting the p11 from the Fast S column, it is absolutely critical that the tubes used to collect the fractions have approx 1 mL of 1 *M* Tris-HCl, pH 7.5

(when an individual fraction volume is 5 mL) in them to readjust the pH of the solution.
7. The p11 is eluted at a volume from the gel-filtration column corresponding to a molecular weight of 27 kDa because it forms a homodimer in solution.
8. In SDS-PAGE analysis of p11, 15% polyacrylamide gel gives the best result. When using a lower-percentage gel, the p11 band tends to run with the dye front, giving unpredictable staining/destaining results.
9. The ceramic hydroxyapatite is much more rigid than the crystalline form. In fact, we have found that the crystalline form is not rigid enough to support a sufficient flow rate. Instead of using the ceramic hydroxyapatite, the dialyzed annexin II *E. coli* lysate supernatant can be applied to a Fast S column equilibrated with 50 mM MES, pH 6.0, 1 mM DTT, and subsequently eluted with a linear NaCl gradient of 0–1 M. However, the supernatant must be adjusted to pH 6.0 with solid MES before application onto the column. Also, when eluting the column, the tubes used to collect the fractions must have 1 mL of 1 M Tris-HCl, pH 7.5 (when an individual fraction volume is 5 mL) to readjust the pH of the solution.
10. It is imperative that the heparin-Sepharose column be regenerated using the method supplied by the manufacturer. A simple wash with high-salt solution (i.e., 1 M NaCl) will not sufficiently clean off some contaminants bound to the column.
11. The N-terminus of annexin II (the p11-binding site) is easily proteolyzed, thus disabling annexin II to bind to p11. Therefore, when incubating recombinant p11 and annexin II on ice overnight to form the recombinant annexin II tetramer (rAIIt), it is advantageous to add protease inhibitors (DIFP, leupeptin, aprotinin, and so on) to the solution to ensure the most complete rAIIt formation.

Acknowledgments

We would like to acknowledge the support of grants from the Medical Research Council (MT 13343), the Alberta Heart and Stroke Foundation and the NIH (1RO1CA78639-01). We also thank Dr. Donald Doering (Massachusetts Institute of Technology) for generously providing the expression vector pAED4.91.

References

1. Kaetzel, M. A. and Dedman, J. R. (1995) Annexins: novel Ca^{2+}-dependent regulators of membrane function. *News Physiol. Sci.* **10,** 171–176.
2. Swairjo, M. A., Concha, N. O., Kaetzel, M. A., Dedman, J. R., and Seaton, B. A. (1995) Ca^{2+}-bridging mechanism and phospholipid head group recognition in the membrane-binding protein annexin V. *Nat. Struct. Biol.* **2,** 968–974.
3. Luecke, H, Chang, B. T., Mailliard, W. S., Schlaepfer, D. D., and Haigler, H. T. (1995) Crystal structure of the annexin XII hexamer and implications for bilayer insertion. *Nature* **378,** 512–515.
4. Favier-Perron, B., Lewit-Bentley, A., and Russo-Marie, F. (1996) The high-resolution crystal structure of human annexin III shows subtle differences with annexin V. *Biochemistry* **35,** 1740–1744.

5. Demange, P., Voges, D., Benz, J., Liemann, S., Gottig, P., Berendes, R., et al. (1994) Annexin V: the key to understanding ion selectivity and voltage regulation? *Trends Biochem. Sci.* **19**, 272–276.
6. Gerke, V. and Weber, K. (1984) Identity of p36K phosphorylated upon Rous sarcoma virus transformation with a protein purified from brush borders; calcium-dependent binding to non-erythroid spectrin and F-actin. *EMBO J.* **3**, 227–233.
7. Nilius, B., Gerke, V., Prenen, J., Szücs, G., Heinke, S., Weber, K., and Droogmans, G. (1996) Annexin II modulates volume-activated chloride currents in vascular endothelial cells. *J. Biol. Chem.* **271**, 30,631–30,636.
8. Nakata, T., Sobue, K., and Hirokawa, N. (1990) Conformational change and localization of calpactin I complex involved in exocytosis as revealed by quick-freeze, deep-etch electron microscopy and immunocytochemistry. *J. Cell Biol.* **110**, 13–25.
9. Senda, T., Okabe, T., Matsuda, M., and Fujita, H. (1994) Quick-freeze, deep-etch visualization of exocytosis in anterior pituitary secretory cells: localization and possible roles of actin and annexin II. *Cell Tiss. Res.* **277**, 51–60.
10. Swairjo, M. A. and Seaton, B. A. (1994) Annexin structure and membrane interactions: a molecular perspective. *Annu. Rev. Biophys. Biomol. Struct.* **23**, 193–213.
11. Waisman, D. M. (1995) Annexin II tetramer: structure and function. *Mol. Cell. Biochem.* **149/150**, 301–322.
12. Kang, H. M., Kassam, G., Jarvis, S. E., Fitzpatrick, S. L., and Waisman, D. M. (1997) Characterization of human recombinant annexin II tetramer purified from bacteria: role of N-terminal acetylation. *Biochemistry* **36**, 2041–2050.
13. Yeatman, T. J., Updyke, T. V., Kaetzel, M. A., Dedman, J. R., and Nicolson, G. L. (1993) Expression of annexins on the surfaces of non-metastatic and metastatic human and rodent tumor cells. *Clin. Exp. Metastasis* **11**, 37–44.
14. Wright, J. F., Kurosky, A., and Wasi, S. (1994) An endothelial cell-surface form of annexin II binds human cytomegalovirus. *Biochem. Biophys. Res. Commun.* **198**, 983–989.
15. Chung, C. Y. and Erickson, H. P. (1994). Cell surface annexin II is a high affinity receptor for the alternatively spliced segment of tenascin-C. *J. Cell Biol.* **126**, 539–548.
16. Cesarman, G. M., Guevara, C. A., and Hajjar, K. A. (1994) An endothelial cell receptor for plasminogen/tissue plasminogen activator (t-PA). II. Annexin II-mediated enhancement of t-PA- dependent plasminogen activation. *J. Biol. Chem.* **269**, 21,198–21,203.
17. Kassam, G., Choi, K. S., Ghuman, J., Kang, H. M., Fitzpatrick, S. L., Zackson, T., Zackson, S., et al. (1998) The role of annexin II tetramer in the activation of plasminogen. *J. Biol. Chem.* **273**, 4790–4799.
18. Choi, K. S., Ghuman, J., Kassam, G., Kang, H. M., Fitzpatrick, S. L., and Waisman, D. M. (1998) Annexin II tetramer inhibits plasmin-dependent fibrinolysis. *Biochemistry* **37**, 648–655.
19. Chung, C. Y., Murphy-Ullrich, J. E., and Erickson, H. P. (1996) Mitogenesis, cell migration, and loss of focal adhesions induced by tenascin-C interacting with its cell surface receptor, annexin II. *Mol. Biol. Cell* **7**, 883–892.

20. Siever, D. A. and Erickson, H. P. (1997) Extracellular annexin II. *Int. J. Biochem. Cell Biol.* **29,** 1219–1223.
21. Hajjar, K. A., Jacovina, A. T., and Chacko, J. (1994) An endothelial cell receptor for plasminogen/tissue plasminogen activator. I. Identity with annexin II. *J. Biol. Chem.* **269,** 21,191–21,197.
22. Thiel, C., Osborn, M., and Gerke, V. (1992) The tight association of the tyrosine kinase substrate annexin II with the submembranous cytoskeleton depends on intact p11- and Ca(2+)-binding sites. *J. Cell Sci.* **103,** 733–742.
23. Jost, M., Weber, K., and Gerke, V. (1994) Annexin II contains two types of Ca^{2+}- binding sites. *Biochem. J.* **298 (Pt 3),** 553–559.
24. Kube, E., Weber, K., and Gerke, V. (1991) Primary structure of human, chicken, and Xenopus laevis p11, a cellular ligand of the src-kinase substrate, annexin II. *Gene* **102,** 255–259.
25. Huang, K. S., Wallner, B. P., Mattaliano, R. J., Tizard, R., Burne, C., Frey, A., et al. (1986) Two human 35 kDa inhibitors of phospholipase A_2 are related to substrate of $pp60^{v\text{-}src}$ and of the epidermal growth receptor/kinase. *Cell* **46,** 191–199.

15

Purification and Characterization of ALG-2

A Novel Apoptosis-Linked Ca^{2+}-Binding Protein

Mingjie Zhang and Kevin W.-H. Lo

1. Introduction

It has been known for decades that Ca^{2+} is actively involved in the processes called programmed cell death or apoptosis (for reviews, *see* **refs.** *1* and *2*). An elevation of cellular Ca^{2+} concentration was observed in glucocorticoid-induced apoptosis of lymphocytes *(3)*, and the apoptotic effects of glucocorticoid can be mimicked by Ca^{2+} ionophores *(4)*. Similarly, release of Ca^{2+} from internal Ca^{2+}- store using thapsigargin (an endoplasmic reticulum Ca^{2+}- ATPase inhibitor) also triggers cell death via apoptotic mechanisms *(5)*. Ca^{2+}-associated apoptosis can be inhibited by extracellular Ca^{2+}-chelators such as glycol-*bis* (β-aminoethyl ether)-*N,N,N,'N'*-tetraacetic acid (EGTA) *(6)* or by the overexpression of a Ca^{2+} buffering protein such as calbindin D_{28K} *(7)*. Although numerous insights have been made with respect to the characterization of Ca^{2+}'s involvement in apoptosis, the detailed role of the ion in these processes remains unclear. The uncertainty that existed regarding the absolute nature of Ca^{2+}'s relationship to apoptosis was largely caused by the lack of Ca^{2+}-dependent molecular targets, which are directly associated with cell death.

Recently, Vito et al. discovered a novel EF-hand Ca^{2+}-binding protein, named apoptosis linked gene-2 (ALG-2), during a functional screening for genes involved in apoptosis. ALG-2 was shown to be required for T-cell receptor, *Fas*-, and glucocorticoid-induced apoptosis as inhibition of protein expression using antisense cDNA resulted in cellular resistance to apoptosis stimuli *(8)*. The gene has been shown to be expressed in all mouse tissues with the highest expression occurring in thymus. Depletion of ALG-2 using antisense RNA does not change the activity of Caspase-9, indicating that ALG-2 is likely to function downstream of the Caspase family of proteases *(9)*. The expression of ALG-2 in T-cell hybridoma is not regulated by apoptosis stimuli *(8)*, suggesting that the apoptotic function of ALG-2 is likely to be regulated by Ca^{2+}-binding to the protein. ALG-2 consists of a single polypepide chain of 191 amino acid residues possessing five EF-hand Ca^{2+}-binding sites *(8)*. Detailed structural characterization has demonstrated that ALG-2 contains two strong Ca^{2+}-binding sites with K_D values in the micromolar range. The protein was found to exist as a dimer in solution, and dimerization occurs via the fifth EF-hand of the protein. The binding of Ca^{2+} induces significant conformational changes of ALG-2 *(10)*. Using a yeast two-hybrid system, a novel ALG-2-binding protein called AIP1/Alix was discovered; the interaction between ALG-2 and AIP1/Alix is indeed Ca^{2+}-dependent *(11,12)*.

In this chapter, we will first describe methods for the efficient purification of large quantities of wild-type and Ca^{2+}-binding deficient mutants of ALG-2. Then, we present a method for studying the dimerization of ALG-2 by means of chemical crosslinking. The method can also be used to probe Ca^{2+}-induced conformational changes of the protein.

2. Materials

2.1. Purification of Wild-Type and Ca^{2+}-Binding Deficient Mutants of ALG-2

1. Bacterial expression plasmid pET3a (Novagen) containing mouse *ALG-2* gene (available from the author).
2. Bacterial strain: *Escherichia coli* BL21(DE3).
3. L-broth (LB) bacterial growth medium (or M9 for isotope labeling).
4. Temperature adjustable orbital shaker.
5. Isopropyl-β-D-thiogalactopyranoside (IPTG).
6. French press cell lysis apparatus (*see* **Note 1**).
7. DEAE-Sepharose column (2.5 cm × 25 cm).
8. Sephacryl-200 gel-filtration column (2.6 cm × 100 cm).

2.2. Chemical Crosslinking of ALG-2

1. HEPES buffer: 20 mM, pH 8.0.
2. Lysine specific chemical crosslinker disuccinimidyl glutarate (DSG, Pierce; spacer length of approx 7.7 D) and 1,5-difluoro-2,4-dinitrobenze (DTNB, Pierce; spacer length approx 3 Å).
3. Triton X-100.

3. Methods

3.1. Expression and Purification of Wild-Type ALG-2

The following expression and purification protocols allow us to obtain about 100 mg of pure ALG-2 from each liter of bacterial culture within 4 d.

3.1.1. ALG-2 Expression

1. Use a single colony of the host BL21(DE3) strain transformed with pET3a containing ALG-2 from a fresh agar plate, to inoculate a culture tube containing 10 mL LB medium and 100 μg/mL of ampicillin. Incubate the tube overnight at 37°C in an orbital shaker (250 rpm).
2. The next morning, transfer the overnight culture to a 2.8-L shake flask containing 1 L LB medium supplemented with 100 μg/mL of ampicillin.
3. Incubate the culture at 37°C until OD_{600} reaches approx 1.0. At this point, take the flask out and cool the culture to 25°C using a cold-water bath or ice bath with gentle shaking (*see* **Note 2**).
4. Induce ALG-2 expression by adding IPTG to the cooled culture with a final concentration of 0.5 mM. The culture was kept incubated at 25°C for approx 12 h with shaking.
5. Harvest the cell by centrifuging the culture at 5000g at 4°C, and store the cell pellet at –80°C for further use.

3.1.2. Purification of Wild Type ALG-2

All of the purification procedures are carried out in a cold room.

1. Thaw the frozen cell pellet from 2 L of culture in 25-mL lysis buffer (50 mM Tris-HCl buffer pH 7.5, containing 2 mM ethylenediaminetetracetic acid (EDTA), 1 mM phenyl methyl sulfonyl fluoride (PMSF), 1 μg/mL leupeptin, and 1 mg/mL antipain). Make sure the cells are completely resuspended (*see* **Note 3**).
2. Lyse cells with three passes through a French press (*see* **Note 1**).
3. Reduce the viscosity of the suspension by sonication (3 min at full power at 50% duty cycle). Keep the cell suspension on ice during the sonication to avoid heating of the mixture.
4. Clarify the suspension by centrifugation at 30,000g, 4°C for 30 min.
5. Pour the supernatant into a 50-mL centrifuge tube, and $CaCl_2$ was added to a final concentration of 3 mM with stirring (*see* **Note 4**). The mixture is incubated on ice for 5–10 min.
6. Centrifuge the suspension at 25,000g for 15 min, and discard the supernatant.
7. Dissolve the pellet with 50 mM Tris-HCl (pH 7.5) containing 2 mM EDTA (*see* **Note 5**).
8. Repeat **steps 4–7** twice.
9. Load the final EDTA-solublized ALG-2 crude mixture to a DEAE-Sepharose column preequilibrated with 50 mM Tris-HCl buffer containing 2 mM EDTA (pH 7.5).
10. Wash the column extensively with the same buffer, and then elute the protein with a linear gradient of 0 to 0.5 M NaCl (*see* **Note 6**).
11. Pool ALG-2-containing fractions, and concentrate the pooled solution using ultrafiltration to a volume of 20 mL or less.
12. Load the protein mixture to a Sephacryl-200 gel-filtration column preequilibrated with 50 mM Tris-HCl buffer containing 1 mM EDTA (pH 7.5), and elute the protein using the same buffer.

13. Combine ALG-2-containing fractions, and dialyze the pure protein against 0.1% NH_4HCO_3 solution before being lyophilized for storage (*see* **Note 7**).

3.1.3. Expression and Purification of Ca^{2+}-Binding Deficient Mutants of ALG-2

In earlier work, we have shown that ALG-2 contains two strong, regulatory Ca^{2+}-binding sites (site I and III). Mutation of the 12th *Glu* residue in these two Ca^{2+}-binding sties can disrupt Ca^{2+}-binding capacity and Ca^{2+}-induced conformational changes of the protein *(10)*. Owing to the lack of Ca^{2+}-induced precipitation property, the mutants could not be purified using **steps 4–8** described in **Subheading 3.1.2.** Instead, these mutant forms of ALG-2 were purified using a modified method.

1. Perform the identical expression procedures for the mutant ALG-2 as described in **Subheading 3.1.1.**
2. Carry out the purification of the mutant ALG-2 at the same manner as described in **Subheading 3.1.2., steps 1–4**.
3. Pour the supernatant into a 50-mL glass beaker on an ice bath, slowly add $(NH_4)_2SO_4$ with constant stirring to a final concentration of 30%. Incubate the mixture on ice with gentle stirring for 15 min, and then clarify the suspension by centrifuge.
4. Pour the supernatant to a clean beaker, and further add $(NH_4)_2SO_4$ to reach a final concentration of 40%. Spin down the precipitant that contains the crude ALG-2 (*see* **Note 7**).
5. Dissolve the pellet with 50 mM Tris-HCl (pH 7.5) containing 2 mM EDTA.
6. Repeat **steps 9–13** in **Subheading 3.1.2.** for the rest of the purification.

3.1.4. Chemical Crosslinking of ALG-2 in the Absence and Presence of Ca^{2+}

Sequence analysis and mutational studies have suggested that ALG-2 is a member of calpain small subunit subfamily protein with five EF-hand Ca^{2+}-binding sites *(10,13)*. However, molecular-weight determination of ALG-2 using analytical gel-filtration chromatography in the presence of EDTA has shown that the protein was eluted as a monomer. We have interpreted this as that the association between the two monomers of ALG-2 is too weak to be detected by gel-filtration chromatography. Therefore, we have used lysine-specific crosslinking reagent, DSG, to probe such weak interactions in solution. The method can also be used to study the Ca^{2+}-induced stabilization of the dimer of ALG-2.

1. Dissolve buffer-free ALG-2 in 20 mM HEPES buffer (pH 8.0) containing 50 mM NaCl, 1 mM DTT to a final protein concentration of 0.5 mg/mL (*see* **Note 8**). To crosslink Ca^{2+}-free ALG-2, one should include 1 mM EDTA in the HEPES buffer.

In contrast, 1 mM of Ca^{2+} and 0.5% of Triton X-100 should be included in the HEPES buffer if one wants to crosslink Ca^{2+}-bound form of ALG-2 (*see* **Note 9**).
2. Add DSG to the protein sample to a final concentration of 0.5 mM. DSG was prepared as a stock solution in dimethyl sulfoxide (0.1 M). To minimize the solvent effect, one should keep the final dimethyl sulfoxide (DMSO) concentration in the reaction mixture below 5% (v/v) (*see* **Note 10**).
3. Incubate the mixture at 25°C for 30 min before quenching the reaction with 0.5 M Tris-HCl buffer (pH 7.5).
4. Analyze the reaction products using 15% sodium dodecyl sulfate-polyacrylamide gel electrophoresis (SDS-PAGE).

4. Notes

1. One can use lysozyme or other cell lysis method to replace French press.
2. We note that essentially all ALG-2 was expressed as inclusion bodies at 37°C. However, the majority of the protein became soluble when the expression temperature was lowered to 25°C. To obtain high-level expression at 25°C, it is necessary to extend the duration of protein expression.
3. It is essential to include 2 mM or more EDTA in the lysis buffer as EDTA keeps ALG-2 in the supernatant.
4. In order to completely precipitate ALG-2 by Ca^{2+}, make sure the final concentration of Ca^{2+} in the supernatant is higher than that of EDTA.
5. The ALG-2-containing pellet can be efficiently resuspended with gentle homogenization using a hand-held homogenizer.
6. We found that pure ALG-2 (>95%) can be obtained after the DEAE-Sepharose column. Small amounts of contaminants can be further removed by gel-filtration column. The ALG-2 sample purified after the gel-filtration column appeared as a single band on a silver-stained SDS-PAGE gel.
7. The wild-type ALG-2 can also be efficiently purified with the $(NH_4)_2SO_4$ fractionation method instead of Ca^{2+}-precipitation/EDTA-solublization procedures. We found that the method described in **Subheading 3.1.2.** gives a slightly higher yield and purity of the protein.
8. Because DSG specifically reacts with free amine groups, one has to make sure that proteins to be crosslinked are devoid of amino group containing buffer such as Tris-HCl. In addition, one should also avoid the use of phosphate buffers for Ca^{2+}-binding proteins as Ca^{2+} will form precipitates with the PO_4^{3-} in the buffer. To avoid any nonspecific crosslinking reactions, the protein concentration should be kept low (0.5 mg/mL). It is advisable to perform a series of crosslinking reaction with different protein concentrations (e.g., 0.1 mg/mL to 5 mg/mL) to ensure the specificity of the reaction. The specificity of the reaction can also be increased by using low molar ratio of DSG to free amino groups of the protein (approx 10:1 in our case).
9. ALG-2 undergoes Ca^{2+}-induced conformational changes in solution. The protein forms soluble aggregates upon the addition of Ca^{2+}, and such aggregation is likely resulted from the association of the solvent-exposed hydrophobic surface of the

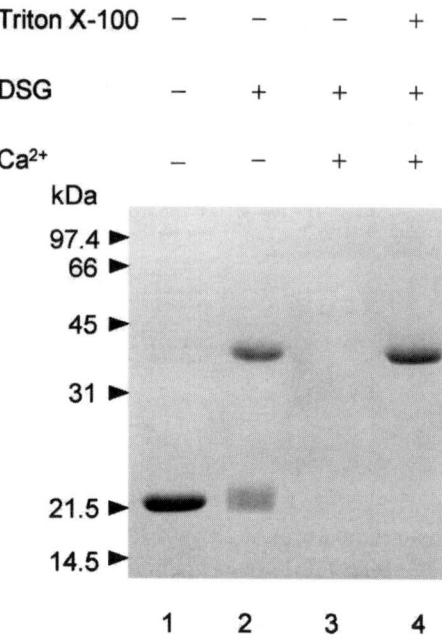

Fig. 1. Chemical crosslinking of lysine residues of ALG-2 with DSG. Ca^{2+}-free ALG-2 was used as the control of the monomeric protein (*lane 1*). Treatment of Ca^{2+}-free ALG-2 with DSG results an extra band at a molecular mass approx 42 kDa, corresponding to the covalently linked ALG-2 dimer (*lane 2*). One can notice that there is a significant portion of ALG-2 remained as a monomer after crosslinking. Cross-linking of Ca^{2+}-bound ALG-2 in the absence of Triton X-100 resulted covalently linked high molecular mass aggregates that were not able to migrate into the separating gel (*lane 3*). Inclusion of 0.5% Triton X-100 in the crosslinking reaction buffer quantitatively shifts ALG-2 into its dimeric species (*lane 4*).

protein *(10,13)*. Crosslinking of Ca^{2+}-bound ALG-2 without Triton X-100 results in covalently linked aggregates (*see* **Fig. 1**, lane 3). However, addition of small amount of Triton X-100 (as low as 0.1%) in the reaction buffer resulted in quantitatively crosslinked dimer (*see* **Fig. 1**, lane 4); whereas the Ca^{2+}-free ALG-2 has a much lower amount of dimer product under the same crosslinking condition (*see* **Fig. 1**, lane 2). The data indicates that the Ca^{2+}-bound form of ALG-2 assumes a much more stable dimer than the Ca^{2+}-free form of the protein. However, analytical gel-filtration chromatography studies showed that Ca^{2+}-bound ALG-2 exists predominantly as a monomer in solution, as the protein was eluted at a molecular mass of approx 21 kDa (*see* **Fig. 2**). A small amount of Triton X-100 should also be included in the protein buffer in order to avoid complications from the nonspecific aggregation of the protein in studying Ca^{2+}-induced conformational changes of the protein using various spectroscopic techniques.

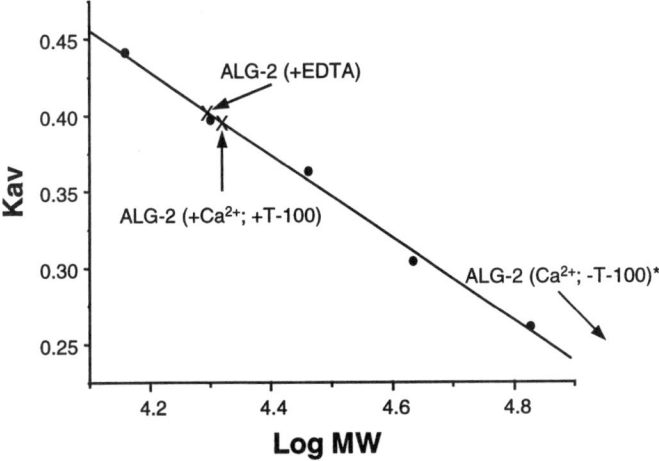

Fig. 2. Molecular mass determination of ALG-2 by analytical gel-filtration chromatography. 100 μL of proteins (1 mg/mL) were injected into a Superdex 200 HR 10/30 column (Amersham Pharmacia Biotech) preequilibrated with 50 mM Tris-HCl, pH 7.5, containing 0.15 M NaCl, 1 mM DTT, 1 mM EDTA (or Ca^{2+}). The column was calibrated using ribonuclease A (14 kDa), Trypsin inhibitor (20 kDa), carbonic anhydrase (29 kDa), ovalbumin (44 kDa), bovine serum albumin (66 kDa), and blue dextran 2000 kDa. K_{av} is defined as $(V_e-V_o)/(V_t-V_o)$, where V_e is the elution volume of a protein, V_o and V_t are the void volume and total volume of the column. Note that without Triton X-100, Ca^{2+}-bound ALG-2 (*) was eluted at a void volume of the column.

10. The distances between the amino groups of the two crosslinked lysine residues (*Lys* 137 in ALG-2, **ref. 10**) is in the range of approx 7 Å; as treatment of the protein under the same condition using another lysine specific crosslinker (DTNB) with a shorter space arm (approx 3 Å) did not result in any covalently linked dimeric protein (data not shown).

References

1. Nicotera, P. and Orrenius, S. (1998) The role of calcium in apoptosis. *Cell Calcium* **23**, 173–180.
2. Trump, B. F. and Berezesky, I. K. (1995) Calcium-mediated cell injury and cell death. *FASEB J.* **9**, 219–228.
3. Kaiser, N. and Edelman, I. S. (1977) Calcium dependence of glucocorticoid-induced lymphocytolysis. *Proc. Natl. Acad. Sci. USA* **74**, 638–642.
4. Kaiser, N. and Edelman, I. S. (1978) Further studies on the role of calcium in glucocorticoid-induced lymphocytolysis. *Endocrinology* **103**, 936–942.
5. Jiang, S., Chow, S. C., Nicotera, P., and Orrenius, S. (1994) Intracellular Ca^{2+} signals activate apoptosis in thymocytes: studies using the Ca^{2+}-ATPase inhibitor thapsigargin. *Exp. Cell Res.* **212**, 84–92.

6. McConkey, D. J., Nicotera, P., Hartzell, P., Bellomo, G., Wyllie, A. H., and Orrenius, S. (1989) Glucocorticoids activate a suicide process in thymocytes through an elevation of cytosolic Ca^{2+} concentration. *Arch. Biochem. Biophys.* **269**, 365–370.
7. Dowd, D. R., MacDonald, P. N., Komm, B. S., Haussler, M. R., and Miesfeld (1992) Stable expression of the calbindin-D28K complementary DNA interferes with the apoptosis pathway in lymphocytes. *Mol. Endocrinol.* **6**, 1843–1848.
8. Vito, P., Lacanà, E., and D'Aamio, L. (1996) Interfering with apoptosis: Ca^{2+}-binding protein ALG-2 and Alzheimer's disease gene ALG-3. *Science* **271**, 521–525.
9. Lacana, E., Ganjei, J. K., Vito, P., and D'Aamio, L. (1997) Interfering with apoptosis: Ca^{2+}-binding protein ALG-2 and Alzheimer's disease gene ALG-3. *J. Immunol.* **158**, 5129–5135.
10. Lo, K. W.-H., Zhang, Q., Li, M., and Zhang, M. (1999) Apoptosis-linked gene product ALG-2 is a new member of the calpain small subunit subfamily Ca^{2+}-binding protein. *Biochemistry* **38**, 7498–7505.
11. Vito, P., Pellegrin, L., Guiet, C., and D'Aamio, L. (1999) Cloning of AIP1, a novel protein that associates with the apoptosis-linked gene ALG-2 in a Ca^{2+}-dependent reaction. *J. Biol. Chem.* **274**, 1533–1540.
12. Missotten, M., Nichols, A., Rieger, K., and Sadoul, R. (1999) Alix, a novel mouse protein undergoing calcium-dependent interaction with the apoptosis-linked-gene 2 (ALG-2) protein. *Cell Death Diff.* **6**, 124–129.
13. Maki, M., Yamaguchi, K., Kitaura, Y., Satoh, H., and Hitomi, K. (1998) Calcium-induced exposure of a hydrophobic surface of mouse ALG-2, which is a member of the penta-EF-hand protein family. *J. Biochem.* **124**, 1170–1177.

16

Crystallization and Structural Details of Ca²⁺-Induced Conformational Changes in the EF-Hand Domain VI of Calpain

Miroslaw Cygler, Pawel Grochulski, and Helen Blanchard

1. Introduction

Calpains are calcium-regulated neutral cysteine proteases that include ubiquitous, as well as tissue-specific, isoforms. The ubiquitous isoforms, µ- and m-calpains are intracellular, nonlysosomal proteases *(1)*. The tissue-specific isoforms include calpain 3, which is found in skeletal muscle, and stomach-specific nCL–2 *(2)*. The calpains catalyze limited proteolysis of substrates involved in cytoskeletal remodeling and signal transduction, although definitive physiological roles are not yet ascertained. They are also thought to contribute to the tissue damage that follows ischemia and reperfusion in conditions such as stroke and cardiac infarct *(3,4)*, stimulating a search for specific and clinically acceptable inhibitors aimed at both the active site and also the Ca^{2+}-binding domains *(5)*. The calpains are heterodimers that consist of an 80-kDa catalytic subunit (the large subunit), and a 30-kDa regulatory subunit (the small subunit).

On the basis of amino acid sequence the large subunit appears to contain four domains (dI–dIV) and the small subunit contains two domains (dV, dVI) *(6)*. Domain dII contains the active-site residues and has some resemblance to cysteine proteases from the papain superfamily *(7)*. Each subunit has a C-terminal Ca^{2+}-binding domain, both of which contain five potential EF-hand motifs with two having a close match to the Ca^{2+}-binding consensus sequence *(8,9)*. These domains play a regulatory role in calpain enzymatic activity. Calpain isoforms differ most significantly in their in vitro Ca^{2+} requirement for activity, which is in the range 5–50 µM for µ-calpain, and approx 350 µM for m-calpain. In contrast, both calpains are activated in vivo at low micromolar Ca^{2+} concentrations, by mechanisms that are not yet fully explained *(10–12)*. Because the large subunits of µ- and m-calpain share approx 60% amino acid sequence identity and the small subunits are identical, then the difference in Ca^{2+} requirement must derive from differences between the large subunits.

The process of activation by calcium appears not to strictly require autolysis because the active site has been shown to be functional before autolysis *(13)*. However, autolysis of the small subunit rapidly follows the addition of calcium, whereas autolysis of the large subunit differs between µ- and m-calpain *(14)*. The half-lives of the autolyzed forms are very short, which limits their time of action in vivo. Calcium is essential for activation and the physiological functions and regulation of the calpains are thought to be linked to Ca^{2+}-regulated signaling pathways. However, because the physiological substrates of the calpains have not been unambiguously identified, these questions remain open. On much indirect evidence, it appears likely that calpains play a regulatory role, rather than simply a degradative role, in cellular metabolism.

Biochemical studies suggest that the association of the two calpain chains occurs mainly through the interactions between the Ca^{2+}-binding domains dIV and dVI *(15)*. The 184 amino acid protein, corresponding to dVI, Met87-Ser270 of the small subunit, forms homodimers in solution *(16)*, which are likely formed in a similar fashion to the physiologically relevant heterodimer association. Recently, the structure of rat *(17)* and porcine *(18a)* small subunit C-terminal domain (dVI) have been determined by X-ray crystallography. In order to investigate calcium-induced conformational changes in dVI we have crystallized this domain without Ca^{2+} (*apo*-dVI), with 200 mM Ca^{2+} (high-Ca^{2+}-dVI), and in the presence of a more physiologically representative 1 mM Ca^{2+} (low-Ca^{2+}-dVI). In addition, a range of mutations in the EF-hands has been introduced to investigate which of the calcium-binding sites have the largest impact on calpain activation *(18b)*.

2. Materials
1. Crystal screen kits from Hampton Research, CA.
2. Highly purified recombinant rat m-calpain dV1 small subunit.

3. Methods
3.1. Crystallization

In order to investigate the mechanism of calpain regulation and the properties of the calpain EF-hands, several domains and fragments of rat m-calpain have been cloned. Domain IV of m-calpain (m-dIV) and heterodimers of m-dIV/dVI can be expressed in *E. coli*, but have so far resisted crystallization.

In contrast, dVI from the small subunit was well expressed (25–50 mg/L of bacterial culture *[19]*) and could be purified and crystallized as a stable homodimer *(16)*. The sequence of the cDNA encodes a protein of 184 residues corresponding to residues Met87-Ser270, with a calculated molecular mass of 21,257 Da. Crystals of dVI were obtained under several conditions and provided a view of the dVI structure at a range of calcium concentrations *(16)*. The initial crystallization conditions were obtained from screening a large set of conditions (Hampton Research, CA) chosen by the factorial method *(20)*. Conditions where small crystals appeared were further refined by changing systematically the concentrations of components of the solution in a narrow range.

3.1.1. Ca^{2+}-Free Form

The crystals of calcium-free rat dVI were obtained using ammonium sulfate as a precipitating agent. The protein concentration used for crystallization was approx 10 mg/mL, with 50 m*M* cacodylate pH 6.5 or phosphate pH 7.5 buffer. The protein solution was mixed in a 1:1 ratio with the reservoir solution that contained 40% saturated ammonium sulfate, 50 m*M* cacodylate pH 6.5 buffer. A hanging drop of the total size of 4 µL was set up over the 1 mL reservoir solution. The small crystals appeared after about 5 d. Macroseeding, by transferring crystals 2–3 times to a fresh crystallization drop, was used to obtain larger crystals suitable for X-ray diffraction data collection. The final crystal size was approx 0.3 × 0.4 × 0.8 mm. These crystals belong to orthorhombic system, space group $C222_1$ with cell dimensions a = 67.5 Å, b = 73.0 Å, c = 156.1 Å, contain two molecules in the asymmetric unit and diffract to 2.3 Å resolution.

3.1.2. Ca^{2+}-Bound Forms

The crystals of calcium-bound rat dVI were obtained using PEG 8K as a precipitant. The protein at 10 mg/mL, 50 m*M* cacodylate pH 6.5, was mixed in a 1:1 ratio with the reservoir solution containing 17% PEG 8K, 50 m*M* cacodylate buffer pH 6.5 buffer, 10% glycerol and 200 m*M* $CaCl_2$. The hanging drop of a total size of 4 µL was set up over a 1 mL reservoir solution. These crystals grew to approx 0.3 mm × 0.3 mm × 0.7 mm after about 7 d. They also belong to an orthorhombic space group $C222_1$ with cell dimensions: a = 74.1, b = 86.4, c = 118.6 Å, contain two molecules in the asymmetric and diffract to 1.7 Å resolution.

The crystals were also obtained at a much lower 1 m*M* $CaCl_2$ concentration under similar conditions as aforementioned except that the reservoir solution contained 13% PEG 8K. These crystals belong to the orthorhombic system, space group $P2_12_12_1$ with cell dimensions: a = 50.3, b = 56.3, c = 141.3 Å. In addition, EF-hand mutants were crystallized under conditions similar to those of the Ca^{2+}-bound wild type dVI. They required the presence of 50–250 m*M*

Ca^{2+}, 6–13% PEG 8K as a precipitant and 7–13% glycerol. Crystals belonging to three different space groups were obtained.

3.2. Data Collection, Structure Solution, and Refinement

The structures were solved by a combination of heavy-atom and multiwavelength anomalous diffraction (MAD) methods. Initial native diffraction datasets for the various forms were collected on a R-Axis IIC area detector (Molecular Structure Corporation, TX). The MAD data sets for Ca^{2+}-free crystals containing selenomethionine or Hg derivative as anomalous scatterers, and Ca^{2+}-bound form soaked in ytterbium compound were collected at three wavelengths each at the Brookhaven National Laboratory synchrotron, Upton, NY. The 1.7 Å resolution dataset for Ca^{2+}-bound form was collected at CHESS facility, Ithaca, NY. The choice of wavelength depended on the heavy atom present in the crystal. The R-Axis data were processed with the Biotex software, whereas the synchrotron data were processed with Denzo/Scalepack *(21)*.

The structures of the Ca^{2+}-free and Ca^{2+}-bound forms were solved independently by either the MAD or multiple isomorphous replacement (MIR) methods. The initial heavy-atom derivative search showed that the crystals were extremely sensitive to mercury compounds and their diffraction deteriorated upon addition of Hg-containing compounds. Because there are only two cysteines in the sequence, we thought that this sensitivity may be related specifically to interactions with one of them. Single Cys-to-Ser mutants were made and indeed showed that this sensitivity was caused by Cys146. The crystals of Ca^{2+}-free Cys146Ser mutant provided a useful Hg derivative. In addition, selenomethionine-substituted dVI was prepared by expressing the protein in a auxotrophic strain of *E. coli* grown in the selenomethionine-supplemented medium and crystals of the Ca^{2+}-free form suitable for the MAD experiment were obtained.

To determine the structure of Ca^{2+}-bound form, we have soaked the crystals in $YbCl_3$, to replace Ca^{2+}, and used anomalous signal from ytterbium in the MAD regime.

The MAD data were treated as a case of isomorphous replacement *(22)* and heavy-atom parameters refined with MLPhare *(23)*. The resulting electron density maps were interpretable and the models of the two forms were built independently with O *(24)*. In each crystal form, there are two independent molecules in the asymmetric unit. They were treated independently in the refinement (X-plor *[25]*). The low-Ca^{2+} form was solved by molecular replacement using the high-Ca^{2+} structure as a model. The 1.7 Å resolution structure of high-Ca^{2+} dVI corresponds to the Cys146Ser mutant because this protein provided the best diffracting crystals. The low-Ca^{2+} dVI and the *apo* dVI correspond to the wild type protein. The final refinement parameters were: high-Ca^{2+}-dVI (50–1.7 Å) R/R_{free} = 0.21/0.25, low-Ca^{2+}-dVI (50–2.5 Å) R/R_{free} = 0.20/0.28, and *apo*-dVI (8–2.3 Å) R/R_{free} = 0.22/0.30 with good geometry.

3.3. Analysis of Structural Differences

The comparison of dVI molecules in various crystal forms was performed initially by their superposition and calculations of interhelical angles, both within and between the EF-hands (CALC_ALL, Yap, K. and Ikura, M.). Further and more-detailed analysis was undertaken with the use of distance difference matrices (DDMs) which were constructed using the DISCOM program *(26a)*. The (i,j) element of the distance difference matrix DD is defined as a difference in the distance between the pair of atoms (i,j) in the first structure and the corresponding distance in the second structure. The DDs for the apo-dVI vs high-Ca^{2+}-dVI are defined as: $DDij = dij(apo)-dij$ (high-Ca^{2+}). To follow the changes in the backbone we have calculated the DD matrices for Cα only. Interresidue contacts were assigned based on the distance between the centers of geometry of each amino acid side-chain in the structure, calculated using CHARM*M (26b)* and two residues are considered to be in contact if their centers of geometry of their side chains are within 8 Å of each other.

3.4. Results and Discussion

3.4.1. The Overall Structure

Amino acid residue numbers refer to the full-length rat calpain small subunit of 270 residues. The 11 N-terminal residues of the bacterially expressed 184-residue dVI were disordered in all structures, so that the X-ray data define residues Glu98 to Ser270 of the whole subunit.

The structure of high-Ca^{2+} dVI is shown in **Fig. 1**. The monomer is compact and predominantly α-helical, with eight α-helices (α1–α8) ranging in length from 2.5–6 turns, the two longest helices extending for approx 30 Å. Their correspondence to E and F helices is as follows: E1 = α1, F1 = α2, E2 = α3, F2 = N-terminal half of α4, E3 = C-terminal half of α4, F3 = α5, E4 = α6, F4 = N-terminal half of α7, E5 = C-terminal half of α7, F5 = α8. Five helix-loop-helix EF-hand motifs (EF1– EF5) are present in each monomer, of which EF1–EF3 bind Ca^{2+} at lower concentrations, EF4 binds Ca^{2+} only at higher concentrations, and EF5 does not bind Ca^{2+} at all. As in many other EF-hand proteins, the calpain dVI EF-hands form pairs (EF1 + EF2; EF3 + EF4), with short antiparallel β-strand interactions between their Ca^{2+}-binding loops. EF5 is unpaired in the monomer, but pairs with its counterpart in the homodimer. EF hands 1 and 2 form the N-terminal Ca^{2+}-binding unit and the EF hands 3 and 4 form the C-terminal Ca^{2+}-binding unit.

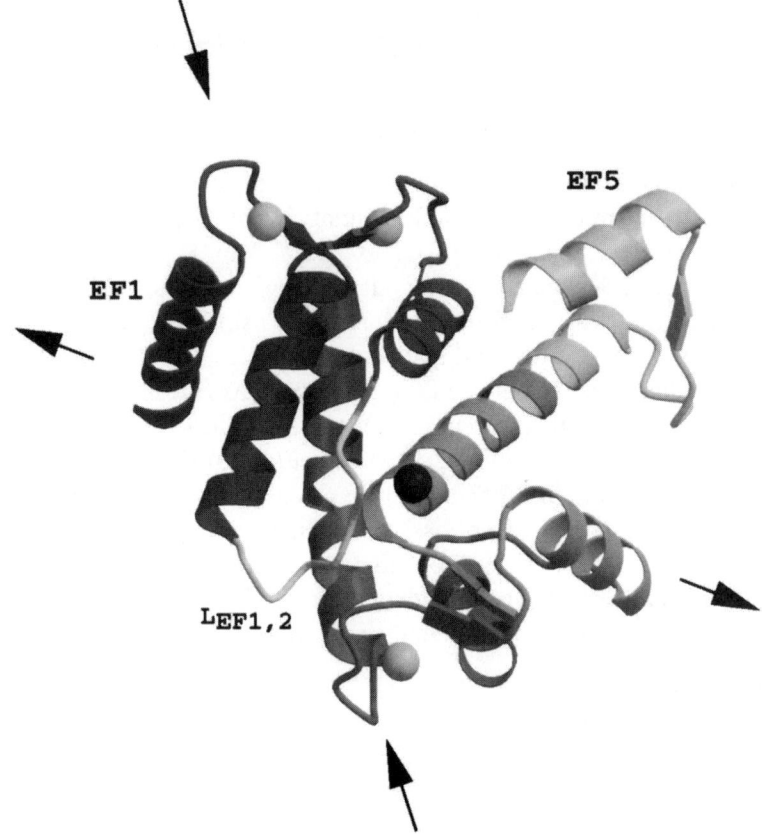

Fig. 1. The schematic representation of calpain domain VI. Helices are shown as springs, short β-strands are shown as arrows. The light spheres represent calcium ions bound to EF1,2,3, whereas the dark sphere shows Ca^{2+} bound to EF4. The gray shading changes gradually from dark at the N-terminus to light at the C-terminus. The arrows mark directions of maximum compression ($\rightarrow \leftarrow$) or stretching ($\leftarrow \rightarrow$) of the molecule upon calcium binding. This and the following figures were prepared with programs Molscript (42) and Raster3D (43).

The N- and C-domains each have a globular form and are interconnected by the long helix, α4, which corresponds to helices F2 and E3. The loop $L_{EF1,2}$ between EF1 and EF2 of the N-domain consists of 15 residues and is atypically long for EF-hand pairs (see Fig. 1). This loop makes contacts with the C-domain, and moves significantly in response to Ca^{2+}, suggesting a mechanism for interdomain communication. A similar-length loop that connects separate Ca^{2+}-binding domains exists also in recoverin (27). The loop connecting EF3 and EF4 is more typical, six residues in length. The second long helix of

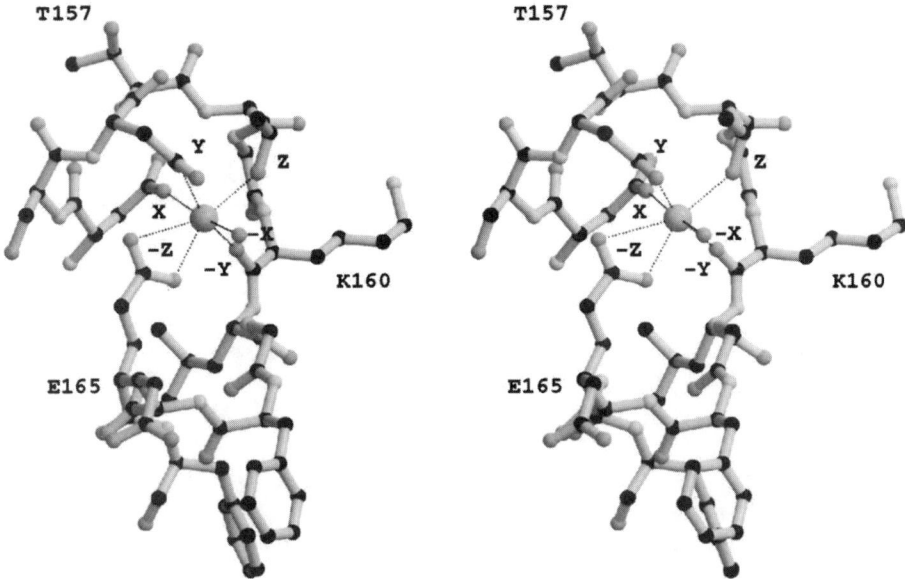

Fig. 2. A stereoview of the Ca^{2+}-binding EF2 hand in ball-and-stick representation showing the ion coordination and the nomenclature for the ligands.

dVI (α7, corresponding to helices F4 and E5) leads from the C-domain into the unpaired EF5. Domain VI is a homodimer both in solution and in the crystal *(17)* and a major part of the binding between the monomers is provided by an intermolecular structural unit composed of EF5 and EF5', which is strikingly similar to a conventional intra-molecular pairs of EF-hands.

3.4.2. EF2,3 Show Typical Calcium Binding

Of the five EF-hands in dVI, only the sequences of EF2 and EF3 conform well to the canonical EF-hand pattern *(28)*. They bind Ca^{2+} with the expected pentagonal bipyramidal coordination involving six ligands from the protein and one water molecule (*see* **Fig. 2**). The protein ligands are: position X, the side-chains of aspartate; position Y, the side-chain of aspartate; position Z, Thr/Ser; position -Z, glutamate; position -Y, mainchain of Lys/Thr; and position -X, a water molecule. The -Z glutamic acid provides the important bidentate coordination.

3.4.3. EF1 Shows a Novel Binding Arrangement

The Ca^{2+}-binding loop of EF1 is one-residue shorter than the canonical EF-hand. The deletion occurs near the N-terminus. In addition, the residue nominally in position X, Leu110, lacks a side chain carboxylate group that

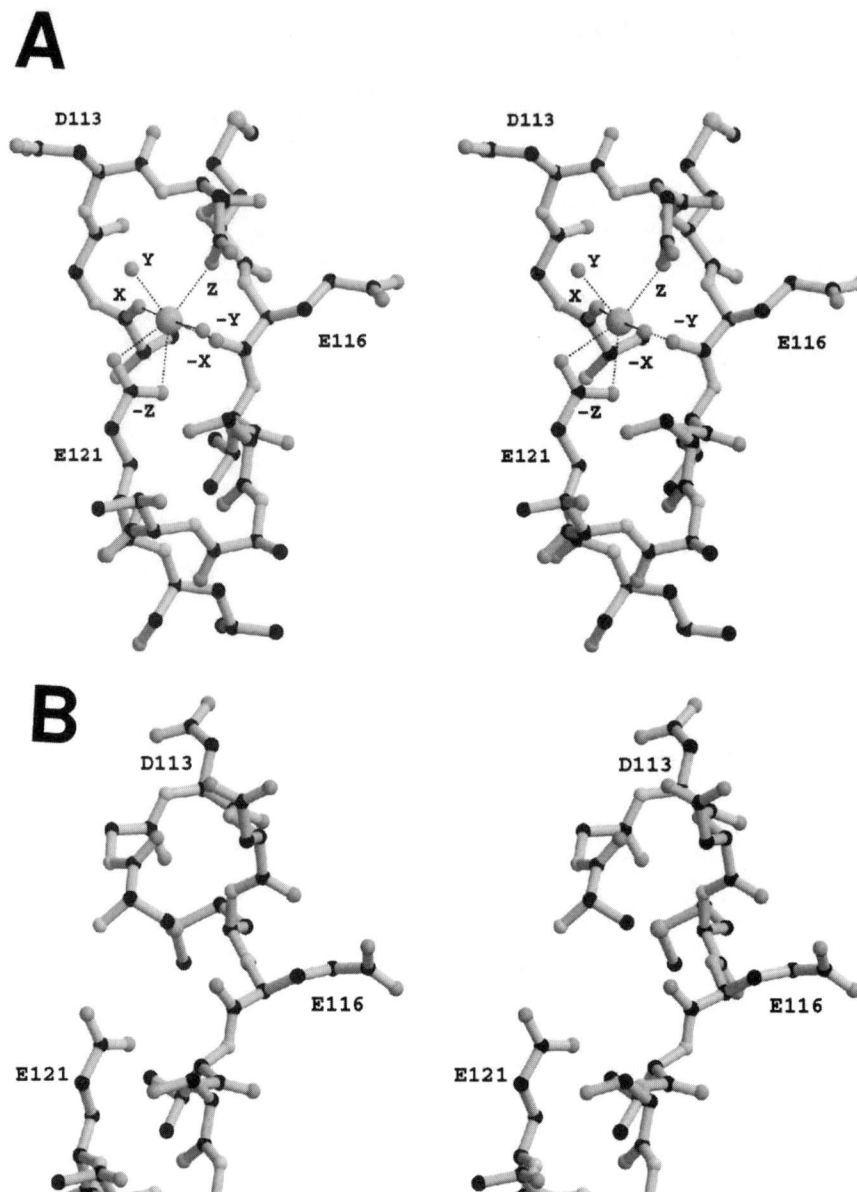

Fig. 3. A stereoview of EF1, an unusual Ca^{2+}-binding hand in ball-and-stick representation. (**A**) conformation of the Ca^{2+}-bound form; (**B**) conformation of the Ca^{2+}-free form in exactly the same orientation. The largest movement is that of Asp113-Asp114. Ca^{2+} ion is shown as a sphere. There are two water molecules in the coordination sphere of the ion (Y and -X).

usually coordinates the Ca^{2+} ion (*see* **Fig. 3**). Despite this, Ca^{2+} is bound to EF1 in a manner very similar to that of a canonical EF-hand and the retention of the ion at 1 m*M* Ca^{2+} concentration indicates a fairly tight binding.

The Ca^{2+} ion coordination normally provided by the side chain at position X is replaced in EF1 by the main-chain carbonyl group of Ala111. This conformation of the main chain causes the potential Y ligand, Asp113, to be too remote to bind the Ca^{2+} ion and instead, a water molecule takes on that role. The remaining ligands in EF1 follow the canonical pattern of interactions: specifically the side chain of Asp114, the carbonyl group of Glu116, the side chain of Glu121 (bidentate), and another water molecule. The EF1 ion coordination sphere, therefore, incorporates two water molecules, whereas typically only one is present *(29,30)*.

At position 15 of the EF-hand consensus sequence there is usually a glycine residue that adopts a conformation unfavorable for other amino acids. This introduces a relatively sharp bend in the loop structure. In EF1 there is a methionine, Met115, in an equivalent position and it adopts a similar main-chain conformation to that of a more common glycine.

3.4.4. Ca^{2+} Binding to EF4

An unusual calcium-binding site was observed in the EF4-hand. Not only is the Ca^{2+} ion bound in the opposite end of the loop to its canonical position, but its coordination sphere contains 8, rather than the usual 7, ligands. Most importantly, one of these ligands, Asp139, is contributed by the $L_{EF1,2}$ loop (*see* **Fig. 4**). This site was occupied by calcium only in the crystals that were grown at relatively high concentrations of Ca^{2+} (more than 20 m*M*) (*see* **Note 1**).

3.4.5. Ca^{2+}-Induced Structural Rearrangements

Fluorescence data suggested that the Ca^{2+}-induced conformational changes of dVI homodimer are relatively small. This has been confirmed by the structural data. The superposition of residues 98–270 for high-Ca^{2+}-dVI and apo-dVI gives an root-mean-square (rms) deviation of 1.82 Å on 173 Cα atom pairs, with the largest distance between the coresponding Cα atoms being 7.5 Å. Comparison of the N-domain and the C-domain separately showed that greater conformational differences are associated with the N-domain. Superposition of 73 Cα atom pairs for the N-domain (residues 101-173, EF1, EF2 and the loop $L_{EF1,2}$) gave an rms deviation of 2.02 Å whereas superposition of 64 Cα atom pairs for the C-domain (residues 175–238, EF3, EF4) gave an rms deviation of 1.07Å (*see* **Note 2**).

The largest differences are within EF1. The calcium-binding loop in the *apo*-dVI has a conformation inconsistent with ion binding. The Asp114 side-chain is turned toward solvent pulling this part of the loop away from the center

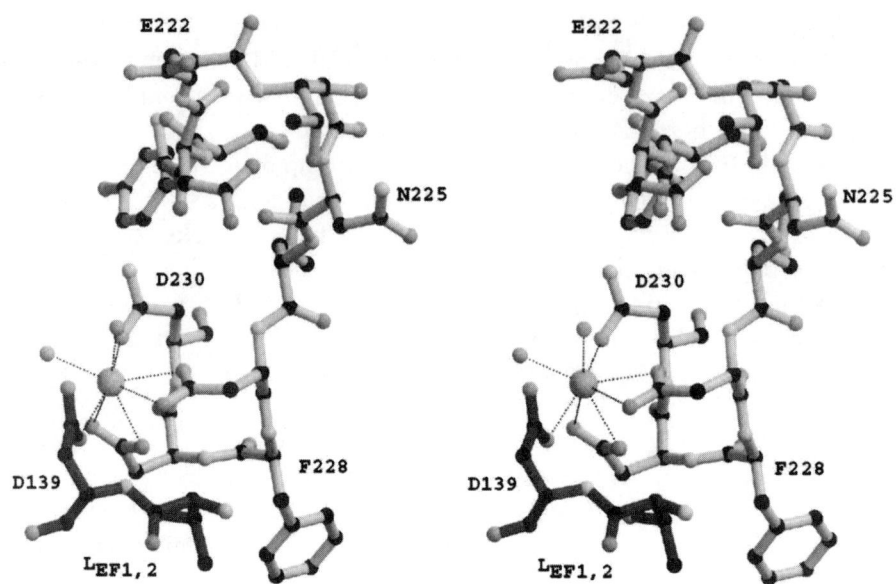

Fig. 4. The coordination of Ca^{2+} in the EF4 hand. The ion sits in an unusual location at the C-terminal end of the loop and is coordinated by eight ligands. The Asp139 from $L_{EF1,2}$ loop (shown in *gray*) is part of the coordination sphere of the ion.

(**Fig. 3**). Out of six hydrogen bonds in EF1 that stabilize the calcium-binding loop, five are conserved between the free and Ca^{2+}-bound form. Three of them are main chain: $O^{Gly112}...N^{Met115}$, $O^{Ser118}...N^{Leu122}$ and $N^{Ser118}...O^{Glu121}$, and two are side chain: $OG^{Ser118}...N^{Glu121}$ and $N^{Ser118}...OE1^{Glu121}$. The hydrogen bond $O^{Ala111}...N^{Asp114}$ in *apo*-dVI is disrupted on Ca^{2+}-binding, but a new hydrogen bond $OE1^{Asp114}...N^{Glu116}$ is formed in its place.

The movements within EF1 can also be described in terms of changes in the interhelical angle and main-chain angles in EF1. The interhelical angle in EF1 decreases from 145° to 123° (by 22°) on Ca^{2+}-binding (i.e., the EF-hand opens, as described later in terms of DDM), which is a greater change than in the other dVI EF-hands, although smaller than the corresponding changes, for example, in calmodulin and TnC (*see* **Table 1**). The proper orientation of the carbonyl group toward the Ca^{2+} is achieved by a readjustment of several main-chain torsion angles within the range of residues 108–117 relative to that in apo-dVI. The most profound change is observed in Ala111, and is likely facilitated by the presence of a following glycine. Superposition of the 28 residues (102 to 129) of the whole of EF1 for apo-dVI and high-Ca^{2+}-dVI gives an rms deviation of 2.42 Å for 28 Cα atom pairs. The major movement takes place in the N-terminus of the loop.

Table 1
Comparison of Interhelical Angles of Calpain dVI, Calmodulin, and Skeletal Muscle Troponin C

	Calpain Apo/Ca	CaM Apo/Ca	TnC Apo/Ca
E1/F1	145/123	134/92	143/99
E1/E2	100/121	90/156	89/146
E1/F2	131/111	128/112	117/115
F1/E2	114/116	128/112	129/114
F1/F2	33/26	50/42	47/37
E2/F2	122/120	129/86	147/97
E3/F3	151/142	132/102	106/103
E3/E4	87/99	80/145	127/125
E3/F4	136/127	145/121	113/125
F3/E4	112/110	142/113	126/127
F3/F4	41/43	33/39	32/51
E4/F4	134/131	131/90	115/109
E5/F5	155/162		

3.4.6. Movement of $L_{EF1,2}$

The binding of Ca^{2+} induces a shift of the linker region $L_{EF1,2}$ (Val130–Asp144), toward the C-domain, with the largest distance of 3.8Å ($C\alpha$-shift) being traversed by Asp139. When the apo-dVI and the high-Ca^{2+}-dVI were superimposed based on 173 pairs of $C\alpha$ atoms, the rms deviation for the 15 $C\alpha$ atom pairs of the $L_{EF1,2}$ loop was 2.1 Å. This reflects predominantly a shift of this loop relative to the rest of the structure. The structural backbone of the EF4 loop is largely unaffected by Ca^{2+}-binding, as shown by a low value of an rms deviation (0.56 Å) for the superposition of only the $C\alpha$ atoms of residues 210–238 in apo-dVI and high-Ca^{2+}-dVI. The most significant change in the loop of EF4 is a rotation of the side chain of Asp229 by 180° around the $C\alpha$-$C\beta$ bond to coordinate Ca^{2+} in EF4. In the apo form, where the distance between $L_{EF1,2}$ loop and EF4 is larger, there are two water molecules bridging their interactions.

3.4.7. The Ca^{2+}-Binding Domains and Ca^{2+}-Induced Conformational Change

In an attempt to classify the structural data on EF-hands from a large number of proteins, the domain formed by a pair of EF-hands was described as either regulatory or structural *(32)*. Structural domains had a high affinity for Ca^{2+} so that the ion is bound at all times, and these domains showed little conformational change on binding Ca^{2+}. In contrast, regulatory domains had a lower affinity for Ca^{2+} so that they could alternate between apo- and Ca^{2+}-bound forms. They showed marked conformational changes upon binding Ca^{2+}, which led to regulatory effects on their target proteins. In these regulatory domains, the two helices in the *apo*-form are closer to antiparallel, with an interhelical vector angle of approx 135°, and are described as closed. On binding Ca^{2+}, they become nearly perpendicular with the interhelical angle to approx 90°, a conformation described as open (*see* **Note 4**).

3.4.8. Ca^{2+}-Induced Implications for Target/Inhibitor Binding

Another common feature of EF-hand proteins is the exposure of a hydrophobic surface in response to Ca^{2+}-binding. For calpain dVI it is known from fluorescence measurements that there is a small, but significant, increase in 6-(p-toluidino)-2-naphthalenesulfonate (TNS) fluorescence in the presence of Ca^{2+}, which suggests an increase in accessible hydrophobic surface. Calpain aggregates in the presence of Ca^{2+}, again implying the exposure of hydrophobic surface. The calpain inhibitor PD150606 is directed to the calcium-binding domains and binds dVI with a higher affinity in the presence of Ca^{2+} *(18a)*. The crystal structure of a Ca^{2+}-bound porcine dVI-PD150606 complex revealed that the inhibitor binds in the hydrophobic pocket located between helices F1 (α2) and F2 (N-terminal of α4) *(18a)*. Comparison of structures of the porcine dVI of calpain with and without inhibitor shows only small conformational changes associated with inhibitor binding. However, the affinity for binding this inhibitor was significantly increased in the presence of calcium *(34)*. In the homologous rat dVI, we find only small Ca^{2+}-induced changes and thus subtle effects must be responsible for such a Ca^{2+}-enhanced inhibitor binding. The hydrophobic pocket in which the inhibitor binds is lined by 17 residues *(18a)*. They are conserved between porcine domain VI and rat calpain dVI and come predominantly from EF1, EF2, and the $L_{EF1,2}$ regions. We find that the binding of calcium causes a significant shift of the α1 helix that changes its environment with respect to α4. Most significantly the Ca^{2+}-induced rearrangement alters the residues in the immediate vicinity of Trp170, which itself is one of the amino acids lining the hydrophobic inhibitor binding pocked *(18a)*. In apo-dVI Phe103 lies close to Trp170 whereas in high-Ca^{2+}-dVI that site is occupied by Leu106 and overall the Trp170 side chain is displaced significantly (CZ3 atom shift by approx 2.5 Å). The helices α1 and α4 move closer together upon Ca^{2+} binding, and this is accompanied by a large increase from 4 to 11 interresidue contacts. Our DDM analysis shows that α2 and α4 are slightly more separated in the calcium-bound form, and that the net effect of Ca^{2+} binding creates a larger space in the region of the reported inhibitor binding site.

3.4.9. Homodimeric dVI and Implications for the Heterodimer dIV–dVI

Heterodimerization is an important aspect in the calpain system with dVI being known to remain bound to the highly homologous domain IV even after extensive autolysis of calpain *(13,35)*, and also that the removal of the last 25 amino acids of dVI (which contribute to EF5) prevents heterodimer formation *(15)*. In the homodimeric dVI structure the fifth EF-hand, that is unpaired in the monomer, interacts with an equivalent region of a second monomer, and is likely representative of the physiological dimer. This mode of dimerization in dVI in which EF-hands from different polypeptide chains embrace, has been observed for the first time in calpain. In general, the interactions between the two monomers are predominantly hydrophobic: 47% of the interface is nonpolar, 23% charged (D, E, H, K, R), and 30% polar (C, S, T, N, Q, Y). The charged and polar homodimer interface incorporates four backbone hydrogen-bonds between the two β-strands of the EF5 of each monomer and additionally 24 hydrogen bonds between donors and acceptors of main-chain and side-chain atoms. An important participant is a side chain of Arg237, which together with its symmetry-related side chain and the side chains of Ser270 participate in an extensive hydrogen-bonded network. Other important hydrogen-bonds are formed between Trp263 and Met241 of a symmetry-related molecule.

The dIV domains of m- and μ-calpain heavy chains display approx 47% sequence identity to dVI indicating that their three-dimensional (3D) structures will be very similar. We have built a homology model for a dIV–dVI heterodimer interface. Our model of the dIV–dVI heterodimer shows that the pattern of extensive hydrogen-bond interactions observed in the homodimer is conserved. The buried surface on the dIV side is mostly composed of nonpolar residues (60%). The charged residues account for 24% of the interface and polar residues for 17%. The heterodimer interface shows an increased nonpolar character relative to that of homodimer, which could explain why the formation of the heterodimer is more likely to happen than the formation of a homodimer.

A further significant observation from the heterodimer model is that in dVI there is no communication between the α8 and Asp154 of the EF2 calcium coordination sphere due to replacement of Tyr269 (dVI) for Leu in dIV, which could affect somewhat different response of these domains to the presence/absence of calcium (*see* **Note 4**).

3.4.10. Relationship to Other EF-Hand Proteins

The Ca^{2+}-binding EF-hand protein superfamily displays a large diversity in tertiary structure, owing to variation in the number of EF-hands, their pairing to form Ca^{2+}-binding domains and the relative orientation of such domains, the length of the polypeptide chain between the paired EF-hands, and the length of interdomain linkers *(29,32)*. Their structures range from the elongated dumbbells of calmodulin *(36)* and troponin C (TnC) *(37)* to much more compact structures such as sarcoplasmic Ca^{2+}-binding protein *(38)* and recoverin *(27)*. Calpain dVI has a unique overall 3D folding motif as ascertained by comparisons to proteins with known structures using the program DALI *(39)* and the classification scheme CATH *(40)*. The calpain C-terminal domains dIV and dVI were originally described as calmodulin-like and it remains true that the highest amino acid sequence identity of calpain dVI to proteins with known 3D structures is to TnC and calmodulin. However, the disposition of the N- and C-domains in dVI relative to the long central α4 helix is different from that in the latter two proteins, leading to a more compact structure. Other proteins like sorcin, grancalcin, and ALG-2 also have high-sequence identity with dVI of calpain and it is predicted that their 3D structures will be similar to dVI *(41)*.

4. Notes

1. The crystal structures of dVI determined in the presence of both 20 m*M* *(18)* and 200 m*M* Ca^{2+} *(17)* revealed one Ca^{2+} ion bound to each of the first four EF-hands (EF1–4). However, the crystals of rat dVI grown at 1 m*M* Ca^{2+} showed only three Ca^{2+} ions, bound at EF1-3 *(17)*. Superposition of the low-Ca^{2+}-dVI and high-Ca^{2+}-dVI structures for residues 98–270 gives an rms deviation of 0.77 Å for 173 Cα atom pairs, indicating that all significant Ca^{2+}-induced shifts of the backbone are invoked prior to binding of Ca^{2+} at EF4. These observations strongly suggest that EF4 is not occupied under physiological conditions and consequently may not be important for the regulatory function of dVI. It is likely, therefore, that the high-resolution structure of high-Ca^{2+}-dVI described here can be considered representative of the physiologically relevant Ca^{2+}-bound form of dVI, with the caveat that EF4 would not contain Ca^{2+}.
2. Inspection of conformational change by means of rms deviations may be biased by the choice of sequences selected for comparison. To obtain a more objective and quantitative description, a distance difference matrix (DDM *[31]*) was constructed for all pairs of Cα atoms in apo-dVI and high-Ca^{2+}-dVI. There were no DD*M* values larger than 7 Å, indicating that relatively small Ca^{2+}-induced changes had occurred. There were, however, 134 DDM values between 5–7 Å, and 624 values between 3–5 Å. The data show the most significant movements occur in the E1 (α1) helix and in $L_{EF1,2}$.

    The largest DDM absolute values are in range 6 Å–7 Å and describe movement of Leu110 and Ala111 in helix E1 toward residues 190–195 in the EF3 coil. The regions that show the largest Ca^{2+}-induced separation are connected with the movement of Glu98, Glu99 at the N-terminus of E1 away from Leu126 in helix F1 and from Asp139, Gly140 in $L_{EF1,2}$.
3. Although this classification held for many EF-hand proteins, examples have recently emerged of EF-pair domains, which move very little on binding Ca^{2+},

but which are clearly regulatory. Cardiac muscle troponin C *(33)* is one such example. In contrast to skeletal muscle troponin C, which opens and closes much like calmodulin, the N-terminal domain of cardiac muscle troponin C binds a Ca^{2+} ion only at the second site with little detectable change, although the molecule clearly regulates cardiac muscle contraction. It is possible that the N-terminal domain of cardiac troponin C does in fact open up more in the presence of calcium and its target proteins than in the presence of Ca^{2+} alone.

The parallels between cardiac troponin C and dVI of calpain are instructive. In dVI of calpain, EF1 in the N-domain opens by 22° and EF3 in the C-domain opens by 9° on binding Ca^{2+}. These values are small and the activity of calpain is entirely dependent on these movements and on similar movements assumed to take place in dIV. It remains an interesting possibility that the movements in calpain EF-hands may be greater in the intact enzyme, where each Ca^{2+}-binding domain is always close to its presumed but presently unknown physiological target peptide in the heterodimer.

4. The structure of intact calpain has been reported recently (*see* Chapter 3, this volume) confirming our predictions,

Acknowledgments

The authors would like to thank Dr. J. S. Elce for a fruitful collaboration and Mrs. Y. Li for excellent technical assistance. H. Blanchard would like to thank M. Nelson and W. Chazin for assistance in the calculations of DDMs and interresidue contact analysis.

References

1. Sorimachi, H., Ishiura, S., and Suzuki, K. (1997) Structure and physiological function of calpains. *Biochem. J.* **328,** 721–732.
2. Sorimachi, H., Saido, T. C., and Suzuki, K. (1994) New era of calpain research. Discovery of tissue-specific calpains. *FEBS Lett.* **343,** 1–5.
3. Bartus, R. T., Elliott, P. J., Hayward, N. J., Dean, R. L., Harbeson, S., Straub, J. A., et al. (1995) Calpain as a novel target for treating acute neurodegenerative disorders. *Neurol. Res.* **17,** 249–258.
4. Iwamoto, H., Miura, T., Okamura, T., Shirakawa, K., Iwatate, M., Kawamura, S., et al. (1999) Calpain inhibitor-1 reduces infarct size and DNA fragmentation of myocardium in ischemic/reperfused rat heart. *J. Cardiovasc. Pharmacol.* **33,** 580–586.
5. Wang, K. K. and Yuen, P. W. (1994) Calpain inhibition: an overview of its therapeutic potential. *Trends. Pharmacol. Sci.* **15,** 412–419.
6. Ohno, S., Emori, Y., Imajoh, S., Kawasaki, H., Kisaragi, M., and Suzuki, K. (1984) Evolutionary origin of a calcium-dependent protease by fusion of genes for a thiol protease and a calcium-binding protein? *Nature* **312,** 566–570.
7. Berti, P. J. and Storer, A. C. (1995) Alignment/phylogeny of the papain superfamily of cysteine proteases. *J. Mol. Biol.* **246,** 273–283.
8. Minami, Y., Emori, Y., Imajoh-Ohmi, S., Kawasaki, H., and Suzuki, K. (1988) Carboxyl-terminal truncation and site-directed mutagenesis of the EF hand struc-

ture-domain of the small subunit of rabbit calcium-dependent protease. *J. Biochem. (Tokyo)* **104,** 927–933.
9. Nakayama, S. and Kretsinger, R. H. (1994) Evolution of the EF-hand family of proteins. *Annu. Rev. Biophys. Biomol. Struct.* **23,** 473–507.
10. Mellgren, R. L. (1997) Specificities of cell permeant peptidyl inhibitors for the proteinase activities of mu-calpain and the 20 S proteasome. *J. Biol. Chem.* **272,** 29,899–29,903.
11. Croall, D. E. and DeMartino, G. N. (1991) Calcium-activated neutral protease (calpain) system: structure, function, and regulation. *Physiol. Rev.* **71,** 813–847.
12. Goll, D. E., Thompson, V. F., Taylor, R. G., and Zalewska, T. (1992) Is calpain activity regulated by membranes and autolysis or by calcium and calpastatin? *Bioessays* **14,** 549–556.
13. Crawford, C., Brown, N. R., and Willis, A. C. (1993) Studies of the active site of m-calpain and the interaction with calpastatin. *Biochem. J.* **296,** 135–142.
14. Saido, T. C., Nagao, S., Shiramine, M., Tsukaguchi, M., Yoshizawa, T., Sorimachi, H., et al. (1994) Distinct kinetics of subunit autolysis in mammalian m-calpain activation. *FEBS Lett.* **346,** 263–267.
15. Elce, J. S., Davies, P. L., Hegadorn, C., Maurice, D. H., and Arthur, J. S. (1997) The effects of truncations of the small subunit on m-calpain activity and heterodimer formation. *Biochem. J.* **326,** 31–38.
16. Blanchard, H., Li, Y., Cygler, M., Kay, C. M., Arthur, J. S. C., Davies, P. L., and Elce, J. S. (1996) Ca2+ binding domain vi of rat calpain is a homodimer in solution: Hydrodynamic, crystallization and preliminary x ray diffraction studies. *Protein Sci.* **5,** 535–537.
17. Blanchard, H., Grochulski, P., Li, Y., Arthur, J. S. C., Davies, P. L., Elce, J. S., and Cygler, M. (1997) Structure of a calpain Ca(2+)-binding domain reveals a novel EF-hand and Ca(2+)-induced conformational changes. *Nat. Struct. Biol.* **4,** 532–538.
18a. Lin, G. D., Chattopadhyay, D., Maki, M., Wang, K. K., Carson, M., Jin, L., et al. (1997) Crystal structure of calcium bound domain VI of calpain at 1.9 Å resolution and its role in enzyme assembly, regulation, and inhibitor binding. *Nat. Struct. Biol.* **4,** 539–547.
18b. Dutt, P., Arthur, J. S. C., Grochulski, P., Cygler, M., and Elce, J. S. (2000) The roles of individual calcium-binding EF-hands in the activation of m-calpain. *J. Biochem.* **15,** 37–43.
19. Graham-Siegenthaler, K., Gauthier, S., Davies, P. L., and Elce, J. S. (1994) Active recombinant rat calpain II. Bacterially produced large and small subunits associate both in vivo and in vitro. *J. Biol. Chem.* **269,** 30,457–30,460.
20. Jancarik, J. and Kim, S.-H. (1999) Sparse matrix sampling: a screening method for crystallization of proteins. *J. Appl. Cryst.* **24,** 409–414.
21. Otwinowski, Z. and Minor, W. (1997) Processing of X-ray diffraction data collected in oscillation mode. *Methods Enzymol.* **276,** 307–326.
22. Ramakrishnan, V. and Biou, V. (1999) Treatment of multiwavelength anomalous diffraction data as a special case of multiple isomorphous replacement, in *Macromolecular Crystallography Part A* (Carter, C. W., Jr. and Sweet, R. M., eds.), Academic, San Diego, CA, pp. 538–557.

23. CCP4 (1998) Collaborative Computational Project Number 4 (1994). The CCP4 suite: programs for protein crystallography. *Acta Crystallogr.* **D50,** 760–763.
24. Jones, T. A., Zou, J. Y., Cowan, S. W., and Kjeldgoard, M. (1991) Improved methods for building models in electron density maps and the location of errors in these models. *Acta. Crystallogr.* **A47,** 110–119.
25. Brünger, A. T. (1993) *X-Plor Version 3.1.* Yale University,
26a. Akke, M., Forsen, S., and Chazin, W. J. (1995) Solution structure of (Cd2+)1-calbindin D9k reveals details of the stepwise structural changes along the Apo—>(Ca2+)II1—>(Ca2+)I,II2 binding pathway. *J. Mol. Biol.* **252,** 102–121.
26b. Brooks, B. R., Bruccoleri, R. E., Olafson, B. D., States, D. J., Swaminthan, S., and Karplus, M. (1983) CHARMM: a program for macromolecular energy minimization and dynamics calculations. *J. Comput. Chem.* **4,** 187–217.
27. Flaherty, K. M., Zozulya, S., Stryer, L., and McKay, D. B. (1993) Three-dimensional structure of recoverin, a calcium sensor in vision. *Cell* **75,** 709–716.
28. Kawasaki, H. and Kretsinger, R. H. (1995) Calcium-binding proteins 1: EF-hands. *Protein Profile* **2,** 297–490.
29. Strynadka, N. C. and James, M. N. (1989) Crystal structures of the helix-loop-helix calcium-binding proteins. *Annu. Rev. Biochem.* **58,** 951–998.
30. McPhalen, C. A., Strynadka, N. C., and James, M. N. (1991) Calcium-binding sites in proteins: a structural perspective. *Adv. Protein Chem.* **42,** 77–144.
31. Nelson, M. R. and Chazin, W. J. (1998) An interaction-based analysis of calcium-induced conformational changes in Ca2+ sensor proteins. *Protein Sci.* **7,** 270–282.
32. Ikura, M. (1996) Calcium binding and conformational response in EF-hand proteins. *Trends Biochem. Sci.* **21,** 14–17.
33. Spyracopoulos, L., Li, M. X., Sia, S. K., Gagne, S. M., Chandra, M., Solaro, R. J. and Sykes, B. D. (1997) Calcium-induced structural transition in the regulatory domain of human cardiac troponin C. *Biochemistry* **36,** 12,138–12,146.
34. Wang, K. K., Nath, R., Posner, A., Raser, K. J., Buroker-Kilgore, M., Hajimohammadreza, I., et al. (1996) An alpha-mercaptoacrylic acid derivative is a selective nonpeptide cell-permeable calpain inhibitor and is neuroprotective. *Proc. Natl. Acad. Sci. USA* **93,** 6687–6692.
35. Nishimura, T. and Goll, D. E. (1991) Binding of calpain fragments to calpastatin. *J. Biol. Chem.* **266,** 11,842–11,850.
36. Babu, Y. S., Bugg, C. E., and Cook, W. J. (1988) Structure of calmodulin refined at 2.2 Å resolution. *J. Mol. Biol.* **204,** 191–204.
37. Herzberg, O. and James, M. N. (1988) Refined crystal structure of troponin C from turkey skeletal muscle at 2.0 Å resolution. *J. Mol. Biol.* **203,** 761–779.
38. Vijay-Kumar, S. and Cook, W. J. (1992) Structure of a sarcoplasmic calcium-binding protein from Nereis diversicolor refined at 2.0 Å resolution. *J. Mol. Biol.* **224,** 413–426.
39. Holm, L. and Sander, C. (1993) Protein structure comparison by alignment of distance matrices. *J. Mol. Biol.* **233,** 123–138.
40. Orengo, C. A., Michie, A. D., Jones, S., Jones, D. T., Swindells, M. B., and Thornton, J. M. (1997) CATH — a hierarchic classification of protein domain structures. *Structure* **5 ,** 1093–1108.

41. Maki, M., Narayana, S. V., and Hitomi, K. (1997) A growing family of the Ca^{2+}-binding proteins with five EF-hand motifs. *Biochem. J.* **328,** 718–720.
42. Kraulis, P. J. (1991) MOLSCRIPT: a program to produce both detailed and schematic plots of protein structures. *J. Appl. Crystallogr.* **24,** 946–950.
43. Merritt, E. A. and Bacon, D. J. (1997) Raster3D: photorealistic molecular graphics. *Methods Enzymol.* **277,** 505–524.

17

Neurocalcin

Role in Neuronal Signaling

Senadhi Vijay-Kumar and Vinod D. Kumar

1. Introduction

The interaction of Ca^{2+} with calcium-binding proteins is one mechanism by which Ca^{2+} is thought to regulate cellular processes *(1)*. In photoreceptors, Ca^{2+} is thought to act as the key modulator of light adaptation, a process whereby absorption of each additional photon is less effective in activating the phototransduction pathway, thus producing smaller alterations in the conductance of the cell. Both biochemical and electrophysiological experiments implicate the enzyme-retinal guanylyl cyclase or rod outer-segment guanylyl cyclase (ROS-GC) as one of the sites of Ca^{2+} action *(2)*. Following illumination, this key enzyme catalyzes the formation of cyclic GMP from GTP, which, in turn, opens cation channels in the outer-segment plasma membrane and reestablishes the dark potential of the cell. The decrease in intracellular Ca^{2+} that occurs upon illumination stimulates cyclase activity *(3)*. It has only recently been established that a calcium-binding protein called Guanylate cyclase activating protein (GCAP) is responsible for this regulation *(4)*.

This is an uncommon mechanism because calcium signals are usually mediated by proteins that are inactive in the calcium-free form and recognize effectors upon binding calcium. The earliest recognized members of this group such as the calmodulin-activated membrane guanylyl cyclases from *Paramecium* and *Tetrahymena* were shown to be activated when the calcium concentration increased from submicromolar to micromolar levels *(5,6)*. However, this unique form of regulation of ROS-GC activation by GCAP appear to fit the physiology of the ROS, i.e., a light flash on dark-adapted ROS leads to hydrolysis of cyclic GMP and reductions in its free concentration and free calcium levels, which triggers activation of the cyclase and restoration of the cyclic GMP concentration, the dark current and the calcium concentration to the dark adapted levels.

Two homologs of GCAP, GCAP-1, and GCAP-2 have been identified in dark-adapted rod outer segments, and both mediate cyclase activity in rod outer segments when the calcium concentrations decreases from 500 nM to 50 nM *(7,8)*. GCAPs belong to a new class of the EF-hand superfamily of calcium-binding proteins that has recently emerged and are expressed primarily in the brain and/or in the photoreceptor cells of various vertebrates *(9)*. One of the most extensively studied members of this new superfamily is a photoreceptor calcium-binding protein called recoverin *(10,11)*. The calcium-bound recoverin has been shown to regulate phototransduction by inhibiting the phosphorylation of activated rhodopsin by rhodopsin kinase *(10,11)*. Numerous vertebrate cognates of recoverin have been identified. They include S-modulin, visinin, frequenin, hippocalcin, vilip, and multiple isoforms of neurocalcin. All of these proteins are typically 190–200 amino acid-residues long, exhibit more than 45% sequence identity, and share some common conserved features: a consensus myristoylation signal at the amino terminus, and two pairs of classical EF-hand calcium-binding loop sequences (EF1, EF2, EF3, EF4), and a conserved *Cys-Pro* pair in the disabled EF1-hand *(10–21)*. Protein-sequence comparisons (*see* **Fig. 1**) indicate that recoverin, S-modulin, and visinin define a distinct subclass that is expressed in photoreceptor cells and have a nonfunctional calcium-binding EF4-hand, whereas, frequenin, hippocalcin, vilip and neurocalcin are members of another sub-class that are primarily expressed in the central and peripheral nervous system and inner retina *(15–19)*. A third subclass includes GCAP-1 and GCAP-2, two mammalian photoreceptor proteins that activate rod outer-segment membrane guanylate cyclase (ROS-GC) *(7,8)*.

In contrast to this well-established negative regulation by calcium, recently it was reported that another Ca^{2+}-binding protein called calcium-dependent guanylate cyclase activating protein (CD-GCAP) whose active moiety is S100b activates ROS-GC in a calcium-dependent manner with little or no influence at nanomolar calcium concentrations *(21)*. Two distinct types of photoreceptor GCs have been cloned from bovine retinal libraries, ROS-GC1 and ROS-GC2 *(22,23)*. These proteins share >40% sequence homology and each contain an extracellular domain, a single transmembrane domain, a conserved kinase homology domain, and a catalytic domain that converts GTP to cGMP. ROS-GC1 responds to three Ca^{2+}-signals. One is modulated by GCAP-1, the second by GCAP-2, and the third is regulated by calcium-dependent GCAP (CD-GCAP). Each signal is transduced by a ROS-GC1 signal-specific domain *(24)*. In contrast, ROS-GC2 is relatively plastic, responding only to Ca^{2+}-signals which are mediated by GCAP2, and the level of stimulation is about 10-fold less

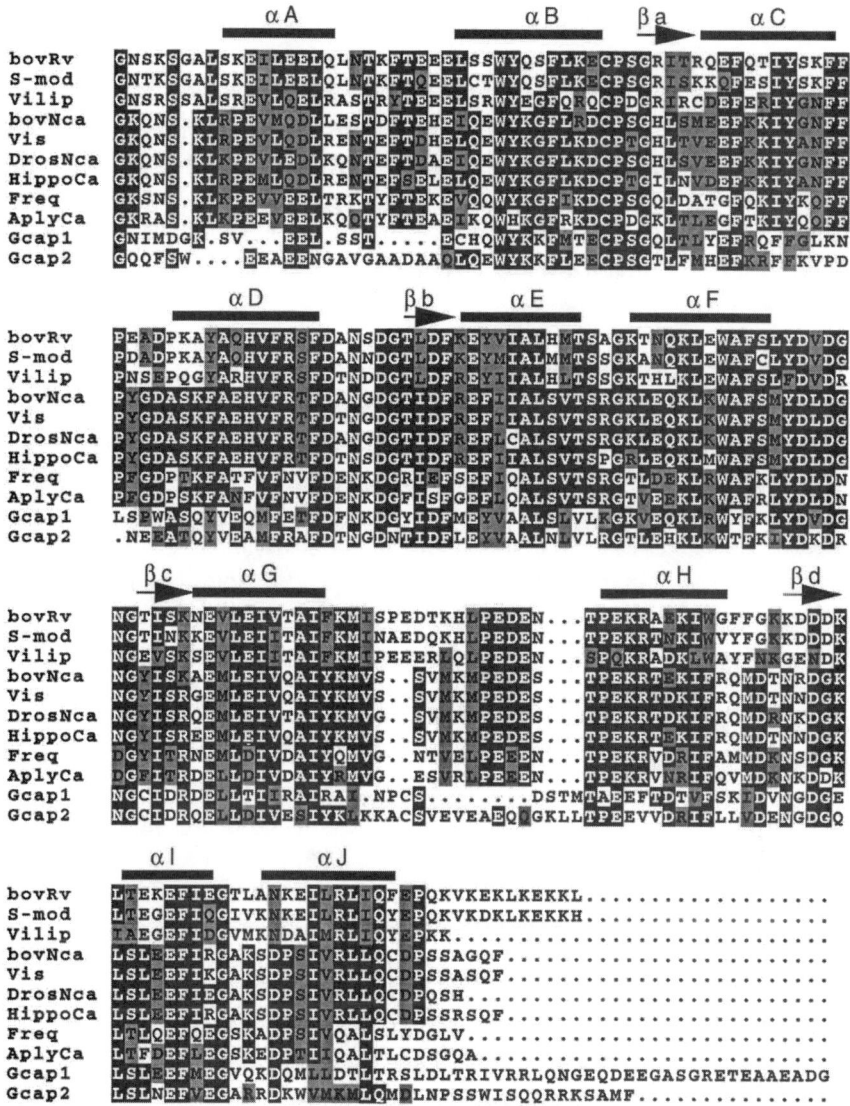

Fig. 1. Amino acid sequence alignment of bovNca (bovine neurocalcin) with recoverin-like proteins. The order in which they are listed: bovRv (bovine photoreceptor recoverin), S-mod (frog photoreceptor S-modulin), Vis (chicken cone visinin), bovNca (Bovine neurocalcin), Vilip (chicken brain villip), DrosNca (Drosophila neurocalcin), HippoCa (rat brain hippocalcin), Freq (Drosophila frequenin), AplyCa (aplycalcin), and Gcap2 (bovine guanylate cyclase activating protein-2). Secondary structural elements of neurocalcin (α-helices and β-sheets) and their notations are shown above the sequences. The numbering used is from bovRv sequence. Conserved residues are shaded in black and similar residues are shaded in gray.

responsive to the CD-GCAP signal *(24,25)*. There is an interesting aspect of the GCAPs and CD-GCAP regulation of ROS-GC1. Ca^{2+} inhibits GCAPs ability to activate the cyclase, but it stimulates CD-GCAP. Ca^{2+}-linkage of the ROS-GC with phototransduction is through GCAPs, and presumably, Ca^{2+}-linkage of the ROS-GC with retinal synaptic activity is through CD-GCAP *(26)*. The present study discloses the identity of a second Ca^{2+}-dependent regulator of ROS-GC1, neurocalcin. Neurocalcin stimulates ROS-GC1 in a Ca^{2+}-dependent fashion. It, however, does not affect ROS-GC2, the other member of ROS-GC subfamily.

Multiple isoforms of neurocalcins have been isolated to date *(18,19)*. The molecular weights of these isoforms range from 23 to 24 kDa. Neurocalcins, unlike other members of this family are differentially distributed. They are expressed mainly in the central nervous system, spinal cord, retina, inner ear, and olfactory epithelium, but they are also found in the zonaglomerulosa of the adrenal gland *(18,19,27–29)*. Both the unmyristoylated and myristoylated forms of neurocalcin bind three calcium ions per molecule at saturating concentrations *(30)*. Unmyristoylated neurocalcin binds calcium with higher affinity than the myristoylated form, but the mode of calcium binding in the myristoylated form is cooperative (Hill Coefficient: 1.8 ± 0.2).

In vitro studies have shown that both recombinant myristoylated and nonacylated forms of Drosophila neurocalcin at high Ca^{2+} concentrations can function as calcium sensors in vision by regulating the phosphorylation of bovine rhodopsin by rhodopsin kinase *(31)*. The potency and inhibitory effect of neurocalcin mimics that of recoverin. Interest in neurocalcin and other members of this family has been heightened by the finding that all members have a covalently attached myristoyl or related N-acyl group at the amino-terminus. Studies on neurocalcin show that the covalent addition of myristoyl group confers on neurocalcin the ability to interact with biological membranes in a Ca^{2+}-dependent manner *(30,32)*. Recent structural studies on recoverin have identified the molecular mechanics of calcium-myristoyl switches *(33,34)*. In the Ca^{2+}-free state, the myristoyl group is sequestered in a deep hydrophobic box, where it is clamped by multiple residues contributed by the three EF-hands. However, upon Ca^{2+}-binding, the myristoyl group is unclamped and extrudes enabling it to interact with a lipid bilayer membrane. This transition is accompanied by a 45° rotation of the N-terminal domain relative to the C-terminal domain, resulting in exposure of many of the hydrophobic residues, which can interact with the target protein. The high degree of conservation of hydrophobic residues among all members of this family of proteins, including neurocalcin suggests that these residues are important in the Ca^{2+}-induced extrusion of the myristoyl group, and that the calcium-myristoyl switch mechanism is almost the same in all family members from yeast to human.

This chapter discusses the crystal structure of neurocalcin, the architecture of Ca^{2+}-binding sites, and provides insights into the structural basis of the protein's broad specificity and its newly defined role in ROS-GC1 regulation of neuronal signaling.

2. Results and Discussion

2.1. Structure

The crystallographic structure of recombinant bovine neurocalcin was determined by multiple isomorphous methods to 2.4 Å resolution *(35)*. The fold of neurocalcin monomer is predominantly α-helical with overall dimensions of approx 35 Å × 61 Å × 41 Å (*see* **Fig. 2A**). Each neurocalcin monomer consists of two pairs of EF hands (N- and C-), the typical 29 residue helix-loop-helix motifs found typically in most calcium-binding proteins like calmodulin *(27)* and other members of EF-hand calcium-binding proteins. The N-terminal domain consists of the EF1 and EF2 hands, and the C-terminal domain comprises of the EF3 and EF4 hands. The EF1 hand includes helixB (24–37), loopI (38–45), and helixC (46–53), followed by a linker region loopII (54–62), to the second EF-hand, which includes helixD (63–72), loopIII (73–81), helixE (82–92). A five residue "hinge" region, loopIV (93–97), connects the N-terminal pair of EF-hands with the C-terminal pair of EF-hands. The C-terminal domains first EF-hand, the EF3 hand, includes helixF (98–108), loopV (109–118), helixG (119–129), followed by an unusually long linker region loopVI (130–145) to the second EF-hand consisting of helixH (146–155), loopVII (156–166), helixI (167–175). The N- and C-terminal pair of EF-hand domains are related by an approximate twofold rotation. The calcium binding loop in each EF-hand domain consists of short β-strands (42–44, β-strand A; 79–81 β-strand B; 115–117 β-strand C; 163–165 β-strand D) which form antiparallel β-sheets. These small β-sheets, when combined with the extensive hydrophobic interactions from the four helices of each pair of EF hands, bring the two Ca^{2+}-binding domains of neurocalcin into close proximity. Beside the four EF-hands, there are two additional helices, N-terminal helix A(11–17), and a short C-terminal helix J(178–183). The short U-shaped linker region (loop IV) connecting the N- and C-terminal pair of EF-hands in neurocalcin is a unique feature of this family of proteins. As a result, the four EF-hands of neurocalcin like recoverin form a compact array on one face of the protein rather than a dumbbell shape as seen with calmodulin.

Gel-permeation chromatography studies indicate that bovine neurocalcin exists as a homodimer in solution. The crystal structure of neurocalcin reveals the presence of two independent neurocalcin molecules in the asymmetric unit. The four EF-hands of each monomer are related to each other by noncrystallographic twofold axis (*see* **Fig. 2B**). In most calcium-binding pro-

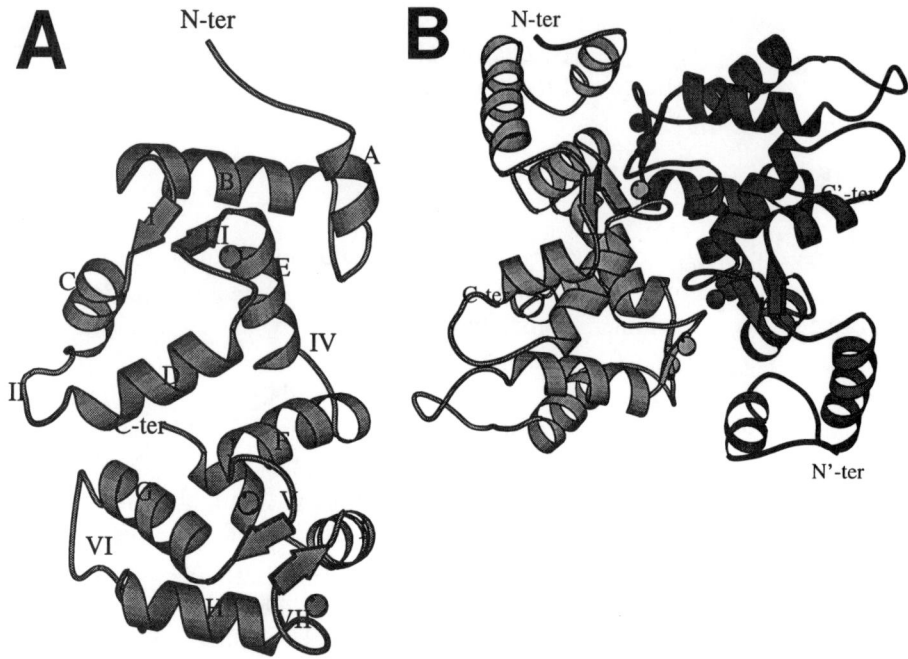

Fig. 2. Ribbon representation of bovine (**A**) neurocalcin monomer. The bound Ca^{2+} atoms are represented as solidspheres. The helices are labeled from A–J and the loops from I–VI. (**B**) neurocalcin dimer: The two monomers are represented as gray and black, respectively. The bound Ca^{2+}-atoms are represented as solid spheres. The noncrystallographic twofold axis relating the two monomers is in the plane of view.

teins, the basic structural/functional unit is a EF-hand pair, rather than a single Ca^{2+}-binding site. Pairing is presumed to stabilize the molecular conformation and influence the Ca^{2+}-binding ability. The interface accessible surface area between the two monomers is 1147 Å^2 is not as extensive compared to other homodimer proteins (37). The interactions between the two monomers are noncovalent interactions, and mostly hydrophobic in nature, involving residues from the EF2, EF3 and EF4 hands. This mode of association, i.e., EF-hand participation in the domain assembly of a protein is rather intriguing, that may be functionally significant for membrane targeting mechanism. It is also possible that such an association may confer upon neurocalcin additional conformational stability.

The crystal structure of neurocalcin revealed four Ca^{2+}-binding EF-hands per molecule. However, calcium is bound only at three of the four EF-hands: EF2, EF3, and EF4. The EF1- hand in neurocalcin is disabled from binding calcium because of the presence of *Cys*38-*Pro*39 sequence at positions 3 and 4 of the 12-residue Ca^{2+}-coordination loop (**Fig. 3A**). The sulfur atom from cysteine is not capable of liganding Ca^{2+}. In addition the presence of *Arg*36 at position 1 is not favorable for liganding Ca^{2+} either, because it lacks a side-chain oxygen. The *Cys-Pro* sequence in EF1-hand is highly conserved in all members of this family (**Fig. 1**). Despite the absence of calcium, the EF1-hand still retains an "open" conformation similar to that of calcium-bound EF-hands observed in other calcium sensors. A disabled EF1-hand is a characteristic feature of this family of calcium sensors and may have been recruited for a different role, i.e., to cradle the myristoyl group in the apo-form *(34)*. The EF2-hand in neurocalcin is in a calcium-bound state, and shows a structural similarity with other calcium-binding protein EF-hands (*see* **Fig. 3A**). However, the EF2-site in the crystal structure of recoverin reveals only a partial occupancy of calcium bound. Despite the partial occupancy, all the residues are in a favorable "open" conformation for calcium-liganding *(38)*. It is presumed this approximates an intermediate between the "on" and "off" states of recoverin *(38)*. Analysis of membrane properties of mutant myristoylated neurocalcin revealed the critical role of EF2-hand in calcium-myristoyl switch mechanism *(30)*. Ca^{2+}-binding to EF2-hand is necessary to induce conformational changes that may lead to the binding of neurocalcin with membranes. The EF3-hand in neurocalcin shows the presence of a calcium bound ion and shows a striking structural similarity with that of calmodulin EF-hands (**Fig. 3C**). The amino acid sequences for the EF3-hand appear to be highly conserved among other members of this subfamily. Recent studies on neurocalcin, show any mutation that results in disabling of the EF3-hand, appears to have profound effects on calcium binding at the other two sites *(30)*. Although any mutation in the EF2-hand had the least effect. These results strongly suggest a strong interdependence of the three-functional EF-hands of neurocalcin in the binding of calcium. The EF4-hand in neurocalcin is in a calcium bound state whereas recoverin is not (*see* **Fig. 3D**) *(38)*. A disabled EF4-hand is a characteristic feature of *rec* subfamily of proteins, which consists of recoverin, S-modulin, and visinin. The EF4-hand in recoverin is disabled from binding calcium because two key residues Gly160 at position 1 and Lys162 at position 3 lack a side-chain oxygen, and are, therefore, unable to bind calcium. Furthermore, Lys161 also forms a salt-bridge with Glu171 at position 12 and acts as a competitive inhibitor of Ca^{2+}-binding *(38)*. As a result, part of the loop that consists of residues 160–165 is pulled away from calcium-binding conformation by approx 5 Å *(35)*. Also, the highly conserved glycine at position 6 is replaced by Asp165.

2.2. Calcium-Myristoyl Switches

The myristoylation consensus sequence at the N-terminus (MGXXXS) is conserved between neurocalcin, recoverin, and other members of this subfam-

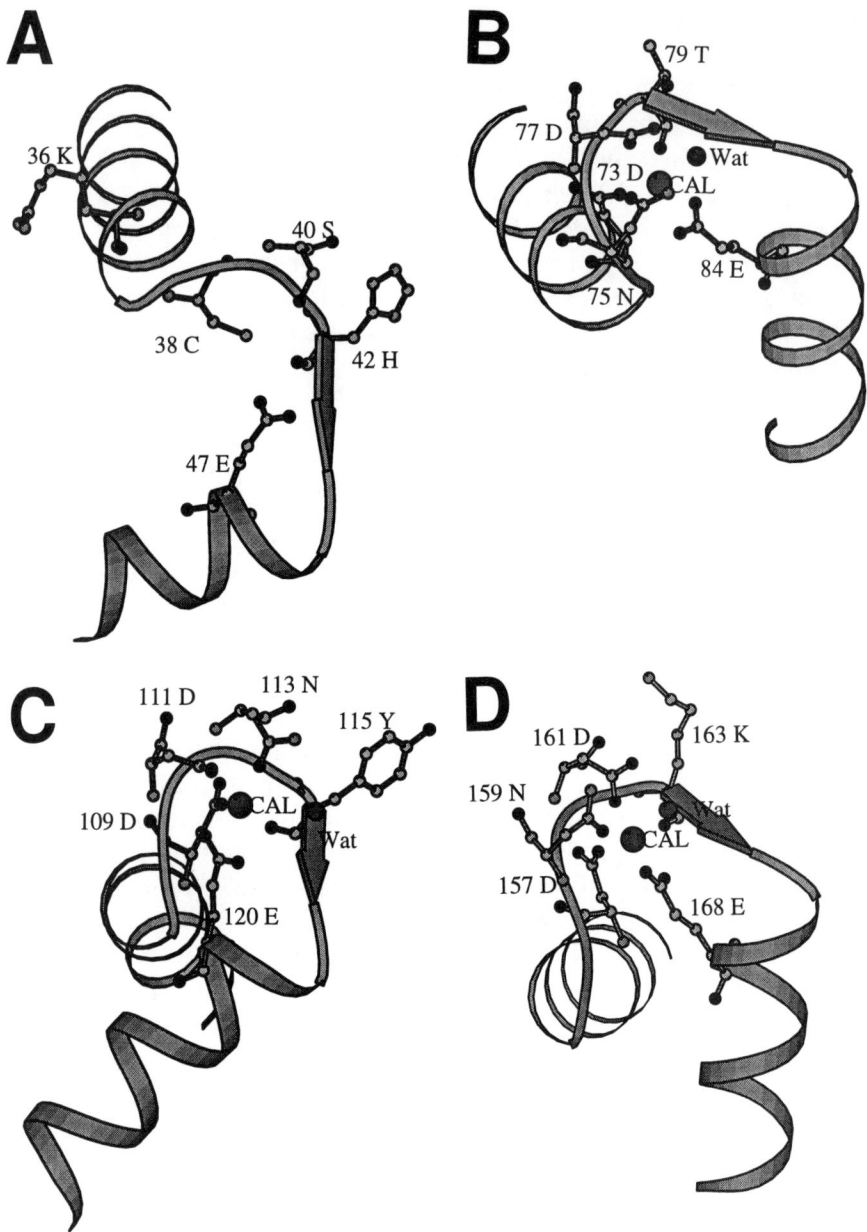

Fig. 3. Representation of the four EF-hands of neurocalcin (EF1, EF2, EF3, EF4). Residues involved in the calcium coordination are shown. (**A**) Disabled EF1-hand, showing the location of lys 36 and *Cys* 38 which prevents the EF1-hand from binding calcium. EF1-hand, however, still retains an open conformation, similar to that of Ca^{2+}-bound EF-hand. (**B**) Ca^{2+}-bound EF2-hand. (**C**) Ca^{2+}-bound EF3-hand. (**D**) Ca^{2+}-bound EF4-hand.

ily (*see* **Fig. 1**). Ten of the hydrophobic residues that clamp the myristoyl group in the Ca^{2+}-free state recoverin (Trp 31, Tyr 32, Phe 49, Ile 52, Tyr 53, Phe 56, and Phe 57 from EF1-hand, Phe 83, Leu 90 from EF2-hand and Trp 104 from EF3-hand) of neurocalcin are conserved with other members of this family suggesting that the structure of the binding pocket is similar in all members of the subfamily. However, these residues in Ca^{2+}-bound recoverin are structurally different from the Ca^{2+}-free recoverin (root-mean-square deviation [rmsd] 9.6 Å) because calcium induces the unclamping of several of these residues and extrusion of myristoyl group outside the protein *(35)*, but structurally conserved with Ca^{2+}-bound neurocalcin (rmsd 1.4 Å). The two glycine residues 42 and 96, which play an important role in the conformational changes from Ca^{2+}-free to Ca^{2+}-bound form, are conserved between recoverin and neurocalcin and most other members of this family suggesting that the calcium-myristoyl switch mechanism is almost the same in all family members, from yeast to humans. Neurocalcin like recoverin has a related concave hydrophobic surface formed by residues Phe 21, Trp 31, Leu 52, Tyr 53, Phe 56, Phe 57, Tyr 86, and Leu 90 (recoverin), which is believed to serve as a possible binding site for protein targets *(39,40)*.

2.3. Guanylate Cyclase Activation

In an attempt to determine if neurocalcin mimics GCAP1 and GCAP2 in stimulating ROS-GC activity, COS cell membranes expressing ROS-GC1 were recombined with purified neurocalcin at different Ca^{2+} concentrations. There was no cyclase stimulation at nanomolar concentrations of Ca^{2+} (*see* **Fig. 4A**) indicating that neurocalcin does not duplicate GCAPs effects; and thus has no role in phototransduction-linked ROS-GC1 activity. However, at a fixed concentration of 1 m*M* Ca^{2+}-levels, Neurocalcin is able to stimulate ROS-GC1 in a concentration-dependent manner; half-maximal activation of the cyclase occurred at approx 1 m*M* and saturation at approx 3 m*M* (*see* **Fig. 4B**). The stimulatory effect of neurocalcin was Ca^{2+}-dependent with its EC50 of approx 20 μ*M* (*see* **Fig. 4A**). In contrast, under similar conditions, neurocalcin does not stimulate ROS-GC2, the other member of ROS-GC subfamily. These observations indicate that neurocalcin is specific for ROS-GC1 regulation, the regulation is Ca^{2+}-dependent, and is unrelated to phototransduction. There is an interesting aspect of the GCAPs and neurocalcin regulation of ROS-GC1. This mode of regulation of ROS-GC1 by neurocalcin is similar to CD-GCAP. The copresence of neurocalcin and ROS-GC1 in amacrine and ganglionic cells suggests that the regulation may be linked to the synaptic activity of these cells. Similar interpretation regarding the role of CD-GCAPs regulation of ROS-GC1 activity has been made previously *(26)*.

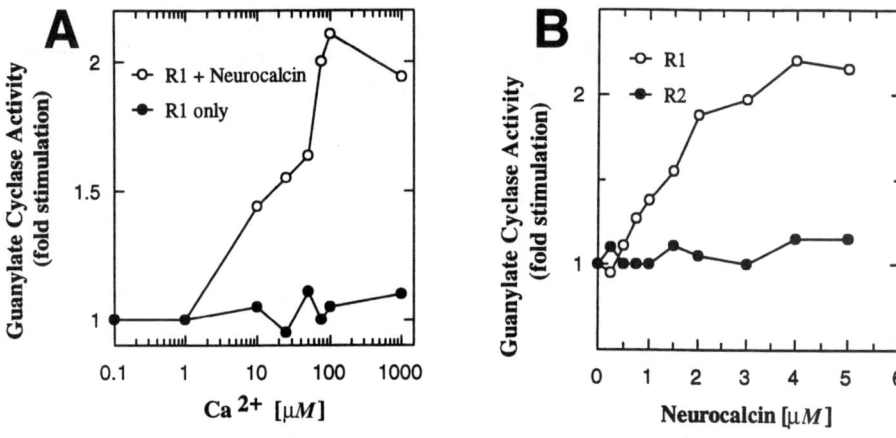

Fig. 4. Effect of neurocalcin and free-Ca^{2+}-concentration on ROS-GC1 and ROS-GC2 activity. Membranes of COS cells expressing ROS-GC1 and ROS-GC2 were prepared as described in **Subheading 3.** These were assayed for guanylate cyclase activity in the presence of **(A)** 1 mM free-Ca^{2+} and incremental concentrations of neurocalcin, and **(B)** 4 mM neurocalcin and indicated free-Ca^{2+}-concentrations.

The neurocalcin-regulated domain of ROS-GC1 has now been identified, and like CD-GCAP resides at the C-terminus, between aa 731–1054. Neurocalcin has little or no effect in ROS-GC1 activity when the GCAP-interacting intracellular domain is mutated *(41)*. These observations suggest possibly two different modes of ROS-GC1 modulating mechanisms; one related to phototransduction-related activity, and second possibly related to retinal synaptic activity. An interesting observation from these studies are the close parallels between neurocalcin and CD-GCAP in their modes of ROS-GC1 regulation. Both are Ca^{2+}-dependent regulators of ROS-GC1 activity, they have no role in phototransduction-related ROS-GC1 activities, and are implicated in ROS-GC1 related retinal synaptic activity. They stimulate ROS-GC1 almost identically in a similar Ca^{2+}-dependent fashion; highly specific manner: neurocalcin has no effect on ROS-GC2, and CD-GCAP is 10-fold more selective than for ROS-GC2; the modulated cyclase domain in both cases reside at the C-terminal region of ROS-GC1 between aa 731–1054 *(42)*. Despite these shared features, there is one notable difference between the two regulators. The achievable cyclase activity is about threefold more with CD-GCAP than neurocalcin *(42)*. The reason for such differences in their cyclase activities is because neurocalcin-regulated domain in aa 731–1054 is quite distinct from that of CD-GCAP regulated cyclase domain. When ROS-GC1 was exposed to the combined saturating concentrations of the two activators (4 mM of neurocalcin and 2 mM of CD-GCAP) at a fixed 1 mM of free-Ca^{2+}-concentration, the resulting stimulation was approximately the sum of those caused by neurocalcin and CD-GCAP alone *(41)*. These observations suggest the neurocalcin-regulated ROS-GC1 domain is distinct from that of GCAP-modulated and CD-GCAP-regulated domain. The physiological implication of these findings are unclear. They predict, however, that under identical conditions, ROS-GC1 will sense both Ca^{2+} signals carried by CD-GCAP and neurocalcin, but the amplitude of response will be far higher with CD-GCAP compared to neurocalcin-mediated Ca^{2+}-signals. The depolarization of neurons cause internal free-Ca^{2+} concentrations to rise to the range of 10–100 mM *(43)*. Given the paired presence of CD-GCAP/ROS-GC1 or neurocalcin/ROS-GC1, the present study provides a conceptual framework for an alternative mechanism by which ROS-GC1 may be linked with neuronal synapse of a variety of retinal cells.

The finding that neurocalcin stimulates ROS-GC1 is intriguing and was unexpected. Structurally, neurocalcin may share stronger identity with GCAP, yet functionally it behaves in altogether an opposite fashion. We have attempted to rationalize these opposite behaviors of neurocalcin and GCAPs toward ROS-GC1. First, the role of calcium-binding EF-hands; recent mutagenesis experiments on GCAP1 and GCAP2 have shown that any mutations within the three calcium-bound EF-hands have significant implications on their regulatory properties during ROS-GC stimulation *(44,45)*. For instance, mutations in the EF2 and EF4 hands of GCAP2 interfere with ROS-GC stimulation, but Ca^{2+} sensitivity remains unaffected. Similarly, when the EF3-hand is disabled in GCAP1, it does not hamper the ability of GCAP1 to stimulate ROS-GC1 activity, but fails to inhibit cyclase activity at high Ca^{2+} levels. Crystal structure of recoverin, despite showing a highly conserved topology with neurocalcin, shows no evidence of activating ROS-GC. Interestingly, recoverin like other rec class of proteins that include S-modulin has a disabled EF4-hand. These results further support the importance of active EF-hands during ROS-GC stimulation.

Second, a closer study between neurocalcin and other members of this subfamily of protein reveals two regions of significant amino acid sequence divergence: the linker region connecting EF3 and EF4 hands; and the C-terminus region. Do these diverse regions undertake a specific conformation at varying Ca^{2+}-concentrations, giving each protein a distinct function? This may indeed be the case supported by the recent mutagenesis studies with GCAPs. A point mutation Y to C in the EF3-hand region results in different phenotypes. The Y99C mutation in GCAP1 decreased the Ca^{2+}-sensitivity toward ROS-GC1 by remaining active at Ca^{2+} concentrations above 1 mM levels. However, a similar mutation Y104C in GCAP2, results in not effecting the Ca^{2+}-sensitivity, but greatly reduced its ability to stimulate ROS-GC1 activity. This is consistent with the authors' suggestion that these mutations may have a profound effect on the surrounding regions of the GCAPs *(46)*, and these regions may play an important role in guanylate cyclase activation. This hypothesis is further supported by Ca^{2+}-dependent studies on GCAP-2 and recoverin *(47)*. Here, the linker region is susceptible to V8 protease hydrolysis only in the Ca^{2+}-bound form of GCAP-2, but has no effect on recoverin. It has been reported that the C-terminal region plays an effective role in electrostatic interactions with phospholipid membranes. It is possible these residues may play a similar role in specific cyclases.

Third, do these proteins need to be in a dimeric or oligomeric state in order for them to play a role in ROS-GC activation? Both neurocalcin and CD-GCAP show evidence of being in a dimeric state, which then poses the question, what role, if any, the dimer interface plays in recognizing the target protein?

3. Methods
3.1. Expression and Purification of Neurocalcin

The cloning, expression, and crystallization of bovine neurocalcin has been described in detail elsewhere *(48)*. The recombinant neurocalcin d gene was expressed in *Escherichia coli*, as a maltose-binding protein fusion protein using pMAL™-c expression vector (New England Biolabs, Beverly, MA). A recombinant neurocalcin d cDNA was amplified by polymerase chain reaction (PCR) from pCalN vector *(19)* using oligonucleotides containing the boundaries of the coding region of neurocalcin d cDNA. The amplified cDNA was inserted into the *Stu*I and *Eco*RI sites of pMAL™-c expression vector. The recombinant plasmid was transformed into *E. coli* strain DH5a, and overexpressed as a maltose-binding fusion protein, according to the manufacturers instructions (New England Biolabs). The purification protocol was modified to obtain material of suitable quality for crystallization experiments. Expression was induced by addition of 0.5 mM isopropylthio-β-D-galactoside. The cell pellet was suspended in 100 mL lysis buffer containing 10 mM phosphate pH 7.2/30 mM NaCl/0.25% Tween-20/10 mM β-mercaptoethanol, and sonicated for 5 min. Sodium chloride was added to the broken cells to a final concentration of 0.5 M, and centrifuged at 9000g for 30 min, and the supernatant was loaded onto a preequilibrated 30 mL amylose resin (New England Biolabs) column. The bound fusion protein was eluted with the 2 vol of column buffer containing 10 mM phosphate pH 7.2/0.5 M NaCl/10 mM β-mercaptoethanol and 10 mM maltose. The fusion protein was subjected to factor *Xa* cleavage and purified on a Mono Q anion-exchange column (Pharmacia). The Mono Q fractions of neurocalcin d protein was dialyzed against 20 mM Tris-Cl, pH 8.0, and 1 mM β-mercaptoethanol, and concentrated by Centricon-10 (Amicon) to 15 mg/mL. The total yield of pure neurocalcin *d* from a 2-L medium is 10–15 mg.

3.2. Crystallographic Analysis

The crystals used for the structure determination had to be macroseeded several times to get suitable diffracting crystals *(49)*. The crystals belong to the monoclinic space group (P21) with cell dimensions a = 42.5Å, b = 93.3Å, c = 50.6 Å, and b = 98.3°, and two monomers in the asymmetric unit. All experimental data were collected at room temperature on a Rigaku R-axis II imaging-plate system, using monochromatized CuKa radiation from rotating anode operated at 5 kW. The diffraction data were processed using RAXIS software *(50)*. The structure was solved by multiple isomorphous replacement (MIR) using a total of five derivatives. Heavy-atom derivatives were prepared by soaking the native crystals in 60% Ammonium sulfate in 50 mM piperazine-N,N'-*bis*-(2-ethane sulfonic acid) (PIPES), pH 6.1, 5 mM calcium chloride, and 0.5 mM β-mercaptoethanol, and the respective heavy-atom reagents at 10–15 mM concentrations for 48 h (**Table 1**). The heavy-atom positions were located from SHELX *(51)*. The calculation and refinement of phases were done with MLPHARE program *(52)* including all derivative data (isomorphous and anomalous) to 3.5 Å resolution. This map was inspected using the graphics display using FRODO *(53)* and showed clear interpretable secondary structures. Subsequently, the map was then improved with density-modification techniques using the program DM, including solvent flattening and histogram matching, and the phases extended to 3.0 Å resolution *(54)*. The map was interpretable enough so that the main chain of regular structural elements could be identified. After many cycles of model building using FRODO and phase combination using SIGMAA-weighted *(55)* omit maps, an initial model was built that consisted of residues 7–129 and 149–185 of each monomer and had an initial R-factor of 0.441–3.0 Å resolution.

Simulated annealing and positional conjugate gradient refinement using NCS restraints in X-PLOR *(56)* was carried out and the resolution was extended slowly to the limiting resolution of 2.4 Å. The R-factor and Rfree for the NCS restrained model was 0.234 and 0.311, respectively. At this point, NCS restraints were removed from the protocol, individual isotropic B-factors were substituted for the grouped ones. The final R-factor value is 0.216 and the corresponding Rfree was 0.308. The model presented herein have good geometry and are at apparent free R-factor minima, values of which are limited by disorder that prevents accurate modeling of residues 134–142 for monomer A and residues 132–145 for monomer B, and the inability to assign density for residues 1–4 and 186–193 of each monomer. The average deviations in bond lengths and angles was 0.008 Å and 1.304°, respectively. The final model consists of residues 5–185 for both the subunits, 46 solvent molecules, and 6 calcium ions.

Table 1
Crystallographic Data Collection, Phasing, and Refinement Statistics

(A) Data collection statistics

Crystal	Resolution Å	Unique reflections	Completeness of data	R_{merge} (%)	Anomalous signal
Native I	2.4	12105	77.20	7.3	
Native II	2.6	10264	81.30	8.7	
Native I and II	2.4	13264	86.52	7.9	
TMLA	2.7	9035	70.60	9.4	Yes
$LuCl_3$ (1)	2.8	7836	80.00	8.1	
$LuCl_3$ (2)	3.0	6262	74.50	10.2	
Mersalyl	2.7	9941	77.70	9.2	
$LuCl_3$ + Mersalyl	2.6	9657	76.20	8.7	

(B) MIRAS phasing statistics for data from 15.0–3.5 Å

Crystal	R_{iso} (%)	Number of sites	Phasing power (a/c)	R_{Cullis} (a/c)	R_{Cullis} (Ano)
TMLA	9.34	2	1.51/0.98	0.85/0.79	0.95
$LuCl_3$(1)	21.0	6	2.82/1.59	0.60/0.62	
$LuCl_3$ (2)	15.7	2	2.24/1.12	0.78/0.94	
Mersalyl	11.94	1	1.25/0.88	0.85/0.77	
$LuCl_3$ + Mersalyl	14.01	3	2.68/1.40	0.68/0.72	

TMLA = Trimethyllead acetate R-factor = $(\Sigma \, ||F_{obs}| - F_{calc}||/\Sigma \, |F_{obs}|$, where F_{obs} and F_{calc} are the observed and calculated structure factor amplitudes, respectively. $R_{iso} = \Sigma \, |I_{ph} - I_p|/\Sigma \, I_p$, where I_p and I_{ph} are the intensities from the protein and the heavy atom derivatives, respectively. Phasing Power = average heavy atom structure factor divided by rms lack of closure. $R_{merge} = \Sigma_{hkl} |I - <I>|/\Sigma \, I$. where I is the observed intensity and $<I>$ is the average intensity from observations of symmetry related reflections, respectively.

(C) Refinement statistics

Resolution (A)	2.4 Å	
Crystallographic R-factor		0.216
Free R-factor	0.308	
Overall average B-factor		35.1 Å2
rmsd bond lengths	0.008 Å	
rmsd bond angles	1.304°	

3.3. ROS-GC Expression Studies

COS7 cells (Simian virus 40-transformed African monkey kidney cells), maintained in Dulbeco's modified Eagles medium (DMEM) with penicillin, streptomycin, and 10% fetal bovine serum (FBS), were transfected with the expression constructs by the calcium-phosphate coprecipitation technique. Transfection with vector alone served as a control. Sixty hours after transfection, cells were washed twice with 50 mM Tris-HCl, pH 7.5, 10 mM MgCl2 buffer, scraped into 2 mL of cold buffer, homogenized, centrifuged for 15 min at 50,000g, and washed several times with the same buffer. This pellet represented the crude membranes.

3.4. Guanylate Cyclase Assay

The crude membranes were assayed for guanylate cyclase activity as described previously (57). Briefly, membranes were preincubated in an ice bath with or without neurocalcin. The assay system consisted of 10 mM theophylline, 15 mM phospho-creatinine, 20 μg of creatinine kinase, and 50 mM Tris-HCl, pH 7.5. This was adjusted to appropriate free-calcium concentrations with precalibrated Ca^{2+}/ethylene glycol-*bis* (β-aminoethyl ether)-$N,N,N,'N'$-tetraacetic acid (EGTA) solutions (Molecular Probes). Total assay volume was 25 μL. The reaction was initiated by the addition of the substrate solution containing 4 mM MgCl2 and 1 mM GTP, and incubated at 37EC for 10 min. The reation mixture was terminated by the addition of 225 μL of 50 mM sodium acetate buffer pH 6.25 followed by heating in boiling water bath for 3 min. The amount of cyclic GMP was determined by radio immunoassay (58).

4. Conclusions

Although the calcium-myristoyl switch mechanism has been well established with the recoverin and other members of this subclass of proteins, the opposite effects GCAP-1 and GCAP-2, compared to neurocalcin on ROS-GC1 stimulation, suggest this family of proteins have rather important roles in coupling calcium cascades and guanine nucleotide-binding cascades (7,8,15). These observations shed new light on this family of proteins and their mode of activation because GCAP1 and GCAP2 activate ROS-GC only in the Ca^{2+}-free form and have an inhibitory effect at high-Ca^{2+} levels. The distinct distribution of neurocalcin in the central and peripheral nervous system suggest that ROS-GC may not be the natural target for neurocalcin. On the other hand, GCs are members of a multigene family with unique tissue-specific expression patterns, it is possible that one of the other GCs from the nervous system may be the natural target for neurocalcin. A question of enormous biological interest is: is it possible these tissues also contain ROS-GCs? If they do, this will represent a general neuronal signaling pathway. Finally, it will be interesting to test the possible interactions of other neurocalcin-like proteins vilip, hippocalcin (calcium-bound EF4-hands) with other GCs to determine whether these proteins may be Ca^{2+} sensitive GC activators.

Acknowledgments

We would like to thank Dr. R. K. Sharma and his laboratory personnel for performing ROS-GC regulation experiments. This work was funded in part by aid from Markey's foundation (Senadhi Vijay-Kumar), and Kimmel Cancer Center (Vinod D. Kumar).

References

1. Kretsinger,R. H. (1985) Hypothesis: Calcium modulated proteins contain EF-hands, in *Calcium Transport in Contraction and Secretion* (Carafoli, E., ed.), North-Holland Publishing, Amsterdam, pp. 469–480.
2. Yarfitz, S. and Hurley, J. B. (1994) Transduction mechanisms of vertebrate and invertebrate photoreceptors. *J. Biol. Chem.* **269,** 14,329–14,332.
3. Koch, K. W. and Stryer, L. (1988) Highly cooperative feedback control of retinal rod guanylate cyclase by calcium ions. *Nature* **334,** 64–66.
4. Gorczyca, W. A., Polans, A. S., Surgucheva, I. G., Subbaraya, I., Baehr, W., and Palczewski, K. (1995) Guanylyl cyclase activating protein. A calcium-sensitive regulator of phototransduction. *J. Biol. Chem.* **270,** 22,029–22,036.
5. Kakiuchi, S., Sobue, K., Yamazaki, R., Nagao, S., Umeki, S., Nozawa, Y., et al. (1981) Ca^{2+}-dependent modulator proteins from tetrahymena pyriformis, sea anemone, and scallop and guanylate cyclase activation. *J. Biol. Chem.* **256,** 19–22.
6. Klumpp, S. and Schultz, J. E. (1982) Characterization of a Ca2+-dependent guanylate cyclase in the excitable ciliary membrane from Paramecium. *Eur. J. Biochem.* **124,** 317–324.
7. Dizhoor, A. M., Olshevskaya, V. V., Henzel, W. J., Wong, S. C., Stults, J. T., Ankoudinova, I., and Hurley, J. B. (1995) Cloning sequencing and expression of 24-kDa Ca^{2+}-binding protein activating photoreceptor guanlyl cyclase. *J. Biol. Chem.* **27,** 25,200–25,206.
8. Goraczniak, R. M., Duda, T., and Sharma, R. K. (1997) Structural and functional characterization of a second subfamily member of the calcium-modulated bovine rod outer segment memberane guanylate cyclase ROS-GC2. *Biochem. Biophys. Res. Commun.* **234,** 666–670.
9. Polans, A., Baehr, W., and Palczewski, K. (1996) Turned on by Ca^{2+}! The physiology and pathology of Ca^{2+}-binding proteins in the retina. *Trends. Neurosci.* **19,** 547–554.
10. Dizhoor, A. M., Ray, S., Kumar, S., Niemi, G., Spencer, M., Brolley, D., et al. (1991) Recoverin: a calcium sensitive activator of retinal rod guanylate cyclase. *Science* **251,** 915–918.
11. Gray-Keller, M. P., Polans, A. S., Palczewski, K., and Detwiler, P. B. (1993) The effect of recoverin-like calcium binding proteins on the photoresponse of retinal rods. *Neuron* **10,** 523–531.
12. Kawamura, S. and Murakami, M. (1991) Calcium-dependent regulation of cyclic GMP phosphodiesterase by a protein from frog retinal rods. *Nature* **349,** 420–423.
13. Kawamura, S., Takamatsu, K., and Kitamura, K. (1992) Purification and characterization of S-modulin, a calcium-dependent regulator on cGMP phosphodiesterase in frog rod receptors. *Biochem. Biophys. Res. Commun.* **186,** 411–417.

14. Yamagata, K., Goto, K., Kuo, C.-H., Kondo, H., and Miki, N. (1990) Visinin: a novel calcium binding protein expressed in retinal cone cells. *Neuron* **2**, 469–478.
15. Pongs, O., Lindemeier, J., Zhu, X. R., Engelkamp, D., Krah-Jentgens, I., Lambrecht, H.-G., et al. (1993) Frequenin: a novel calcium-binding protein that modulates synaptic efficacy in the Drosophila nervous system. *Neuron* **11**, 15–28.
16. Kobayashi, M., Takamatsu, K., Saitoh, S., Mirra, M., and Noguchi, T. (1992) Molecular cloning of hippocalcin, a novel calcium-binding protein of the recoverin family exclusively expressed in hippocampus. *Biochem. Biophys. Res. Commun.* **189**, 511–517.
17. Lenz, S. E., Henschel, Y., Zopf, D., Voss, B., and Gundelfinger, E. D. (1992) VILIP, a cognate protein of the retinal calcium binding proteins visinin and recoverin, is expressed in the chicken brain. *Mol. Brain Res.* **15**, 133–140.
18. Terasawa, M., Nakano, A., Kobayashi, R., and Hidaka, H. (1992) Neurocalcin: a novel calcium binding protein from bovine brain. *J. Biol. Chem.* **267**, 19,596–19,599.
19. Okazaki, K., Watanabe, M., Ando, Y., Hagiwara, M., Terasawa, M., and Hidaka, H. (1992) Full sequence of neurocalcin, a novel calcium binding protein abundant in the central nervous system. *Biochem. Biophys. Res. Commun.* **185**, 147–153.
20. Dyer, J. R., Sossin, W. S., and Klein, M. (1996) Cloning and characterization of aplycalcin and aplysia neurocalcin, two new members of the calmodulin superfamily of small calcium-binding proteins. *J. Neurochem.* **67**, 932–942.
21. Pozdynakov, N., Goraczniak, R., Margulis, A., Duda, T., Sharma, R. K., Yoshida, A., and Sitarammaya, A. (1997) Structural and functional characterization of retinal calcium-dependent guanylate cyclase activator protein (CD-GCAP): identity with S100b protein. *Biochemistry* **36**, 14,159–14,166.
22. Goraczniak, R. M., Duda, T., Sitaramayya, A., and Sharma, R. K. (1994) Structural and functional characterization of the rod outer segment memberane guanylate cyclase. *Biochem. J.* **302**, 455–461.
23. Goraczniak, R. M., Duda, T., and Sharma, R. K. (1997) Structural and functional characterization of a second subfamily member of ROS-GC2. *Biochem. Biophys. Res. Commun.* **234**, 666–670.
24. Krishnan, A., Goraczniak, R. M., Duda, T., and Sharma, R. K. (1998) Third calcium-modulated rod outer segment membrane guanylate cyclase transduction mechanism. *Mol. Cell. Biochem.* **178**, 251–259.
25. Duda, T., Goraczniak, R., and Sharma, R. K. (1996) Molecular characterization of S100A1-S100B protein in retina and its activation mechanism of bovine photoreceptor guanylate cyclase. *Biochemistry* **35**, 6263–6266.
26. Cooper, N. G. F., Liu, L., Pozdnyakov, N., Margulis, A., and Sitarammaya, A. (1995) The bovine rod outer segment guanylate cyclase, ROS-GC is present in both outer segment and synaptic layers of the retina. *J. Mol. Neurosci.* **6**, 211–222.
27. Nakano, A., Terasawa, M., Watanabe, M., Okazaki, K., Inoue, S., Kato, M., et al. (1993) Distinct regional localization of neurocalcin, a Ca^{2+}-binding protein, in the bovine adrenal gland. *J. Endocrinol.* **138**, 283–290.
28. Ilino, S., Kobayashi, S., Okazaki, K., and Hidaka, H. (1995) Immunohistochemical localization of neurocalcin in the rat inner ear. *Brain Res.* **680**, 128–134.
29. Ilino, S., Kobayashi, S., Okazaki, K., and Hidaka, H. (1995) Neurocalcin-immunoreactive receptor cells in the rat olfactory epithelium and vomeronasal organ. *Neurosci. Lett.* **191**, 91–94.

30. Ladant, D. (1995) Calcium and membrane binding properties of bovine neurocalcin d expressed in Escherichia coli. *J. Biol. Chem.* **270**, 3179–3185.
31. Furobert, E., Chen, C.-K., Hurley, J. B., and Teng, D. H.-F. (1996) Drosophila neurocalcin, a fatty acylated Ca^{2+}-binding protein that associates with membranes and inhibits in vitro phosphorylation of bovine rhodopsin. *J. Biol. Chem.* **271**, 10,256–10,262.
32. Teng, D. H.-F., Chen, C.-K., and Hurley, J. B. (1994) A highly conserved homologue of bovine neurocalcin in Drosophila melanogaster is a Ca^{2+}-binding protein expressed in neuronal tissues. *J. Biol. Chem.* **269**, 31,900–31,907.
33. Tanaka, T., Ames, J. B., Harvey, T. S., Stryer, L., and Ikura, M. (1995) Sequestration of the membrane-targeting myristoyl group of recoverin in the calcium-free state. *Nature* **376**, 444–447.
34. Ames, J. B., Ishima, R., Tanaka, T., Gordon, J. I., Stryer, L., and Ikura, M. (1997) Molecular mechanics of calcium-myristoyl switches. *Nature* **389**, 198–202.
35. Vijay-Kumar, S. and Kumar, V. D. (1999) Crystal structure of recombinant bovine neurocalcin. *Nat. Struct. Biol.* **6**, 80–88.
36. Babu, Y. S., Bugg, C. E., and Cook, W. J. (1988) Structure of calmodulin refined at 2.2 Å resolution. *J. Mol. Biol.* **204**, 191–204.
37. Jones, S. and Thornton, J. M. (1996) Principles of protein–protein interactions. *Proc. Natl. Acad. Sci. USA.* **93**, 13–20.
38. Flaherty, K. M., Zozulya, S., Stryer, L., and McKay, D. B. (1993) Three-dimensional structure of recoverin, a calcium sensor in vision. *Cell* **75**, 709–716.
39. Resh, M. D. (1994) Myristylation and palmitylation of Src family members: the fats of the matter. *Cell* **76**, 411–413.
40. McLauglin, S. and Aderem, A. (1995) The myristoyl-electrostatic switch: a modulator of reversible protein-membrane interactions. *Trends Biochem. Sci.* **20**, 272–276.
41. Kumar, V. D., Vijay-Kumar, S., Krishnan, A., Duda, T., and Sharma, R. K. (1999) A second calcium regulator of rod outer segment memberane guanylate cyclase, ROS-GC1: neurocalcin. *Biochemistry*, submitted.
42. Duda, T., Goraczniak, R. M., and Sharma, R. K. (1996) Molecular characterization of S100A1-S100B protein in retina and its activation mechanism of bovine photoreceptor guanylate cyclase. *Biochemistry* **35**, 6263–6266.
43. Augustine, G. J. and Neher, E. (1992) Calcium requirements for secretion in bovine chromaffin cells. *J. Physiol.* **450**, 247–251.
44. Dizhoor, A. M. and Hurley, J. B. (1996) Inactivation of EF-hands make GCAP-2 (p24) a constitutive activator of photoreceptor guanylyl cyclase by preventing a Ca^{2+}-induced activator-to-inhibitor transistion. *J. Biol. Chem.* **271**, 19,346–19,3450.
45. Rudznicka-Nawrot, M., Surgucheva, I., Hulmes, J. D., Heseleer, F., Sokal, I., Crabb, J. W., et al. (1998) Changes in biological activity and folding of guanylate cyclase-activating protein as a function of calcium. *Biochemistry* **37**, 248–257.
46. Dizhoor, A. M., Boikov, S. C. G., and Olshevskaya, E. (1998) Constitutive activation of photoreceptor guanylate cyclase by Y99C mutant of GCAP-1. *J. Biol. Chem.* **273**, 17,311–17,314.
47. Hughes, R. E., Brzovic, P. S., Dizhoor, A. M., Klevit, R. E., and Hurley, J. B. (1998) Ca^{2+}-dependent conformational changes in bovine GCAP-2. *Prot. Sci.* **7**, 2675–2680.

48. Kumar, V. D., Hidaka, H., Okazaki, K., and Vijay-Kumar, S. (1996) Crystallization and preliminary X-ray crystallographic studies of recombinant Bovine neurocalcin d. *Protein Struct. Funct. Genet.* **25,** 261–264.
49. Thaller, C., Weaver, L. H., Eichele, G., Wilson, E., Karlsson, R., and Jansonius, J. N. (1981) Repeated seeding technique for growing large single crystals of proteins. *J. Mol. Biol.* **147,** 465–469.
50. Higashi, T. (1990) PROCESS: a program for indexing and processing RAXIS IIC imaging plate data, Rigaku Corp., Japan.
51. Sheldrick, G. M. and Schneider, T. R. (1997) SHELXL:high resolution refinement. *Meth. Enzymol.* **277,** 319–343.
52. Collaborative Computational Project Number 4 (1994) The CCP4 suite. Programs for protein crystallography. *Acta Crystallogr.* **D50,** 760–763.
53. Jones, T. A. (1985) Interactive computer graphics: FRODO. *Meth. Enzymol.* **115,** 157–171.
54. Cowtan, K. D. and Main, P. (1993) Improvement of macromolecular electron density maps by the simultaneous application of real and reciprocal space constraints. *Acta. Crystallogr.* **D49,** 148–157.
55. Read, R. J. (1986) Improved fourier coefficients for maps using phases from partial structures with errors. *Acta. Crystallogr.* **A42,** 140–149.
56. Brunger, A. T. (1992) XPLOR version 3. **1,** A system for x-ray crystallography and NMR. Yale University Press, New Haven, CT.
57. Paul, A. K., Marala, R. B., Jaiswal, R. K., and Sharma, R. K. (1987) Coexistence of guanylate cyclase and atrial natiuretic factor receptor in a 180 kD protein. *Science* **235,** 1224–1226.
58. Nambi, P., Aiyar, N. V., Roberts, A. N., and Sharma, R. K. (1982) Relationship of calcium and membrane guanylate cyclase in adrenocortcotropin-induced steroidogenesis. *Endocrinology* **111,** 196–200.
59. Kraulis, P. (1991) MOLSCRIPT — a program to produce detailed and schematic plots for protein structures. *J. Appl. Crystallogr.* **24,** 946–950.

18

Crystallization and Structure–Function of Calsequestrin

ChulHee Kang, William R. Trumble, and A. Keith Dunker

1. Introduction

X-ray crystallography provides one of the most important tools for understanding structure–function relationships in proteins. Over the last several years, the scientific community has accumulated a great deal of experience making crystals and solving protein structures by X-ray diffraction. However, often the X-ray approach cannot be used because of the failure to obtain crystals, which is still considered as the single greatest experimental problem in crystallography. Currently, there is no method for predicting the conditions needed to obtain crystals for a given protein. Researchers simply screen a large number of conditions hoping to find crystals in any of the trials. Given crystals under one set of conditions, even crystals of low quality, conditions in the neighborhood of the successful trial are again surveyed in hopes of finding conditions for obtaining improved crystals. Even though crystals are obtained, some are not suitable for structure determination by X-ray diffraction. The first requirement is that a crystal must contain a sufficiently high degree of order relating one molecule to another over a sufficient number of molecules. In addition, a crystal must have an appropriate size and shape. Needle-like crystals are of insufficient size in two dimensions and may require application of seeding techniques using initial crystals as nuclei for larger crystals, but it is often not a straightforward procedure.

Calsequestrin from rabbit skeletal muscle sarcoplasmic reticulum (SR) is the prototype of an important family of calcium regulatory protein with members in all types of eukaryotic cells (for review, *see* **refs.** *1* and *2*). This protein presented calcium-induced needle-like crystals more than 10 yr ago *(3–5)*.

During the intervening period, the structure of this molecule was not determined, evidently because of the failure to find altered conditions that yielded suitable, rather than needle-like, crystals.

Based on structure–function aspects of calsequestrin determined from correlations among various studies, both in vivo and in vitro, we formed an hypothesis regarding the reasons for the development of the observed needle-like crystal morphology induced by calcium. This hypothesis, in turn, suggested alternative approaches for crystallization, approaches designed to yield suitable, rather than needle-like, crystals. Using these alternative approaches, we were able to obtain suitable crystals and then to solve the structure of calsequestrin by X-ray crystallography *(6)*.

In this brief chapter, we will first discuss the background leading to our hypothesis, and then will discuss the alternative crystallization approaches that were indicated by this hypothesis. Next, we will present the resulting structure for calsequestrin. Finally, we will discuss the details of the structure that support our original hypothesis for the origin of needle-like crystals. All together these data provide insight into the mechanisms underlying the dependence of the calcium regulatory function on the structure of calsequestrin.

1.1. Function of Calsequestrin

The calsequestrin protein regulates calcium levels in the lumen of the SR membrane of muscle cells. The SR membrane is a specialized elaboration of the endoplasmic reticulum (ER) and, like the ER, the SR serves as a calcium storage/release system for regulating calcium ion concentration. The ER from other types of animal and plant cells contain very acidic proteins, which are reminiscent of the C-terminal region of calsequestrin and bind calcium ion with a relatively high capacity, but low affinity (for review, *see* **ref. 7**).

In muscle cells, the regulation of calcium ion by the SR, in turn, controls the state of the actin-myosin fibrils, with release of calcium from the SR bringing about muscle contraction and uptake of calcium by the SR bringing about relaxation *(2,8,9)*. For the fastest muscles, a slow step in the contraction/relaxation cycle is calcium pumping into the SR membrane *(10)*. In this pump-storage-release of calcium by the SR, calsequestrin plays a key role through buffering the calcium levels within the lumen of the SR. Calsequestrin is not, however, a passive calcium buffer, but rather plays an active role in regulating calcium levels and facilitating its release from the SR lumen *(11–13)*.

Calsequestrin is not uniformly distributed within the lumen of the SR membrane, but rather is bound at the calcium release channel *(11–13)*, also known as the ryanodine receptor *(14)*. Binding at this receptor appears to be mediated by two additional proteins, junctin and triadin *(11–13)*. Diffusion times depend on the square of the distance. Thus, by localizing calsequestrin and its bound calcium at the ryanodine receptor, diffusion times for calcium release are drastically reduced.

In vitro calcium titration studies suggest that initial calcium or other divalent cation binding at lower concentrations causes calsequestrin to fold into a more compact, globular form *(3,12,15–21)*, whereas low affinity binding at higher calcium concentrations is accompanied by calcium-induced protein aggregation *(18,22)*. A fibrous or needle-like character is typically observed for these calcium-induced aggregates *(22)*. During purification and storage in the presence of calcium ion, calsequestrin readily forms needle-like crystals *(3–5)*. In vivo, ultrastructure studies reveal that calsequestrin forms arrays of linear polymers *(23–25)*. Crosslinking studies indicate substantial calsequestrin–calsequestrin interactions within the SR membrane *(26)*. In summary, these data suggest that, not only is calsequestrin localized at the calcium-release channel, but, in addition, calsequestrin forms calcium-induced linear polymers. This calcium-induced polymerization likely contributes to the localization and function of calsequestrin, and furthermore the needle-like crystal morphology probably reflects the morphology of the natural polymer.

1.2. Model for Calcium Binding and Release

Each calsequestrin molecule binds and releases large numbers of calcium ions during a contraction–relaxation cycle, estimated to be 40 to 50 ions per molecule. For the binding and release of so many divalent cations, calsequestrin needs to have a large number of negatively charged side chains. Consistent with its cation binding function, calsequestrin, for example, of rabbit skeletal muscle, has a PI near 3.7, a total of 110 carboxylate groups, and an excess of 80 carboxyl side chain groups over the sum of the positively charged ones *(2)*.

This large excess of negatively charged side chains is not uniformly distributed over the entire molecule. Rather, the amino terminal half of calsequestrin is slightly negative, whereas the residues in the carboxyl terminal half are more than 35% aspartates or glutamates. There is further asymmetry within the carboxyl terminal half. For example, 16 of the last 21 residues, including the last 14 in a row, are aspartic acids. Frog skeletal calsequestrin contains an even longer run of consecutive negative residues, 45, at the carboxyl terminus *(2)*.

The asymmetric charge distribution in calsequestrin led to the view that the amino terminal half folds into a typical globular protein, whereas the negatively charged carboxy terminus forms a largely disordered tail. Furthermore, various studies, including our own, suggested that, when in low-salt environment (especially in the absence of calcium), calsequestrin is largely unfolded. Raising calcium concentrations first leads to the formation of globular structure as identified by protection against protease digestion and increased helical content by circular dichroism. Raising the calcium levels further leads to concomitant low affinity binding of large numbers of calcium ions and protein precipitation.

The binding of an inhibitor that prevents certain protein–protein interactions was shown to block both folding and calcium-induced precipitation of calsequestrin, which suggested that the folding of the amino terminal half is required for subsequent low-affinity calcium binding. This led to a model in which the calcium-induced aggregation of calsequestrin requires both interactions among the folded, globular amino termini and a low-affinity interaction, possibly involving cross-bridges between the negatively charged, mostly disordered carboxyl terminal regions *(18)*.

1.3. Ionic Strength Induced Folding of Calsequestrin

Lowering or raising the pH to very low or high values frequently causes order-to-disorder transitions in proteins, leading to either molten globule-like forms or random coil-like forms, depending on the protein and the ionic strength. At lower ionic strength, the asymmetric charge brought about by the pH extreme often leads to the random coil form, whereas raising the ionic strength at a pH extreme frequently induces partial folding into a molten globule-like form. Thus, proteins can undergo interconversions among well-ordered structures, molten globule-like forms, and random-coil-like chains, depending to a large degree on the net charge of the side chains and the ionic strength of the solution *(27–29)*.

The large negative charge of calsequestrin results in a molecule that has a significant tendency to be disordered, even at neutral pH. As indicated above, calcium ions induce calsequestrin to fold into a more compact form. However, this could be as much the general effects of ionic strength on a highly charged protein, rather than the result of specific calcium interactions.

The general tendency of charged proteins to be disordered suggests that raising the ionic strength with salts other than those containing calcium or other divalent cation could likewise lead to the folding of calsequestrin. Circular dichroism and trypsin digestion studies confirm that monovalent cations at sufficiently high concentrations can lead to folded calsequestrin molecules that are very similar to those induced by calcium. The major difference is that monovalent ion-induced folding does not lead to protein–protein aggregation and precipitation.

1.4. Hypothesis

Calcium-induced needle-like crystals were reported over 10 yr ago *(3–5)*. Although these crystals were unsuitable for structure determination, they were important for demonstrating that calsequestrin could be crystallized. Given the ionic-strength dependent, large-scale structural changes in the calsequestrin molecule, it seemed possible that the calsequestrin molecule might undergo significant conformational change upon crystallization. Thus, a very important observation, made by Raman spectroscopy, was that calsequestrins structure in the crystals is extremely similar if not identical to its structure in solution *(3)*.

Based on all these observations, we formed the hypothesis that the growth of calcium-induced needle-like crystals was caused by a two step process:

1. High-ionic strength-induced the folding of calsequestrin into a globular structure, perhaps with disordered regions;
2. Calcium-induced the cross-bridging of adjacent calsequestrin molecules by means of the very negatively charged carboxyl terminal part of the protein.

The prior folding was shown to be a prerequisite for protein aggregation, and, therefore, perhaps also for crystallization. The proposed cross-bridging could account for the needle-like crystals by accelerating crystal growth in one direction. Of course this accelerated growth would be related to the observation that calsequestrin forms linear polymers in its functional state in vivo. Given this hypothesis, the obvious approach for obtaining suitable crystals would be to crystallize calsequestrin in the absence of calcium.

2. Materials
2.1. Calsequestrin

Calsequestrin was purified from the skeletal muscle of New Zealand white rabbits as described in **Subheading 3**.

3. Methods
3.1. Purification of Calsequestrin

SR microsomal membranes were isolated from the white skeletal muscle of rabbits by the method described before *(18,24)*. Both back and leg muscles were ground in a meat grinder and about 50 g of ground meat in 250 mL of 0.3 M sucrose, 5 mM imidazole-HCl, pH 7.4, and 0.23 mM phenylmethylsulfonyl fluoride (PMSF) were homogenized in a Waring blender three times for 30 s each at the maximal speed. The homogenates were centrifuged for 10 min at 7700g, followed by filtration of the supernatants through 6–8 layers of cheesecloth. Microsomal pellets were obtained by centrifuging the supernatants for 30 min at 110,000g. Pellets were resuspended in 0.3 M sucrose, 5 mM imidazole-HCl, pH 7.4, 0.23 mM PMSF, and 1.1 mM leupeptin. SR microsomes were then further centrifuged for 30 min at 110,000g, and the pellets were resuspended with ice-cold 100 mM Na$_2$CO$_3$ at a protein concentration of 1–2 mg/mL. The suspension was incubated for 30 min on ice and recentrifuged as aforementioned. The supernatants obtained were first equilibrated with buffer A (20 mM MOPS, pH 7.0, 0.5 M NaCl, and 0.1 mM EGTA), then applied to a phenyl-Sepharose column prewashed with buffer A. After the column was washed with buffer A, calsequestrin was eluted with buffer A containing 20 mM CaCl$_2$. Purity of the calsequestrin was checked by SDS gel electrophoresis. For a typical protein preparation, with a loading of approx 20 mg, essentially one band is observable by normal Coomassie staining. Purified calsequestrin was concentrated with an Amicon ultrafiltration cell using a 30-kDa molecular mass cutoff membrane and stored at –20°C. No particular effort was made to exclude divalent cations except for extensive dialysis. Calcium and other divalent cation levels remained at the ambient levels determined by their trace amounts in the various salts and distilled water used in these experiments.

3.2. Crystallization

About 500 separate crystallization trials were surveyed using the vapor diffusion method to find methods to grow suitable crystals of rabbit skeletal muscle CSQ. Initial crystallization experiments were done by the hanging drop vapor diffusion method *(30)* with an incomplete factorial approach *(31)* to screen the widest possible range of crystallization conditions. With these techniques, each crystallization condition can be tested with as little as 1 µL (5–10 µg) of sample. These crysallization trials involved 16 different precipitating agents. Use of high-ionic-strength solutions with monovalent cations was emphasized. All of the trials were carried out at both 4°C and room temperature. Good-sized crystals (non-needle) were found when 2-methyl-2,4 pentane diol (MPD) was used as the precipitating agent. The best crystals were grown from a solution containing 10% (v/v) MPD, 0.1 M Na$_3$Citrate, 0.05 M Na cacodylate, pH 6.5 and 5 mg/mL protein, for a nominal initial Na$^+$ concentration of 0.35 M. Drops of 4 µL in volume were equilibrated against a reservoir having twice the concentration of all the reagents except for the protein. Rectangular crystals appeared within 1 wk. By 2 wk, the crystals achieved a suitable size: typically 0.2 × 0.2 × 0.8 mm.

3.3. Crystallography

The crystals were of the orthorhombic space group C222 with one molecule per unit cell. The dimensions of the crystal axes were a = 59.74, b = 145.56, c = 111.79 Å. The crystals diffracted to 2.3 Å and complete diffraction data were collected to that resolution using a Rigaku R-axis 11 imaging plate. Heavy atom derivatives were prepared by soaking the crystals in solutions containing 2-chloromecuri-4-nitrophenol, p-chloromecuriphenylsulfonate, or potassium tetrachloroplatinum. Complete 3.0 Å diffraction data sets, including anomalous data, were collected for all derivatives. Heavy-atom postions were determined by analyzing difference Patterson maps and confirmed by anomolous difference Patterson maps. Multiple isomorphous replacement was used in combination with anomalous data and the solvent flattening technique. For 14,500 reflections above the 2s level (based on F) with Bragg spacings between 10–2.4 Å (data completeness is 78.3% for this set of reflections), refinement yielded a structure at 2.4 Å resolution with an R-factor of 18.8%. The root-mean-square (rms) deviations from ideality are 0.013 Å for bond lengths and 3.0' for bond angles. The current model also contains 58 water molecules.

3.4. Structure and Packing of Calsequestrin

The resulting structure *(6)*, to our surprise, exhibited three nearly identical tandem domains; each of these has the thioredoxin protein fold (*see* **Fig. 1**). These three domains interact to form a disk-like shape with an approximate radius of 32 Å and a thickness of 35 Å. The overall center of the protein, which corresponds to the interfaces of the three domains, is hydrophilic rather than hydrophobic. The entire exterior surface of calsequestrin is dominated by negative charges on all sides.

The individual monomers stack in the crystal lattice into ribbon-like polymers *(6)*. The few positive charges in calsequestrin are mostly involved in salt bridges between domains within a monomer or between monomers; thus, the surfaces of the ribbon polymers are almost entirely negative. The monomers stack with alternate orientations in each ribbon polymer. These alternating orientations define two distinct twofold axes that are perpendicular to the long axis of the ribbon. This leads to two types of packing interfaces within the polymer *(6)*.

The first type of interface, which we call front-to-front, involves the fitting of the convex globule from Domain II on one subunit into a concave depression on the other, and this interface also involves "arm exchange or domain swapping" between the extended amino terminal ends of the two adjacent molecules. Each extended arm comprises 10 N-terminal residues from domain I. This extended arm binds along a groove between two β-strands of Domain II in the neighbor (*see* **Fig. 2**). Also, the front-to-front interface contains a number of negatively charged groups *(6)*.

The second type of interface, which we call back-to-back, involves bringing together two jaw-like openings that lie between domains I and III, thus creating a substantial cavity within this interface. On two sides of this cavity, a symmetry-related pair of helix–helix interactions stabilizes the subunit–subunit dimerization. Each helix–helix interface contains a pair of lysine/glutamate salt bridges and a glutamate from one helix that associates with amino terminal (e.g., the positive dipole) end of its neighbor; it is as if a side chain from one helix acts as an N-cap residue for its neighbor (*see* **Fig. 3**). The last observed

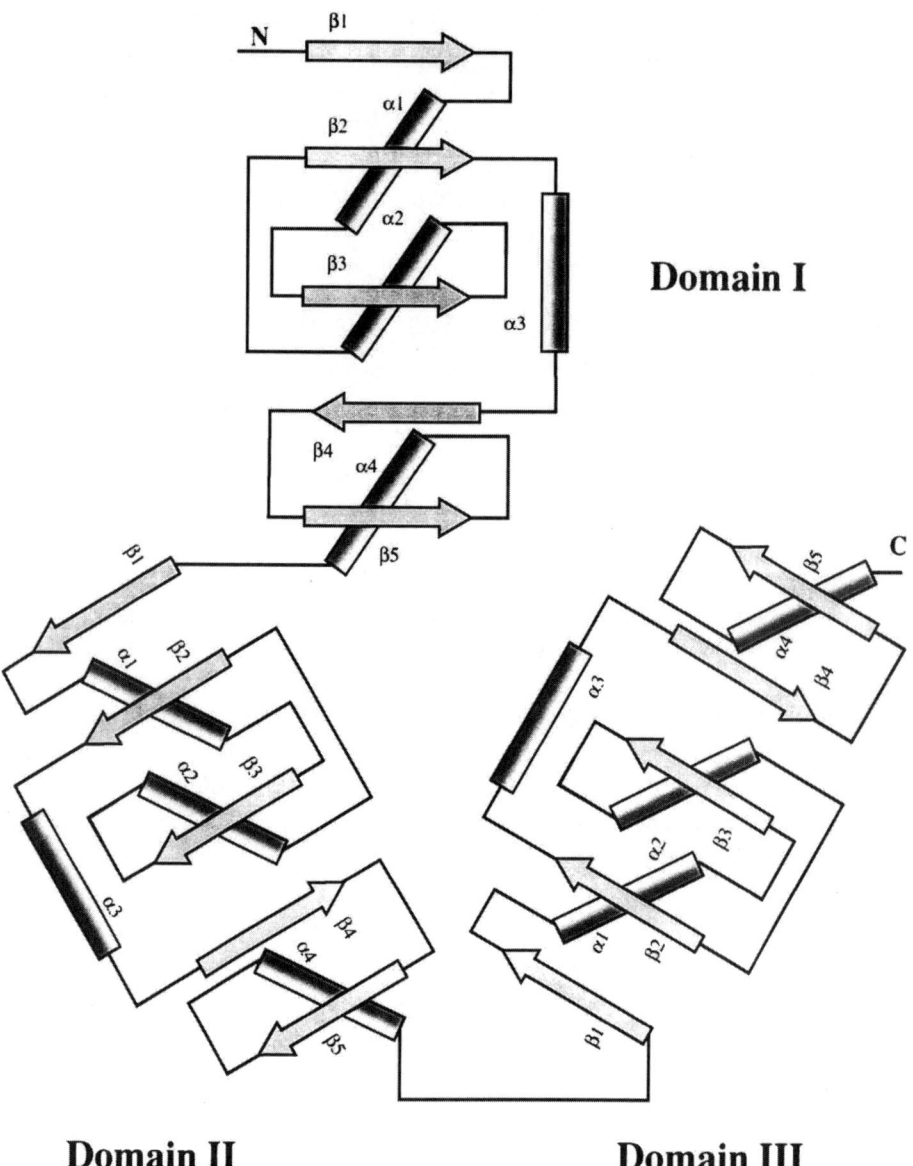

Fig. 1. Topology diagram of the calsequestrin structure. The topology of an individual domain is identical with that of *E. coli* Thioredoxin (**domain I**, 12–124; **domain II**, 125–228; **domain III**, 229–352). Circles denote predicted locations of possible binding clefts from the strand topology.

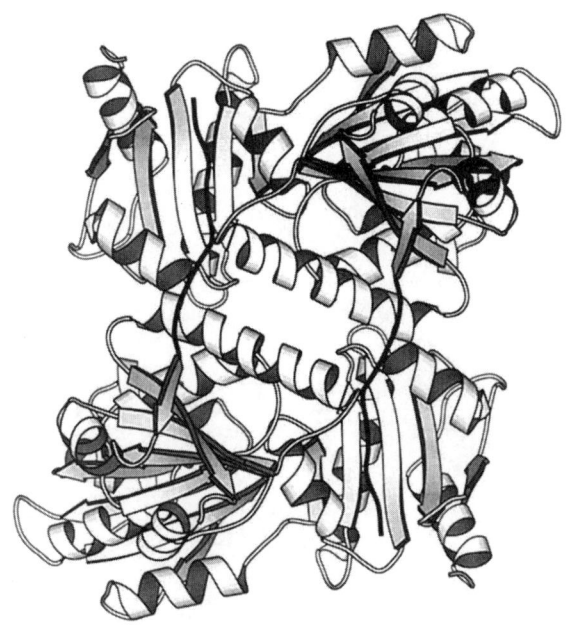

Fig. 2. Ribbon representation of the front-to-front interaction. Two molecules are related by a crystallographic twofold axis. The extended N-terminal residues (^{3}GLDFPEYDGVDRV^{15}) of one monomer, depicted with black, become inserted into the other monomer.

residue at the two carboxyl terminal ends lie within the cavity; each subunit contains an additional polyanionic tail that is unobserved due to protein disorder *(6)*; it is unclear how many of these 20 additional residues reside within the cavity and how many extend to the exterior of this structural feature.

4. Notes
4.1. Expected and Observed Calsequestrin Structure

Sequence analysis of this protein family prior to the structure determination gave no hint of the three-domain structure *(32,33)*. The expectation was that calsequestrin would contain a globular N-terminal domain and a C-terminal disordered region *(18)*. Indeed, the last 130 residues in this protein are more negative than any stretch of comparable length of any protein previously deposited in the Protein Data Bank. The finding of three domains in tandem was totally unexpected. Even when armed with the three-domain information, our more recent sequence analysis gives no hint of the structural repeat (work in progress). The sequence similarities among the domains are not distinguish-

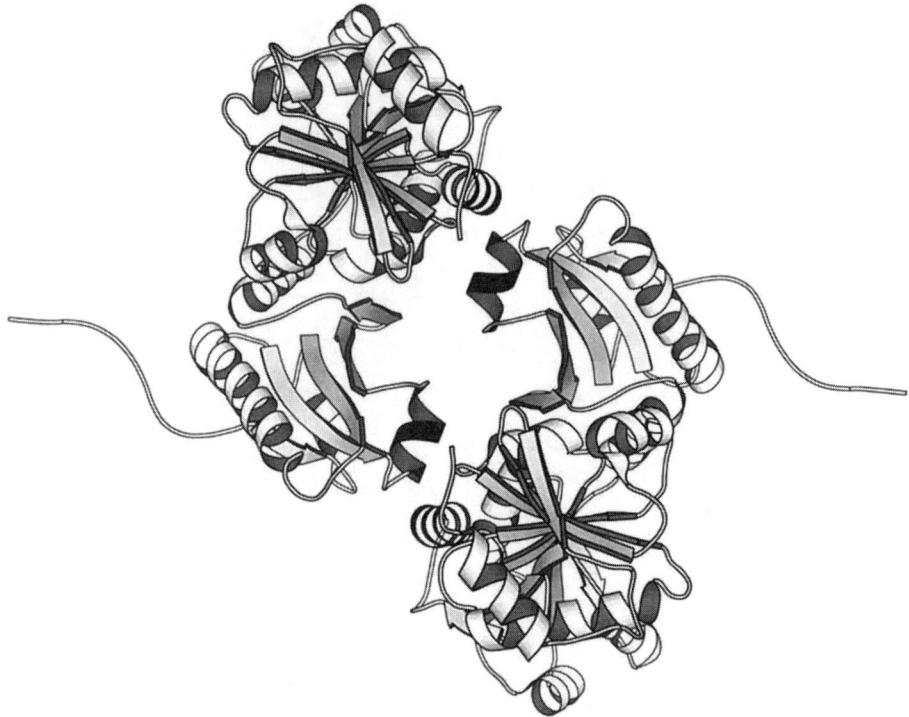

Fig. 3. Ribbon representation of the back-to-back interaction. Two molecules are related by a crystallographic twofold axis. The extended N-terminal residues (^{79}KDAAVAKKLG^{87}) and SAH site (^{214}SEEEIVNFV^{222}) are depicted with black.

able from random by any method tested so far. As for the expectation that the C-terminal region would be disordered, this expectation was at least partially realized. Residues 327–333, a polyanionic loop, and 348–368, the very highly negative C-terminal end, were unobserved in the calsequestrin structure, probably because of their structural disorder *(6)*.

4.2. Low Divalent Cation Concentrations and High Ionic Strength

Our original hypothesis contained two aspects:

1. Monovalent cations induce calsequestrin folding without aggregation by neutralizing the high negative charge without providing the divalent ions for cross-bridging.
2. Divalent cations induce needle-like crystals from rapid crystal growth in one dimension because of the divalent-ion-dependent cross-bridging. Thus, our strategy was to try to crystallize calsequestrin using conditions having high-ionic strength and low-divalent cation concentrations.

With regard to the first point, the starting conditions contained 0.35 M Na$^+$ and the reservoir contained twice this amount as aforementioned with only ambient amounts of divalent ions. Small crystals were visible under the microscope after 1 wk, at which time the protein-containing drop was probably approaching equilibrium with the reservoir and is well above that needed to induce calsequestrin folding as measured by protease protection and CD. Regarding the second point, extensive dialysis against distilled water was used to lower the free calcium. In addition, a high concentration of citrate buffer (0.1 M initial, approaching 0.2 M final) during crystallization further insured that free calcium would be too low to induce needle formation. That is, from the solubulity of calcium citrate *(34)*, which leads to a solubility product of about 1.9×10^{-12}, we estimate that the free calcium levels would be too low to occupy the low affinity calcium-binding sites to any significant degree. Because binding to the low affinity sites is proposed to induce needles, the citrate buffer insures that calcium would be too low to induce needle formation by this means.

4.3. Formation of Crystalline Needles

Protein crystal formation is a complex process, so we cannot be sure that the low divalent cation concentration was truly of critical importance in our success in obtaining useful crystals of calsequestrin. However, the details of the structure of the polymers observed within the crystal appear to be able to account for the formation of needle-like crystals in the presence of calcium. That is, both the front-to-front and the back-to-back interfaces require the close juxtaposition of negatively charged residues. In the case of the back-to-back interface, around 45 negative charges from each protein are localized at least partially within the cavity bounded by this contact. Divalent cations would be expected to cross-bridge these neighboring negative charges. Such divalent ion cross-bridging, in turn, would accelerate the formation of these interfaces, speed up crystal growth along this one direction, and thereby favor the formation of needles. In the absence of divalent cations, these interfaces would be expected to form much more slowly because of negative charge repulsion.

4.4. Important Structural Features for the Mechanism of Calsequestrin

An important aspect of our initial hypothesis was that the needle crystals of calsequestrin were related to its structure and function in vivo. As aforementioned, the crystals we obtained contained a polymeric organization that could account for the observed needles. However, one can frequently identify linear arrangements of molecules within a crystal. So, it is uncertain whether the indicated polymer is actually related to the functional form of calsequestrin or is merely an arbitrary identification of a polymer within the crystal, especially because the crystallization conditions insured the virtual absence of calcium from the crystals.

More experiments need to be carried out to definitively resolve this uncertainty. However, several observations support the physiological relevance of the indicated polymer. Both interfaces involve extensive fitting between the two proteins. The front-to-front interface not only fits a convex surface into a concave one, but also utilizes interactions via arm-exchange, in which an extended chain of one protein fits into a groove on the surface of its neighbor. The back-to-back interface brings together two pairs of helices that not only make two salt bridges each, but also include a glutamate residue from one helix that serves as an N-cap residue for its neighbor. These interactions are more extensive and more complex than those typically observed in crystal contacts; rather these interfaces are similar to the interfaces in proteins that function as oligomers. For these reasons, the indicated polymer is likely to be the functional form of calsequestrin.

If indeed the indicated polymer in the crystal represents the functional form of the calsequestrin molecule, considerable insight is provided regarding the role of calsequestrin structure in the function of calcium regulation. First, at the ion and protein concentrations within the SR, it seems likely that calsequestrin remains polymerized during a contraction/relaxation cycle. Each polymer has a negatively charged surface. An array of such polymers would create long, narrow, negative-lined solvent channels leading to the calcium release channel. One-dimensional diffusion of calcium along the surface of the polymer *(9)* and restricting the solvent to long, narrow tubes would work together to speed up the diffusion of calcium from its binding sites in calsequestrin to its release channel. Furthermore, the lack of fixed structure of the calcium-binding sites means that binding affinities would be low, and diffusion-limited on-rates and off-rates would be as fast as possible. Clearly, the solved structure for calsequestrin *(6)* suggests a very efficient machine for calcium binding and release in the SR.

References

1. Maclennan, D. H. and Wong, P. T. (1971) Isolation of a calcium-sequestering protein from sarcoplasmic reticulum. *Proc. Natl. Acad. Sci. USA* **68,** 1231–1235.
2. Yano, K. and Zarain-Herzberg, A. (1994) Sarcoplasmic reticulum calsequestrins: structural and functional properties. *Mol. Cell Biochem.* **135,** 61–70.
3. Williams R. W. and Beeler, T. J. (1986) Secondary structure of calsequestrin in solutions and in crystals as determined by Raman spectroscopy. *J. Biol. Chem.* **261,** 12,408–12,413.
4. Maurer, A., Tanaka, M. Ozawa, T., and Fleischer, S. (1985) Purification and crystallization of the calcium binding protein of sarcoplasmic reticulum from skeletal muscle. *Proc. Nat. Acad. Sci. USA* **82,** 4036–4040.

5. Hayakawa, K., Swenson, L., Baksn, S., Wei, Y., Michalak, M., and Derewenda, Z. S. (1994) Crystallization of canine cardiac calsequestrin. *J. Mol. Biol.* **235,** 357–360.
6. Wang, S., Trumble, B., Liao, H., Dunker, K., and Kang, C. (1998) Crystal structure of calsequestrin from rabbit skeletal muscle sarcoplasmic reticulum. *Nat. Struct. Biol.* **5,** 476–483.
7. Krause, K. H. and Michalak, M. (1997) Calreticulin. *Cell* **88,** 439–443.
8. Maclennan, D. H., Campbell, K. P., and Reithmeier, R. A. F. (1983) Calsequestrin. *Calcium and Cell Function* **4,** 151.
9. MacLennan, D. H. and Reithmeier, R. A. (1998) Ion tamers. *Nat. Struct. Biol.* **5,** 409–411.
10. Rall, J. A. (1996) *News Physiol. Sci.* **11,** 249–255.
11. Lytton, J. and MacLennan, D. H. (1992) Sarcoplasmic reticulum, in *The Heart and Cardiovascular System.* (Fozzard, H. A., Harber, E., Jennings, R. B., Katz, A. M., and Morgan, H. E., eds.), 2nd ed., Raven, New York, pp. 1203–1222.
12. Mitchell, R. D., Simmerman, H. K. B. and Jones, L. R. (1988) Ca2+ binding effects on protein conformation and protein interactions of canine cardiac calsequestrin. *J. Biol. Chem.* **263,** 1376–1381.
13. Guo, W. and Campbell, K. P. (1995) Association of triadin with the ryanodine receptor and calsequestrin in the lumen of the sarcoplasmic reticulum. *J. Biol. Chem.* **270,** 9027–9030.
14. Inui, M., Saito, A., and Fleischer, S. (1987) Purification of the ryanodine receptor and identity with feet structures of junctional terminal cisternae of sarcoplasmic reticulum from fast skeletal muscle. *J. Biol. Chem.* **262,** 1740–1747.
15. Ikemoto, N., Nagy, B., Bhatnagar, G. M., and Gergely, J. (1974) Studies on a metal-binding protein of the sarcoplasmic reticulum. *J. Biol. Chem.* **249,** 2357–2365.
16. Ostwald, T. J. MacLennan, D. H., and Dorrington, K. J. (1974) Effects of cation binding on the conformation of calsequestrin and the high affinity calcium-binding protein of sarcoplasmic reticulum. *J. Biol. Chem.* **249,** 5867–5871.
17. Aaron, B. M., Oikawa, K., Reithmeier, R. A., and Sykes B. D. (1984) Characterization of skeletal muscle calsequestrin by 1H NMR spectroscopy. *J. Biol. Chem.* **259,** 11,876–11,881.
18. He, Z., Dunker, A. K., Wesson, C. R., and Trumble, W. R. (1993) Ca(2+)-induced folding and aggregation of skeletal muscle sarcoplasmic reticulum calsequestrin. The involvement of the trifluoperazine-binding site. *J. Biol. Chem.* **268,** 24,635–24,641.
19. Ikemoto, N., Bhatnagar, G. M. Nagy, B., and Gergely, J. (1972) Interaction of divalent cations with the 55,000-dalton protein component of the sarcoplasmic reticulum. Studies of fluorescence and circular dichroism. *J. Biol. Chem.* **247,** 7835–7837.
20. Cozens, B. and Reithmeier, R. A. (1984) Size and shape of rabbit skeletal muscle calsequestrin. *J. Biol. Chem.* **259,** 6248–6252.
21. Ohnishi, M. and Rethmeier, R. A. (1987) Fragmentation of rabbit skeletal muscle calsequestrin: spectral and ion binding properties of the carboxyl-terminal region. *Biochemistry* **26,** 7458–7465.
22. Tanaka, M., Ozawa, T., Maurer, A. M., Cortese, J., and Fleischer, S. (1986) Apparent cooperativity of Ca^{2+} binding associated with crystallization of Ca^{2+}-binding protein from sarcoplasmic reticulum. *Arch. Biochem. Biophys.* **251,** 369–378.

23. Franzini-Armstrong, C., Kenney, L. J., and Varriano-Marston, E. (1987) The structure of calsequestrin in triads of vertebrate skeletal muscle: a deep-etch study. *J. Cell Biol.* **105,** 49–56.
24. Saito, A., Seiler, S., Chu, A., and Fleischer, S. (1984) Preparation and morphology of sarcoplasmic reticulum terminal cisternae from rabbit skeletal muscle. *J. Cell Biol.* **99,** 875–885.
25. Maurer, A., Tanaka, M., Ozawa, T., and Fleischer, S. (1985) Purification and crystallization of the calcium binding protein of sarcoplasmic reticulum from skeletal muscle. *Proc. Natl. Acad. Sci. USA* **82,** 4036–4040.
26. Maguire, P. B., Briggs, F. N., Lennon, N. J., and Ohlendieck, K. (1997) Oligomerization is an intrinsic property of calsequestrin in normal and transformed skeletal muscle. *Biochem. Biophys. Res. Commun.* **240,** 721–727.
27. Ohgushi, M. and Wada, A. (1983) Molten-globule state: a compact form of globular proteins with mobile side-chains. *FEBS Lett.* **164,** 21–24.
28. Dolgikh, D. A., Gilmanshin, R. I., Brazhnikov, E. V., Bychkova, V. E., Semisotnov, G. V., Venyaminov, S. Yu., and Ptitsyn, O. B. (1981) Alpha-Lactalbumin: compact state with fluctuating tertiary structure? *FEBS Lett.* **136,** 311–315.
29. Ptisyn, O. B. (1995) Molten globule and protein folding. *Adv. Protein Chem.* **47,** 83–229.
30. McPherson, A. (1990) Current approaches to macromolecular crystallization. *Eur. J. Biochem.* **189,** 1–23.
31. Jancarik, J. and Kim, S.-H. (1991) Sparse matrix sampling: a screening method for crystallization of proteins. *J. Appl. Cryst.* **24,** 409–411.
32. Fliegel, L., Ohnishi, M., Carpenter, M. R., Khanna, V. K., Reithmeier, R. A. and MacLennan, D. H. (1987) Amino acid sequence of rabbit fast-twitch skeletal muscle calsequestrin deduced from cDNA and peptide sequencing. *Proc. Natl. Acad. Sci. USA.* **84,** 1167–1171.
33. Scott, B. T., Simmerman, H. K., Collins, J. H., Nadal-Ginard, B., and Jones, L. R. (1988) Complete amino acid sequence of canine cardiac calsequestrin deduced by cDNA cloning. *J. Biol. Chem.* **263,** 8958–8964.
34. Freier R. K. (1976) *Aqueous Solutions: Data for Inorganic and Organic Compounds*, vol. 1. Walter de Gruyter, New York.

19

Use of Fluorescence Resonance Energy Transfer to Monitor Ca^{2+}-Triggered Membrane Docking of C2 Domains

Eric A. Nalefski and Joseph J. Falke

1. Introduction

The C2 domain is a membrane-targeting domain present in numerous signaling proteins that carry out a wide variety of biochemical functions *(1–3)*. In many of these proteins, Ca^{2+} triggers docking of the C2 domain to phospholipid vesicles. Several methods have been utilized to measure the Ca^{2+}-triggered docking of C2 domain proteins, or their isolated C2 domains, to natural or artificial phospholipid membranes in vitro. These include sedimentation of proteins bound to natural membranes *(4)*, synthetic vesicles *(5)* or sucrose-loaded vesicles *(6)*; copurification of radiolabeled vesicles with affinity-tagged proteins bound to agarose beads *(7)*; isolation by size-exclusion chromatography of protein-vesicle complexes *(8)*; monitoring intrinsic fluorescence changes in the C2 domain in response to membrane docking *(9)*; and fluorescence resonance energy transfer (FRET) between protein and vesicles *(10–12)*. Of these approaches, only fluorescence techniques allow measurement of protein docking to membranes in real time. This feature enables equilibrium experiments that: (1) monitor membrane docking in response to incremental changes in Ca^{2+} concentration; (2) test the reversibility of the docking process; and (3) assess the Ca^{2+}-independent component of membrane docking, all in a single sample. Moreover, FRET may be used to directly monitor the kinetics of membrane docking and release in stopped-flow fluorescence measurements *(12)* (*see* **Note 1**). Thus, FRET can measure both the equilibrium and rate constants for membrane association.

FRET arises from the nonradiative transfer of energy from a fluorescent donor to an acceptor molecule *(13)*. FRET requires overlap between donor emission and acceptor absorbance spectra and is exquisitely sensitive to the distance between the donor and the acceptor. The efficiency of FRET (E) is given by the equation:

$$E = R_0^6 / R^6 + R_0^6$$

where R is the distance between the donor and acceptor and R_0 is the characteristic donor-acceptor distance resulting in half-maximal FRET. Typical R_0 values are on the order of 10–50 Å, which enables FRET to be useful for many biochemical applications. The standard FRET assay for docking of the C2 domain to phospholipid membranes utilizes one or more intrinsic tryptophans in the C2 domain as the fluorescence donor (λ_{em}^{max} = 320 to 350 nm) and N-(5-dimethylaminonapthalene-1-sulfonyl) (DANSYL) groups covalently attached to the head group of phospholipids as the fluorescence acceptor (λ_{ex}^{max} ~330 nm). The R_0 value of 21 Å for this pair *(13)* guarantees that FRET is only observed when the domain is docked to the membrane and not when free in solution. This assay is generally useful in C2 domain systems, because most C2 domains contain a highly conserved tryptophan *(2)*, located approx 12 Å from the Ca^{2+}-binding cleft, at position 71 in $cPLA_2$ and 223 in PKCβ. Even C2 domains that lack the conserved tryptophan can serve as fluorescence donors if they possess another tryptophan at an appropriate location: rat synaptotagmin I C2A, for instance, gives a detectable signal in the FRET assay even though it contains a single tryptophan at position 259, which is distant from the conserved site. In addition, tryptophans outside the C2 domain —in a different domain of the native protein or in the fusion partner linked to the C2 domain for expression—may serve as suitable donors. Finally, this FRET assay is not restricted to the tryptophan-DANSYL combination: numerous fluorophores attached to phospholipid head groups are available commercially that are suitable acceptors for various extrinsic donor fluorophores, such as fluorescein, that can be attached to native or engineered cysteines in the C2 domain via alkylating reagents *(14)* (*see* **Note 2**).

2. Materials

1. A fluorescent spectrophotometer capable of UV excitation and emission measurements is required. Because fluorescence signals are highly temperature dependent, sample temperatures should be controlled by thermostatted jacketed cuvet holders to maintain assay temperature (4–37°C). Multiple sample holders allow controlled manipulation of separate samples simultaneously.
2. Standard four-sided fluorescence quartz cuvets commercially available from many suppliers are preferred for UV wavelengths in the standard FRET, although

disposable, low UV-absorbing methacrylate cuvets from Fisher Scientific (Pittsburgh, PA) can be substituted at longer wavelengths. Glass or disposable polystyrene cuvets are suitable for visible wavelengths in other FRET applications. Standard cuvets allow the use of sample volumes as low as 1.5–2.0 mL and teflon cell stir bars for continuous mixing of samples. See **Subheading 3.2.** for decalcification and **Note 3** for cleaning.

3. Assay solutions should be made up with double-distilled water (ddH$_2$O), be adjusted to physiological ionic strength and pH, and contain pH buffers that bind Ca^{2+} poorly and are free of contaminants that may absorb or fluoresce in the wavelength ranges used. Suitable buffers, which include piperazine-*N*-*N'*-*bis*[2-ethanesulfonic acid] (PIPES), *N*-[2-hydroxyethyl]piperazine-*N'*-[2-ethanesulfonic acid] (HEPES), and 3-[*N*morpholino]propanesulfonic acid (MOPS), but not the Ca^{2+}-binding buffer tris(hydroxymethyl)aminomethane (Tris), may be decalcified (*see* **Subheading 3.2.**). Concentrated stocks of Ca^{2+} should be made up gravimetrically in assay buffer using the highest commercial grade of CaCl$_2$ available.

4. Pure synthetic phospholipids, as well as fluorescent phospholipid acceptors, can be purchased from a variety of sources, including Molecular Probes (Eugene, OR), Avanti Polar Lipids (Alabaster, AL), and Sigma (St. Louis, MO). For the standard FRET assay, DANSYL-labeled phosphatidylethanolamine (DANSYL-PE) serves as the fluorescence acceptor from protein tryptophans. Phospholipids should be stored as powders or as chloroform stocks at –20°C. Fluorescent phospholipids should be kept in the dark to minimize photobleaching.

5. Protein is purified, concentrated, and stored under conditions that prevent inactivation by degradation or aggregation. Such conditions must be established for each protein. For example, concentrated stocks of the cPLA$_2$ C2 domain remain monomeric when stored in the presence of assay buffer containing 1 m*M* Ca^{2+}, but aggregate when stored in the presence of excess ethylenediaminetetraacetic acid (EDTA) *(9)*. In contrast, the rat synaptotagmin I C2A domain remains monomeric even in the absence of Ca^{2+}. Where appropriate, Ca^{2+} used for storage may be removed immediately prior to assay (*see* **Subheading 3.2.**).

3. Methods
3.1. Protein and Phospholipid Preparation

Phospholipid vesicles are prepared by mixing chloroform stocks of unlabeled and fluorescently labeled phospholipids and drying them down under a gentle stream of N$_2$ to form a thin film. Acceptor DANSYL-PE concentrations are kept low (2.5–5% mol/mol) to ensure that the dominant interactions are between the protein and the unlabeled phospholipid. Assay buffer is added, and small unilamellar vesicles are prepared by sonication using a probe microtip on the highest allowed setting for a total of 5 min. To prevent overheating of the sample, the sample container is suspended in an NaCl-ice water bath and the sample is subjected to cavitation pulses (0.6 s on, 0.4 s off). Using a device such as the SONICATOR® Ultrasonic Liquid Processor (Misonix, formerly Heat Systems, Farmingdale, NY), we obtain uniform transparent suspensions of phospholipid vesicles.

Protein is purified, concentrated, and dialyzed or exchanged into assay buffer. Contaminating Ca^{2+} may be removed if necessary prior to assay (**Subheading 3.2.**).

3.2. Ca^{2+} Decontamination from Solutions and Containers

Levels of contaminating Ca^{2+} arising from glassware or in ordinary laboratory solutions may be relatively high (1–10 μM) and thus may interfere with experiments investigating the low-μM Ca^{2+} concentrations that trigger membrane docking by many C2 domains. Two approaches may be used to deal with contaminating Ca^{2+}. First, contaminating Ca^{2+} may be neglected when the desired Ca^{2+} levels are generated by adding known amounts of Ca^{2+} to mM concentrations of suitable Ca^{2+} chelators such as EDTA or ethylene glycol-bis(β-aminoethyl ether)-$N,N,N,'N'$-tetraacetic acid (EGTA) in the FRET assay *(15)*. Free Ca^{2+} levels are calculated using a computer program such as CHELATOR *(16)* and should be confirmed with use of a Ca^{2+} electrode or a spectroscopic Ca^{2+} indicator. Alternatively, Ca^{2+} contamination may be reduced to submicromolar levels by using no glassware, by decalcifying all reusable plasticware to which all solutions come into contact (by soaking successively in 33% nitric acid and 100 mM EDTA and extensive washing with ddH_2O), by decalcifying all protein, phospholipid, and buffer solutions (by passing over Ca^{2+}-chelating resins such as Chelex 100 [Bio-Rad, Hercules, CA]), and by decalcifying cuvets and stir bars (by soaking in 100 mM EDTA and washing extensively with ddH_2O). In general, new microfuge tubes and micropipet tips of high quality do not require decalcification. These rigorous procedures have allowed us to lower contaminating Ca^{2+} levels to < 300 nM.

3.3. Instrument Settings

Excitation wavelength is set at 280 nm to excite tryptophans and the emission is recorded at 520 nm to monitor DANSYL emission (I) (*see* **Note 4**). Excitation and emission slits are adjusted to minimize photobleaching of samples and to maximize sample signal (e.g., 4 and 8 nm, respectively, on our SLM 48000S spectrometer). Fluorometer voltage and gain settings are adjusted to ensure that the maximal FRET signal lies within the linear range of detection. To avoid sample photobleaching, throughout the assay the excitation shutter is opened for only the few seconds required for taking readings and is closed during addition of stock aliquots and the ensuing mixing and temperature equilibration period, typically at least two minutes. Multiple readings are recorded and averaged for each data point. All readings are corrected for changes in volume accompanying additions.

3.4. Data Collection

1. Phospholipid (50–150 µM total) is equilibrated in a total volume of 1.5–2.0 mL in assay buffer. Fluorescence readings are taken prior to addition of protein to determine the baseline fluorescence (I_0) caused by weak direct excitation of DANSYL by 280 nm light. The equilibration period should be long enough to establish a stable baseline fluorescence, which may take up to 15 min.
2. Microliter amounts of concentrated protein are added to a final concentration of 0.1–1 µM. The fluorescence signal in the absence of added Ca^{2+} is recorded once it has stabilized, which may take up to 5 min. An increased DANSYL signal at this step represents FRET caused by membrane docking triggered by contaminating Ca^{2+} or Ca^{2+}-independent membrane docking (*see* **Note 5** for important control experiments).
3. Microliter aliquots of concentrated stocks of Ca^{2+} are added to raise the total Ca^{2+} concentration incrementally through the submicromolar to hundreds of micromolar range, and stabilized fluorescence readings are recorded. Increased DANSYL signals represent FRET arising from Ca^{2+}-triggered membrane docking or slow Ca^{2+}-independent membrane docking that accumulates during the experiment.
4. Once the DANSYL signal has plateaued over a several-fold range of Ca^{2+} concentrations, the added Ca^{2+} is chelated with addition of excess EDTA from a 0.5 M stock. Ideally, the resulting DANSYL signal should return to the initial baseline reading in order to demonstrate complete reversibility of the docking process and lack of Ca^{2+}-independent binding.

3.4. Data Analysis

The fluorescence baseline (I_0) is subtracted from all readings, and the difference (ΔI) is plotted as a function of total Ca^{2+} (*see* **Fig. 1**). If the measured Ca^{2+}-apparent dissociation constant is significantly greater than the concentration of protein, which is frequently the case for C2 domains in the FRET assay described here, then the amount of Ca^{2+} bound by the protein during the titration can be neglected and the total Ca^{2+} concentration may be taken as the free concentration. Alternatively, when a Ca^{2+}-buffering system is used to control the Ca^{2+}, the free Ca^{2+} concentration provided by CHELATOR or a similar computer program can be used directly. The plot is fit with different suitable binding equations *(17)*, the most useful of which for the C2 domain is the modified Hill equation:

$$\Delta I = \Delta I_{max} (x^H / x^H + K^H)$$

where ΔI_{max} is the maximal calculated FRET signal induced by Ca^{2+}, x is the free Ca^{2+} concentration, K is a constant representing the concentration of Ca^{2+} that induces half-maximal protein docking to membranes, and H is the Hill coefficient, providing information on the number of sites and the degree of cooperativity between them (*see* **Note 6**).

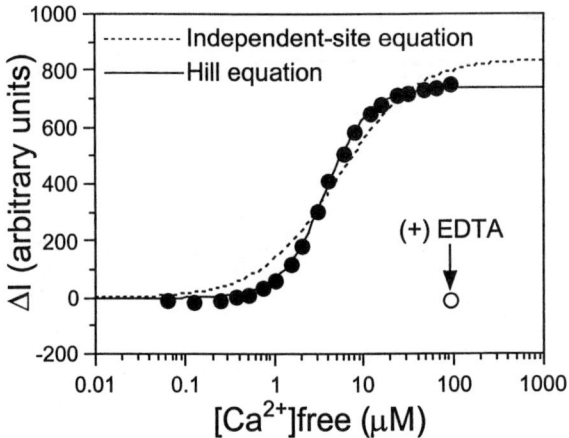

Fig. 1. Ca^{2+}-triggered docking of the $cPLA_2$ C2 domain to phosphatidylcholine vesicles measured by FRET. ΔI was measured by subtracting baseline fluorescence from each I value, and data were fit using the independent-site equation (*dashed line*) or the modified Hill equation (*solid line*). Superior fit was provided by the Hill equation, yielding parameters: $\Delta I_{max} = 739 \pm 5$ arbitrary units, $K = 3.71 \pm 0.07$ μM, and $H = 1.77 \pm 0.05$, indicating membrane docking is triggered by multiple Ca^{2+} ions that bind with positive cooperativity. Adding C2 alone caused no increase in DANSYL emission, demonstrating no Ca^{2+}-independent membrane docking and little Ca^{2+} contamination. Open symbol represents ΔI value after completion of Ca^{2+} titrations upon addition of excess EDTA [(+) EDTA], demonstrating complete reversibility of membrane docking. Experimental conditions: 25°C; 100 mM KCl, 20 mM PIPES, pH 7.0, 60 μM vesicles (95% phosphatidylcholine, 5% DANSYL-PE mol/mol), 0.5 μM $cPLA_2$ C2 domain; $\lambda_{ex} = 280$ nm, $\lambda_{em} = 520$ nm.

4. Notes

1. FRET may be carried out in a stopped-flow fluorescence spectrometer to measure rapid kinetics of protein release from membranes triggered by Ca^{2+} chelation (*9*). In the apparatus, the contents of one syringe (protein and saturating concentrations of Ca^{2+} [100 μM] and labeled phospholipid vesicles) are rapidly mixed with another syringe containing a large excess of EDTA (5 mM) in assay buffer. Tryptophan is excited ($\lambda_{ex} = 280$ nm) and DANSYL emission may be monitored by suitable cutoff filters. Time-courses are best fit with various binding equations.
2. Intrinsic cysteines in the protein may be labeled with fluorescent probes via thiol-specific alkylating agents (*14*), many examples of which are commercially available (*18*). Fluorescein (λ_{em} max ~520 nm) coupled to an engineered cysteine on the surface of the $cPLA_2$ C2 domain serves as a FRET donor upon Ca^{2+}-triggered membrane docking to vesicles containing Texas Red (λ_{ex} max ~582 nm) or tetramethylrhodamine (λ_{ex} max ~540 nm) acceptors (*14*).

3. Cuvets and stir bars should be washed successively with ddH$_2$O and methanol, soaked for at least 30 min in 100 mM EDTA, exhaustively washed with ddH$_2$O, and dried.
4. In principle, membrane docking could be monitored by the *decrease* in tryptophan emission caused by FRET to the DANSYL group, rather than by the sensitized *increase* in DANSYL-emission. However, in practice, tryptophan-emission intensities may actually *increase* as a result of environmental changes in the C2 domain induced by membrane docking to phospholipid vesicles lacking acceptor molecules, as occurs for the cPLA$_2$ C2 domain *(9)*. In such cases, the net tryptophan intensity changes will be the sum of the gain caused by environmental changes and the loss caused by FRET. Thus, the signal-to-noise of a FRET signal recorded as such may be reduced compared to that measured by monitoring the gain in DANSYL emission.
5. Control experiments must always be carried out to confirm that observed increases in DANSYL emission are a result of FRET from protein to membrane and not caused by effects of Ca^{2+} on membranes, i.e., by carrying out parallel Ca^{2+} titrations in the absence of added protein. Samples should be visually inspected for the presence of particulates or membrane aggregates, which may form in the presence of high Ca^{2+} concentrations (in the millimolar range).
6. ΔI_{max} represents the calculated FRET at saturating Ca^{2+} concentrations, which is influenced by the intensity of the intrinsic tryptophan emission and its distance from the lipid surface, the concentration and density of both donor and acceptor molecules, as well as physical features of the phospholipids such as composition and method of preparation. ΔI values may be divided by ΔI_{max} for comparing separate Ca^{2+} titrations. K from the Hill equation determines the midpoint in the titration, a qualitative indicator of the average binding affinity. The Hill coefficient may be regarded as an indicator of positive cooperativity under certain circumstances (e.g., for a two-site binding system when $1.4 < H < 2$) *(19)* and suggests the minimal number of cooperating Ca^{2+} ions that trigger membrane docking. To a first approximation, when membrane docking is controlled by a single, independent Ca^{2+}-binding site, H is unity. Alternatively, if multiple Ca^{2+} sites with positive cooperativity trigger docking, H exceeds unity and K is an average dissociation constant for the various sites. It should be emphasized, however, that in cases where successive Ca^{2+}-binding events yield different FRET changes, the Hill coefficient will be distorted. For more qualitative Hill analysis, equilibrium dialysis should be carried out to measure ^{45}Ca^{2+}-binding.

Acknowledgments

Financial support was provided by NIH RO1-48203 (to J. J. Falke) and NIH Post-Doctoral Fellowship GM–18303 (to E. A. Nalefski).

References

1. Ponting, C. P. and Parker, P. J. (1996) Extending the C2 domain family: C2s in PKCs δ, ε, η, θ, phospholipases, GAPs, and perforin. *Prot. Sci.* **5,** 162–166.

2. Nalefski, E. A. and Falke, J. J. (1996) The C2 domain calcium-binding motif: structural and functional diversity. *Prot. Sci.* **5,** 2375–2390.
3. Rizo, J. and Südhof, T. C. (1998) C2-domains, structure and function of a universal Ca^{2+}-binding domain. *J. Biol. Chem.* **273,** 15,879–15,882.
4. Clark, J. D., Lin, L.-L., Kriz, R. W., Ramesha, C. S., Sultzman, L. A., Lin, A. Y., et al. (1991) A novel arachidonic acid-selective cytosolic PLA_2 contains a Ca^{2+}-dependent translocation domain with homology to PKC and GAP. *Cell* **65,** 1043–1051.
5. Yamaguchi, T., Shirataki, H., Kishida, S., Miyazaki, M., Nishikawa, J., Wada, K., et al. (1993) Two functionally different domains of rabphilin-3A, Rab3A p25/smg p25A-binding and phospholipid- and Ca^{2+}-binding domains. *J. Biol. Chem.* **268,** 27,164–27,170.
6. Mosior, M. and Epand, R. M. (1993) Mechanism of activation of protein kinase C: roles of diolein and phosphatidylserine. *Biochemistry* **32,** 66–75.
7. Davletov, B. and Südhof, T. C. (1993) A single C2 domain from synaptotagmin I is sufficient for high affinity Ca^{2+}/phospholipid binding. *J. Biol. Chem.* **268,** 26,386–26,390.
8. Grobler, J. A. and Hurley, J. H. (1998) Catalysis of phospholipase Cδ1 requires that Ca^{2+} bind to the catalytic domain, but not the C2 domain. *Biochemistry* **37,** 5020–5028.
9. Nalefski, E. A., Slazas, M. M., and Falke, J. J. (1997) Ca^{2+}-signaling cycle of a membrane-docking C2 domain. *Biochemistry* **36,** 12,011–12,018.
10. Bazzi, M. D. and Nelsestuen, G. L. (1987) Association of protein kinase C with phospholipid vesicles. *Biochemistry* **26,** 115–122.
11. Brose, N., Petrenko, A. G., Südhof, T. C., and Jahn, R. (1992) Synaptotagmin: a calcium sensor on the synaptic vesicle surface. *Science* **256,** 1021–1025.
12. Nalefski, E. A., Sultzman, L. A., Martin, D. M., Kriz, R. W., Towler, P. S., Knopf, J. L., and Clark, J. D. (1994) Delineation of two functionally distinct domains of cytosolic phospholipase A_2, a regulatory Ca^{2+}-dependent lipid-binding domain and a Ca^{2+}-independent catalytic domain. *J. Biol. Chem.* **269,** 18,239–18,249.
13. Wu, P. and Brand, L. (1994) Resonance energy transfer: methods and applications. *Anal. Biochem.* **218,** 1–13.
14. Nalefski, E. A. and Falke, J. J. (1998) Location of the membrane-docking face on the Ca^{2+}-activated C2 domain of cytosolic phospholipase A_2. *Biochemistry* **37,** 17,642–17,650.
15. Nalefski, E. A., McDonagh, T., Somers, W., Seehra, J., Falke, J. J., and Clark, J. D. (1998). Independent folding and ligand specificity of the C2 calcium-dependent lipid binding domain of cytosolic phospholipase A_2. *J. Biol. Chem.* **273,** 1365–1372.
16. Schoenmakers, T. J. M., Visser, G. J., Flik, G., and Theuvenet, A. P. R. (1992) CHELATOR: an improved method for computing metal ion concentrations in physiological solutions. *BioTechniques* **12,** 870–879.
17. Eftink, M. (1997) Fluorescence methods for studying equilibrium macromolecule-ligand interactions. *Methods Enzymol.* **278,** 221–257.

18. Haugland, R. P. (1996) Thiol-reactive probes, in *Handbook of Fluorescent Probes and Research Chemicals*, Molecular Probes, Eugene, OR, pp. 47–62.
19. Grabarek, Z. and Gergely, J. (1983) On the applicability of Hill type analysis to fluorescence data. *J. Biol. Chem.* **258,** 14,103–14,105.

20

Ca²⁺-Binding Mode of the C₂A-Domain of Synaptotagmin

Josep Rizo, Josep Ubach, and Jesús García

1. Introduction

The ability of neurons to communicate with each other lies at the center of brain function. Communication between neurons occurs primarily by synaptic transmission and is mediated by neurotransmitters that are released by synaptic vesicle exocytosis. This process is tightly regulated by a complex protein machinery and is triggered by Ca^{2+} influx into a presynaptic terminal when an action potential causes opening of voltage-gated Ca^{2+}-channels *(1)*. The synaptic vesicle protein synaptotagmin I is an essential component of this machinery *(2)*, acting most likely as a Ca^{2+} sensor in neurotransmitter release *(3)*. The Ca^{2+}-binding properties of synaptotagmin I are thus likely to be key for the mechanism of interneuronal communication.

Most of the cytoplasmic region of synaptotagmin I is formed by two consecutive domains homologous to the C_2-domain of protein kinase C (PKC) *(4)*. C_2-domains are widespread Ca^{2+}-binding modules of approx 15 kDa that are particularly abundant in proteins involved in signal transduction and membrane traffic *(5,6)* (*see* also Chapter 19 in this volume). The first C_2-domain of synaptotagmin I (C_2A-domain) is the most extensively studied among these protein modules, and characterization of its structural and binding properties not only has yielded insights into how synaptotagmin I functions, but has also provided a framework to predict and understand the properties of C_2-domains in general. The crystal structure of the Ca^{2+}-free C_2A-domain revealed for the first time the basic architecture of C_2-domains, which consists of a compact β-sandwich formed by two four-stranded β-strands *(7)*. The C_2A-domain was shown by NMR spectroscopy to bind three Ca^{2+} ions in a tight cluster at the tip

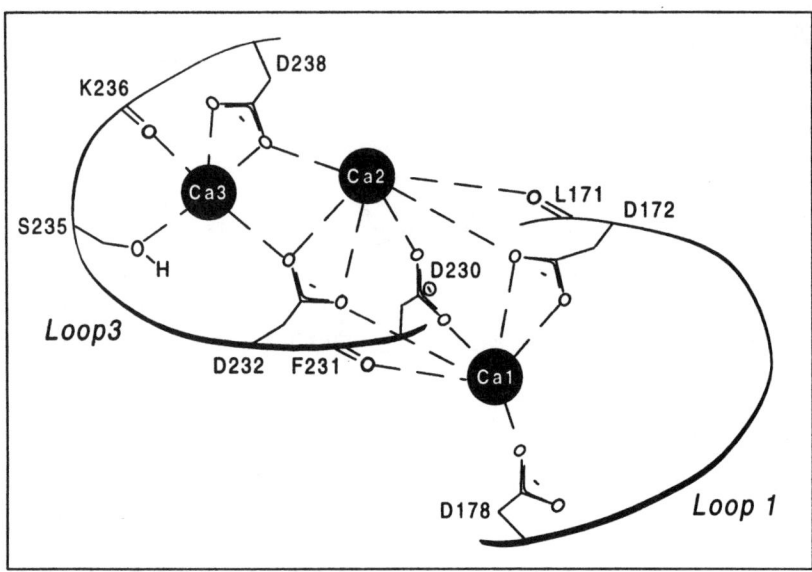

Fig. 1. Ca^{2+}-binding mode of the C_2A-domain of synaptotagmin I. The diagram summarizes the three Ca^{2+}-binding sites (*solid circles labeled Ca1–Ca3*) and their ligands. All ligands correspond to two loops (*labeled loops 1 and 3*).

of the domain *(8,9)*. The Ca^{2+}-binding motif is very different from the EF-hand motif and involves ligands from two loops that are distant in the sequence (referred to as loops 1 and 3); the Ca^{2+} ligands include three backbone carbonyl groups, as well as five aspartate and one serine side chains (*see* **Fig. 1**). The conservation of these residues in many C_2-domains indicates that the three Ca^{2+}-binding sites are widespread in this family of protein modules (*see* sequence alignment in **ref. 6**).

The solution structure of the Ca^{2+}-saturated C_2A-domain revealed that Ca^{2+}-binding does not cause significant conformational changes in the C_2A-domain *(10)*. This result shows that the C_2A-domain functions via a fundamentally different mechanism from that observed in EF-hand proteins such as calmodulin, where Ca^{2+}-induced conformational changes expose hydrophobic surfaces that bind to target molecules *(11)*. In contrast, NMR analysis of the Ca^{2+}-dependent interaction between the C_2A-domain and syntaxin, an essential component of the synaptic membrane fusion machinery, showed that binding to syntaxin is triggered by the drastic change in the electrostatic potential of the C_2A-domain caused by binding of multiple Ca^{2+} ions to a small region of the domain *(8,12,13)*. This observation led to an electrostatic switch model of neurotransmitter release whereby repulsion between negatively charged surfaces of the C_2A-domain and syntaxin prevents exocytosis in the absence of

Ca^{2+}; upon Ca^{2+} influx, Ca^{2+}-binding to the C_2A-domain reverses its electrostatic potential, inducing binding to syntaxin and initiating exocytosis.

From a methodological point of view, the most interesting aspect of the NMR studies of the C_2A-domain summarized above was the determination of its complete Ca^{2+}-binding mode, both because of the complexity of the motif and because of the scarcity of information available from X-ray crystallography. Ca^{2+} ions have been commonly included in the NMR structures of EF-hand proteins based on the characteristic Ca^{2+}-binding mode observed in the crystal structures of the same or other EF-hand proteins. However, the C_2A-domain could not be cocrystallized with Ca^{2+}, and its low Ca^{2+}-affinity hindered saturation by diffusion of Ca^{2+} salts into crystals of the apo-C_2A-domain. Thus, the crystals cracked when soaked with more than 100 µM Ca^{2+}, and at this concentration only the highest affinity Ca^{2+}-binding site could be identified *(7)*. On the other hand, the high solubility of the C_2A-domain in the absence and presence of saturating Ca^{2+} concentrations allowed elucidation of its complete Ca^{2+}-binding mode in solution by a combination of NMR spectroscopy and site-directed mutagenesis *(8–10)*. This chapter will concentrate on the approach used for these studies. The approach should be applicable to the study of the Ca^{2+}-binding mode of proteins or protein domains with low Ca^{2+} affinities and with molecular weights amenable to high resolution NMR analysis (currently less than 40,000). Particular considerations for the analysis of high affinity Ca^{2+}-binding sites are discussed in **Subheading 4**.

2. Materials

1. High-field NMR spectrometer (400 MHz or higher) equipped with at least three channels, a 5-mm triple resonance probe, and preferably with pulsed-field gradients.
2. Computer workstation with software to process and analyze NMR spectra (e.g., NMRPipe *[14]* or NMRView *[15]*), as well as to calculate protein structures using NMR restraints (e.g., CNS *[16]*).
3. Buffer for protein, Ca^{2+} and Mn^{2+} solutions. The buffer should not precipitate nor bind these cations. The buffer concentration should be high enough (e.g., 40 mM) to minimize pH changes upon addition of Ca^{2+} or Mn^{2+}. The solvent should contain 5% D_2O for frequency lock in the NMR experiments.
4. Protein samples (*see* **Note 1**):
   a. One 1-mM sample of uniformly $^{15}N,^{13}C$-labeled, Ca^{2+}-saturated protein for structure determination (0.3 mL).
   b. Two 100-µM samples of uniformly ^{15}N-labeled, Ca^{2+}-free protein for Ca^{2+} titrations (0.6 mL each).
   c. Two 200-µM samples of uniformly ^{15}N-labeled protein, one Ca^{2+}-free and another Ca^{2+}-saturated, for Mn^{2+} titrations (0.6 mL each).
   d. One 100-µM sample of uniformly ^{15}N-labeled, Ca^{2+}-free protein for each mutant designed to test the Ca^{2+}-binding model developed initially (0.6 mL).

5. Concentrated solution of $CaCl_2$ (e.g., 1 M).
6. Concentrated solution of $MnCl_2$ (e.g., 1 M).
7. Several high-quality 5-mm NMR tubes such as 528-PP or 535-PP from Wilmad (Buena, NJ) and one NMR microtube such as BMS-005 from Shigemi (Allison Park, PA) (*see* **Note 2**).

3. Methods

Elucidation of the Ca^{2+}-binding mode of the C_2A-domain involved an iterative approach that combined high-resolution structure determination with site-directed mutagenesis, as well as Ca^{2+}- and Mn^{2+}-titrations of the C_2A-domain monitored by NMR spectra *(8,10)*. Because of the iterative nature of the methodology, the order of the steps indicated below is only orientative and the results obtained in each step often need to be reanalyzed according to the data obtained in other steps.

1. Determine optimal conditions to analyze the protein under study by NMR spectroscopy. This is a general requirement for any protein NMR analysis and involves testing the solubility of the protein under an array of conditions with different pH, ionic strength, and buffer composition. Because structure elucidation is only necessary for the Ca^{2+}-bound form of the protein and the NMR experiments used for this purpose require high protein concentrations (approx 1 mM), the conditions should be optimal for the protein solubility in the presence of saturating Ca^{2+} concentrations. The minimum concentration of Ca^{2+} required for saturation can be determined from the initial Ca^{2+} titration described in **step 2**. The protein also needs to be soluble at 100–200 µM concentrations under the same buffer conditions in the absence of Ca^{2+} (*see* **Note 3**).
2. Perform a preliminary Ca^{2+} titration monitored by ^1H–^{15}N heteronuclear single quantum correlation (HSQC) spectra using a 100-µM ^{15}N-labeled protein sample (*see* **Note 4**). These spectra contain one cross-peak for each nonproline residue of a protein, providing multiple probes that are sensitive to Ca^{2+}-binding. Thus, the Ca^{2+} titration causes progressive shifts in the HSQC cross-peaks from NH groups in residues at or near the Ca^{2+}-binding region, whereas cross-peaks from distant NH groups remain unaffected (*see* **Fig. 2A**) (*see* **Notes 5 and 6**). The titration should be performed adding increasing Ca^{2+} concentrations in an exponential fashion until no changes are observed in the ^1H–^{15}N HSQC spectrum, indicating saturation (*see* **Note 7**). The results will reveal the range of Ca^{2+} concentrations where the different Ca^{2+}-binding sites are titrated.
3. Determine the three-dimensional (3D) structure of the Ca^{2+}-saturated protein using a 1 mM ^{15}N,^{13}C-labeled sample. (For the methods used for this purpose, *see* Chapters 20 and 21 in Volume 2.) No Ca^{2+} ions should be included at this stage. It is critical to obtain as many interproton distance restraints as possible to define accurately the conformation of the Ca^{2+}-binding region because proper orientation of the potential Ca^{2+} ligands will facilitate their final identification. In the process followed to solve the structure of the protein, assignments for the

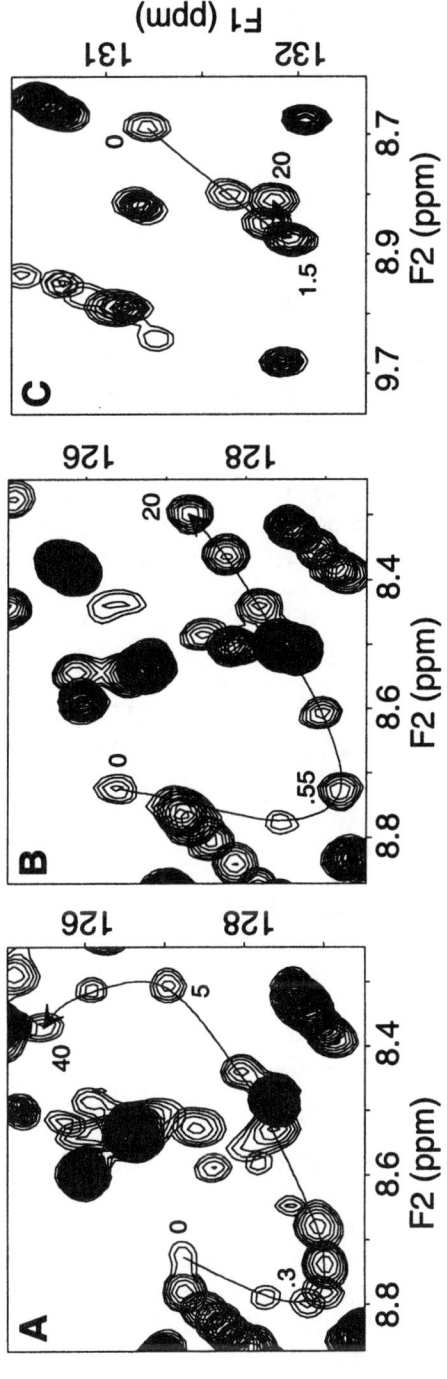

Fig. 2. Ca^{2+}-titrations of wild-type and mutant C$_2$A-domains monitored by ^{1}H–^{15}N HSQC spectra. The plots represent superpositions of expansions from ^{1}H–^{15}N HSQC spectra acquired at different Ca^{2+}-concentrations for the wild type C$_2$A-domain (A) and for mutant C$_2$A-domains containing a Ser235Ala (B) or Asp238Ala (C) subtitution. A curved arrow has been drawn to indicate the position of the cross-peak corresponding to Asp238 NH as a function of Ca^{2+} concentration. The spectra represent a subset of those acquired during the titrations. The total Ca^{2+} concentrations used for each superimposed spectrum in millimolar units were (some are indicated next to the cross-peaks): (A) 0, 0.1, 0.2, 0.3, 0.45, 0.6, 1.2, 2, 5, 20 and 40; (B) 0, 0.075, 0.25, 0.55, 1, 3, 6 and 20; (C) 0, 0.2, 0.4 0.8, 1.5 and 20. Note that some cross-peaks do not shift significantly upon addition of Ca^{2+}.

^1H–^{15}N HSQC cross-peaks of the Ca^{2+}-saturated protein are obtained. ^1H–^{15}N HSQC cross-peak assignments for the Ca^{2+}-free protein can be obtained by following the cross-peak movements in the Ca^{2+} titrations (*see* **Note 8**).

4. Perform an additional Ca^{2+} titration monitored by ^1H–^{15}N HSQC spectra with a 100-μM ^{15}N-labeled protein sample, using at least 15 different Ca^{2+} concentrations within the range where HSQC cross-peak changes were observed in the first titration (**step 2**). This titration will allow to better define the number of components in the titration curves, which are reflected by the different directions of movement of the cross-peaks and by the Ca^{2+}-concentrations required for such movement. This is exemplified for the cross-peak of D238 of the C_2A-domain in **Fig. 2A**. The different components are caused by binding of Ca^{2+} to different sites in the protein. Thus, the titration yields the number of Ca^{2+}-binding sites and at the same time the affinity of each site (*see* **Notes 9** and **10**; *see* also **Note 6**).

5. Develop an initial model for the Ca^{2+}-binding mode of the protein. The model is developed by analysis of the Ca^{2+}-induced chemical shift changes observed for the ^1H–^{15}N HSQC cross-peak from each residue in the protein, and from close examination of the structure of the protein in the region(s) where the largest changes are observed. Although there is no strict correlation between the shift induced in an NH group by binding of one Ca^{2+} ion and the NH/Ca^{2+}-ion distance, larger shifts are usually observed for NH groups close to the Ca^{2+} ion. Analysis of which NH groups have larger shifts for a particular component of the titration provides information on the order in which the different Ca^{2+}-binding sites are titrated. For instance, the NH group of D238 is closest to site Ca2 of the C_2A-domain; because the largest changes for this NH group are observed in the second titration component (*see* **Fig. 2A**), this component is likely to correspond to Ca^{2+}-binding to site Ca2.

6. Perform two titrations with Mn^{2+} monitored by ^1H–^{15}N HSQC spectra, one with a Ca^{2+}-free 200-μM ^{15}N-labeled protein sample and another with a Ca^{2+}-saturated 200-μM ^{15}N-labeled protein sample. Mn^{2+} is a paramagnetic ion that substitutes well for Ca^{2+} in protein-binding sites and causes strong relaxation effects on nuclei in its vicinity, resulting in severe broadening of the corresponding resonances. Thus, selective broadening of ^1H–^{15}N HSQC cross-peaks from NH groups near the metal-binding sites are expected upon addition of Mn^{2+} (*see* **Fig. 3**). These broadening effects can be used initially in a qualitative manner to identify metal binding sites in the structure of the protein, and later in a more quantitative manner (*see* **step 9**). Because this quantitative analysis requires measurement of cross-peak intensities, substoichiometric concentrations of Mn^{2+} are required to avoid broadening beyond detection of all cross-peaks from NH groups close to the metal (*see* **Notes 11** and **12**). The Mn^{2+} titration performed in the absence of Ca^{2+} is expected to sample primarily the highest affinity-binding sites. In the Ca^{2+}-saturated protein, the Mn^{2+} titration will sample preferentially sites where the relative affinity for Mn^{2+} vs Ca^{2+} is highest (*see* **Note 13**). These experiments provide complementary information to that obtained from the Ca^{2+} titrations and

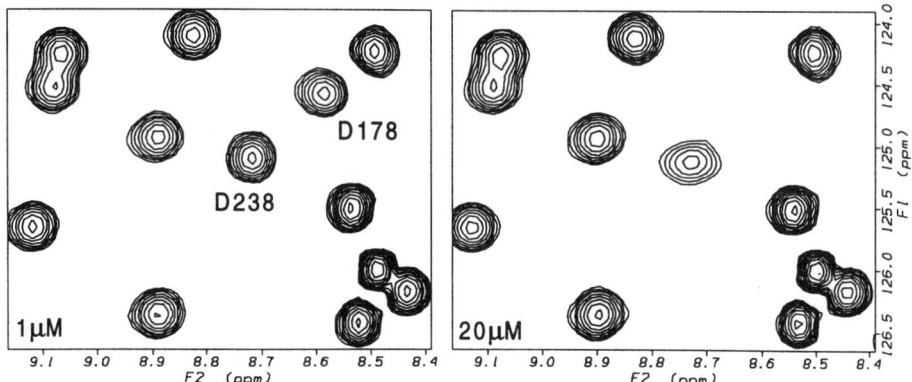

Fig. 3. Mn^{2+}-induced broadening effects in ^1H–^{15}N HSQC spectra of the C_2A-domain. Expansions of ^1H–^{15}N HSQC spectra of the C_2A-domain acquired in the absence of Ca^{2+} and presence of 1 μM or 20 μM Mn^{2+} are shown on the left and right pannels, respectively. Note that only the cross-peaks from D238 and D178 NH are significantly broadened with 20 μM Mn^{2+}, whereas no broadening is observed for all other cross-peaks, which correspond to NH groups that are far from the metal-binding sites. The D178 NH cross-peak is broadened more severely that of D238 NH, in agreemet with the fact that D178 NH is closer to site Ca1 than D238 NH (Ca1 is the only site being significantly populated by Mn^{2+} under these conditions).

are used to refine the preliminary Ca^{2+}-binding model and/or identify binding sites that were not evident from the Ca^{2+} titrations.

7. Based on the working Ca^{2+}-binding model, design single point mutations to disrupt specific Ca^{2+}-binding sites. Perform Ca^{2+} titrations monitored by ^1H–^{15}N HSQC with a 100-μM ^{15}N-labeled sample of each mutant. The specific component of the titration corresponding to the Ca^{2+}-binding site intended to be disrupted should not be observed if the model is correct. For instance, mutation of *Ser*235 to alanine in the C_2A-domain removed the component corresponding to site Ca3 (*see* **Fig. 2B**) as expected from the Ca^{2+}-binding motif shown in **Fig. 1**. Similarly, mutation of Asp238 to alanine resulted in loss of the components corresponding to sites Ca2 and Ca3 (although slight residual binding to site Ca2 remained) (*see* **Fig. 2C**).

8. Determine the final Ca^{2+}-binding mode with iterative calculations of the structure of the protein including Ca^{2+} ions and incorporating ligand/Ca^{2+}-ion restraints progressively. Initially, the only restraints used for the Ca^{2+} ion in each site force them to be near ligands that have been ascertained to participate in binding to that site by the mutagenesis experiments. These initial calculations immediately position the ions approximately in their binding sites and most of the ligands for each ion can be deduced from their location within less than 3 Å from the correspond-

ing Ca^{2+} ion in a large percentage of the structures calculated. Additional calculations are then performed introducing new restraints between the new ligands identified and the Ca^{2+} ions. This increases the convergence of the structure calculations and facilitates identification of other ligands that were not evident before. For the C_2A-domain, this iterative process allowed unambiguous identification of all ligands except the carbonyl group of Leu171 (see **Fig. 1**), which was at a distance of less than 3 Å from site Ca2 in only about 50% of the structures calculated including restraints for all other ligands. Leu171 CO was accepted as a bona fide ligand of site Ca2 because these structures exhibited lower interproton distance violations than those with long distances between this ligand and site Ca2. The structure of the Ca^{2+}-saturated C_2-domain of PKC-β determined independently by X-ray crystallography *(17)* revealed an identical Ca^{2+}-binding mode.

9. Perform a quantitative analysis of the Mn^{2+}-induced broadening effects to check the consistency of these effects with the final Ca^{2+}-binding mode obtained. The analysis is performed by measuring the intensities of cross-peaks that exhibit significant broadening in a ^1H–^{15}N HSQC spectrum acquired at a given Mn^{2+} concentration. The intensities are normalized with those observed in a ^1H–^{15}N HSQC spectrum acquired for the same sample in the absence of Mn^{2+}. Then, the differences between the inverse of the normalized intensities in the absence and presence of Mn^{2+} (Δi^{-1}) are calculated. On the other hand, average distances (r) between the different metal binding sites and the NH protons corresponding to the broadened cross-peaks are calculated form the ensemble of final structures obtained. Logarithmic plots of Δi^{-1} vs r are constructed independently for each metal-binding site. A linear correlation is expected for the site sampled preferentially by Mn^{2+} at the given Mn^{2+} concentration (e.g., see **Fig. 4**) whereas no correlation is expected for the plots corresponding to other metal-binding sites (see **Note 14**). The observation of good correlations serves as an independent measure of the quality of the final Ca^{2+}-binding mode determined in **step 8**.

4. Notes

1. The minimum number of protein samples is indicated. Additional samples may be required to repeat some of the experiments, and a $^{15}N,^{13}C$-labeled sample with partial or full perdeuteration may be required to determine the structure of proteins larger than 20 kDa. The protein concentrations indicated are orientative. Concentrations around 1 mM (sample a) are usually required to elucidate a high resolution protein structure at 500 MHz. The concentrations of samples b–d are appropriate to yield high-quality NMR data with a 500 MHz NMR spectrometer using the experimental parameters indicated in **Notes 4** and **11**.
2. The microtube is used for sample a to reduce the total amount of protein needed because $^{15}N,^{13}C$-labeling is quite expensive.
3. Protein solubility is most often minimal near its isoelectric point. Thus, the isoelectric point should be calculated for the apo-protein and for the Ca^{2+}-saturated protein. Because the number of Ca^{2+} ions that bind to the protein is unknown *a priori*, isoelectric points should be calculated assuming different numbers of

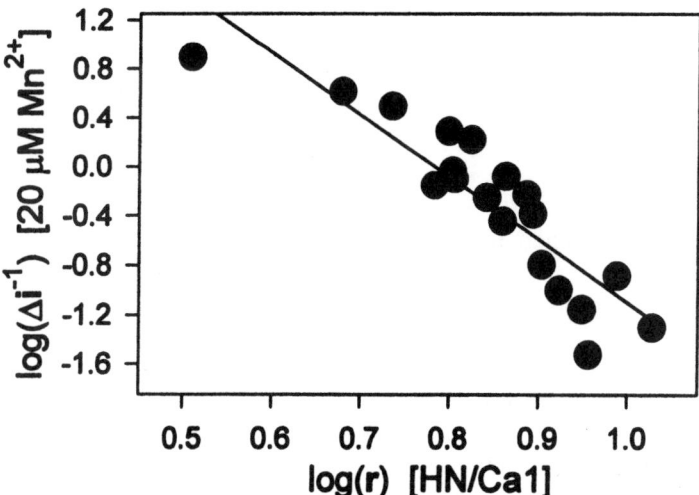

Fig. 4. Logarithmic plot of Δi^{-1} observed for ^1H-^{15}N HSQC cross-peaks of the C_2A-domain at 20 µM Mn^{2+} in the absence of Ca^{2+} vs the NH/Ca1 distances calculated from the solution structure of the Ca^{2+}-saturated C_2A-domain.

bound Ca^{2+} ions. Whenever possible, the pH should be under 7.5, to avoid fast exchange of amide protons with the solvent, and above 5 to prevent protonation of acidic side chains that may participate in Ca^{2+}-binding.

4. ^1H–^{15}N HSQC experiments should preferably be performed using a sensitivity-enhanced pulse sequence incorporating pulsed-field gradients and a water flip-back pulse such as that described in **ref. 18**. We usually use acquisition times of 100 ms in both the ^1H and the ^{15}N dimensions, with spectral widths of 8000 Hz and 1000–1500 Hz, respectively. The number of transients per free-induction decay is set to a multiple of the phase cycle to yield a total time of 1 h for acquisition of the spectrum. It is advisable to extend this time to 2 or 4 h if the protein concentration is less than 100 µM, the field is less than 500 MHz, or broadening because of chemical exchange is observed in the midpoints of the titration.

5. Ca^{2+}-induced shifts for residues far from the binding sites may be observed if there is an overall Ca^{2+}-induced conformational change. In this case, the Ca^{2+} titrations are still useful to determine the Ca^{2+}-concentration required for saturation and the number of Ca^{2+} ions that bind to the protein. However, information on the location of metal-binding sites may be more easily obtained from the Mn^{2+}-induced broadening effects (**steps 6** and **9**) than from the Ca^{2+}-induced chemical shift changes.

6. For proteins with intermediate or high Ca^{2+}-affinities (k_D below 10 µM) progressive disappearance of HSQC cross-peaks with concomitant appearance of new HSQC cross-peaks will be observed. In this case, the number of Ca^{2+}-binding sites corresponds to the number of Ca^{2+} equivalents required to achieve saturation.

7. The titration starts with a sample that does not contain Ca^{2+}. This can be easily accomplished for proteins with low-Ca^{2+} affinity through extensive dialysis or successive dilution/concentration with Ca^{2+}-free buffer. Increasing Ca^{2+} concentrations (e.g., 25 µM, 50 µM, 100 µM, and so on) are achieved by adding a small volume (e.g., 3 µL) of buffer with the appropriate Ca^{2+} concentration. All these solutions are prepared from a mother solution of concentrated $CaCl_2$ (1 M).
8. In crowded regions of the spectrum it may be difficult to follow all the cross-peak movements throughout the titration. In such cases, multidimensional NMR data should also be acquired on the $^{15}N,^{13}C$-labeled sample in the absence of Ca^{2+} to obtain resonance assignments for the apo-protein.
9. The direction of movement for a given cross-peak caused by Ca^{2+}-binding to two different sites may be the same by coincidence, which may thus appear as binding to only one site. However, because many cross-peaks are expected to exhibit Ca^{2+}-induced shifts, it is highly unlikely that this coincidence will occur for every cross-peak. A single component for multiple Ca^{2+}-binding sites may result for all cross-peaks in cases of high cooperativity or if all sites have very similar affinities by mere coincidence. In both cases, the number of binding sites should then be deduced from the Mn^{2+}-induced broadening effects (**steps 6** and **9**).
10. Accurate affinities can be obtained by fitting the Ca^{2+}-dependence of each particular chemical shift to standard protein-ligand equilibrium equations *(19)*. However, good estimates of the affinity associated to each titration component can also be obtained from the midpoint of the component. For instance, if the midpoint of the second component occurs at 600 µM Ca^{2+} (total Ca^{2+} added) and the protein concentration is 100 µM, the kD of the corresponding binding site corresponds to the free Ca^{2+} concentration, i.e., 450 µM. For high-affinity Ca^{2+}-binding (k_D below 10 µM), accurate affinities are better obtained with fluorescence-based assays, which can be performed with protein concentrations in the low micromolar range *(20)*.
11. The Mn^{2+} titration is started adding a small concentration of Mn^{2+} (e.g., 1 µM) after acquiring a reference ^1H–^{15}N HSQC spectrum in the absence of Mn^{2+}. Increasing concentrations of Mn^{2+} are then added until most of the cross-peaks from protons near the Ca^{2+}-binding region are broadened beyond detection. The most useful spectra are those where differential broadening is observed for a large number of cross-peaks. For the C_2A-domain, these concentrations were 20 µM and 50 µM Mn^{2+} for the Ca^{2+}-free sample, and 100 µM and 200 µM Mn^{2+} for the Ca^{2+}-saturated sample. ^1H–^{15}N HSQC spectra are acquired with analogous parameters to those used for the Ca^{2+} titrations (*see* **Note 4**), but with longer total acquisition times (e.g., 4–8 h) to allow reliable quantitation of cross-peak intensities.
12. The observation of significant broadening at substoichiometric Mn^{2+} concentration requires a fast-exchange regime between the Mn^{2+}-free and Mn^{2+}-bound forms of the protein, which is expected for low-affinity binding sites. For sites with k_D below approx 10 µM, it is preferable to use stoichiometric concentrations of a paramagnetic Ca^{2+} analog that only induces moderate resonance broadening (e.g., lanthanides such as samarium).

13. For the Ca^{2+}-free C_2A-domain, Mn^{2+} concentrations up to 20 μM sampled mostly the highest affinity site (Ca1 in **Fig. 1**), and concentrations of 50 μM and above started to sample the site with the second highest affinity (Ca2). In the presence of Ca^{2+}, site Ca3 was primarily sampled.
14. The increase in the transverse relaxation rate of a nucleus induced by proximity to a Mn^{2+} ion is proportional to r^{-6} *(21)*. The difference in the inverted signal intensity for the nucleus is approximately proportional to the increase in transverse relaxation rate. Hence, $\log(\Delta i^{-1})$ is expected to depend linearly on $\log(r)$. In the analysis of the C_2A-domain, linear correlations were obtained for NH/Ca1 distances in experiments performed with 20 μM Mn^{2+} and no Ca^{2+}, and for NH/Ca3 distances in experiments performed with 100 μM Mn^{2+} and 20 μM Ca^{2+}. At 50 mM Mn^{2+} in the absence of Ca^{2+}, deviations from linearity in the $\log(\Delta i^{-1})/\log[r(NH/Ca1)]$ plots for the cross-peaks from the two NH groups closest to site Ca2 indicated partial sampling of this site. These results show the importance of performing the analysis at different Mn^{2+} concentrations, and for the Ca^{2+}-free and Ca^{2+}-bound forms of the protein.

References

1. Südhof, T. C. (1995) The synaptic vesicle cycle: a cascade of protein–protein interactions. *Nature* **375,** 645–653.
2. Geppert, M., Goda, Y., Hammer, R. E., Li, C., Rosahl, T. W., Stevens, C. F. and Südhof, T. C. (1994) Synaptotagmin I: a major Ca^{2+} sensor for transmitter release at a central synapse. *Cell* **79,** 717–727.
3. Südhof, T. C. and Rizo, J. (1996) Synaptotagmins: C_2-domain proteins that regulate membrane traffic. *Neuron* **17,** 379–388.
4. Perin, M. S., Fried, V. A., Mignery, G. A., Jahn, R., and Südhof, T. C. (1990) Phospholipid binding by a synaptic vesicle protein homologous to the regulatory region of protein kinase C. *Nature* **345,** 260–263.
5. Rizo, J. and Südhof, T. C. (1998) C_2-domains, structure and function of a universal Ca^{2+}-binding domain. *J. Biol. Chem.* **273,** 15,879–15,882.
6. Nalefski, E. A. and Falke, J. J. (1996) The C2 domain calcium-binding motif: structural and functional diversity. *Protein Sci.* **5,** 2375–2390.
7. Sutton, R. B., Davletov, B. A., Berghuis, A. M., Südhof, T. C., and Sprang, S. R. (1995) Structure of the first C_2 domain of synaptotagmin I: a novel Ca^{2+}/phospholipid-binding fold. *Cell* **80,** 929–938.
8. Ubach, J., Zhang, X., Shao, X., Südhof, T. C., and Rizo, J. (1998) Ca^{2+} binding to synaptotagmin: how many Ca^{2+} ions bind to the tip of a C_2-domain? *EMBO J.* **17,** 3921–3930.
9. Shao, X., Davletov, B. A., Sutton, R. B., Südhof, T. C., and Rizo, J. (1996) A bipartite Ca^{2+}-binding motif in C_2-domains of synaptotagmin and protein kinase C. *Science* **273,** 248–251.
10. Shao, X., Fernandez, I., Südhof, T. C., and Rizo, J. (1998) Solution structures of the Ca^{2+}-free and Ca^{2+}-bound C_2A-domain of synaptotagmin I: does Ca^{2+} induce a conformational change? *Biochemistry* **37,** 16,106–16,115.

11. Zhang, M., Tanaka, T., and Ikura, M. (1995) Calcium-induced conformational transition revealed by the solution structure of apo calmodulin. *Nat. Struct. Biol.* **2**, 758–767.
12. Shao, X., Li, C., Fernandez, I., Zhang, X., Südhof, T. C., and Rizo, J. (1997) Synaptotagmin-syntaxin interaction: the C_2-domain as a Ca^{2+}-dependent electrostatic switch. *Neuron* **18**, 133–142.
13. Fernandez, I., Ubach, J., Dulubova, I., Zhang, X., Südhof, T. C., and Rizo, J. (1998) Three-dimensional structure of an evolutionary conserved N-terminal domain of syntaxin 1A. *Cell* **94**, 841–849.
14. Delaglio, F., Grzesiek, S., Vuister, G. W., Zhu, G., Pfeifer, J., and Bax, A. (1995) NMRPipe: a multidimensional spectral processing system based on UNIX pipes. *J. Biomol. NMR* **6**, 277–293.
15. Johnson, B. A. and Blevins, R. A. (1994) NMRView: a computer program for the visualization and analysis of NMR data. *J. Biomol. NMR* **4**, 603–614.
16. Brunger, A. T. et al. (1998) Crystallography and NMR System (CNS): a new software suite for macromolecular structure determination. *Acta Crystallogr. D* **54**, 901–921.
17. Sutton, R. B. and Sprang, S. R. (1998) Structure of the protein kinase C-b phospholipid-binding C_2-domain complexed with Ca^{2+}. *Structure* **6**, 1395–1405.
18. Zhang, O., Kay, L. E., Olivier, J. P., and Forman-Kay, J. (1994) Backbone ^1H and ^{15}N resonance assignments of N-terminal SH3 domain of drk in folded and unfolded states using enhanced-sensitivity pulsed field gradient NMR techniques. *J. Biomol. NMR* **4**, 845–858.
19. Williams, T. C., Corson, D. C., Sykes, B. D., and MacManus, J. P. (1987) Oncomodulin. *J. Biol. Chem.* **262**, 6248–6256.
20. Ubach, J., Garcia, J., Nittler, M. P., Südhof, T. C., and Rizo, J. (1999) Structure of the Janus-faced C_2B-domain of rabphilin. *Nature Cell Biol.* **1**, 106–112.
21. Taylor, A., Sawan, S., and James, T. L. (1982) Structural aspects of the inhibitor complex formed by N-(leucyl)-o-aminobenzenesulfonate and manganese with Zn^{2+}-Mn^{2+} leucine aminopeptidase (EC 3. 4. 11. 1). *J. Biol. Chem.* **257**, 11,571–11,576.

21

Study of Calcineurin Structure by Limited Proteolysis

Seun-Ah Yang and Claude Klee

1. Introduction

The use of limited proteolysis to study protein structure was pioneered by Linderstrom–Lang *(1)*. During the past 20 years, limited proteolysis has proven to be a powerful tool to demonstrate that most of the calmodulin-regulated enzymes contains an inhibitory domain and that their activation by calmodulin involves the displacement of this domain *(2–6)*. The sequential limited proteolysis of the Ca^{2+}/calmodulin-regulated protein phosphatase, calcineurin, is a good example of how this method can be used to dissect the functional domains of a protein and to identify conformational changes accompanying ligand binding and interaction with regulatory proteins.

Calcineurin is a heterodimer of a 59-kDa catalytic subunit, calcineurin A, tightly bound ($K_d \leq 10^{-14}$ *M*) to a 19 kDa Ca^{2+}-binding regulatory subunit, calcineurin B *(7,8)*. A schematic representation of the functional-domain organization of calcineurin A determined by limited proteolysis is shown in **Fig. 1**. In the absence of calmodulin, proteases rapidly degrade the regulatory domain of calcineurin A, whereas the catalytic domain is preserved yielding a constitutively active enzyme that no longer binds or requires calmodulin for activity *(8)*. The activated enzyme remains bound to calcineurin B, and is still dependent on Ca^{2+}, but it requires ten times less Ca^{2+} than the native enzyme for activity suggesting that in the native protein the regulatory domain modulates the affinity of calcineurin B for Ca^{2+} by a mechanism not yet identified *(9)*. In the presence of calmodulin, the calmodulin-binding domain is protected from proteolysis, but removal of a 6-kDa carboxyl-terminal fragment is sufficient to render the *p*-nitrophenyl phosphatase activity of the enzyme independent of

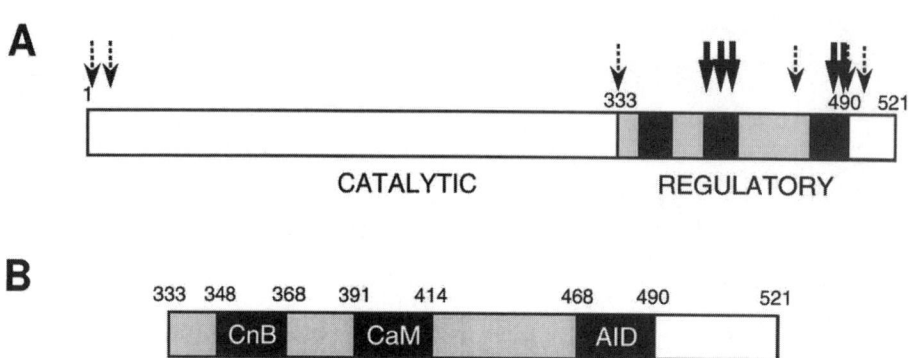

Fig. 1. Functional domain organization of calcineurin A. (**A**) Schematic representation of the catalytic and regulatory domains of calcineurin A identified by limited proteolysis. The arrows indicate the major sites of cleavage by clostripain in the presence of calmodulin (25 min digest, *dashed arrows*) and in the absence of calmodulin (0.5 min digest, *plain arrows*). (**B**) Extended representation of the regulatory domain and identification of the calmodulin-binding, calcineurin B-binding and autoinhibitory domains identified by limited proteolysis and synthetic peptides.

calmodulin, while preserving its ability to bind calmodulin. Thus, the regulatory domain of calcineurin contains two distinct subdomains, a calmodulin-binding and an inhibitory domain. The amino-terminal sequences of the calcineurin fragments generated during the proteolysis were used to identify the cleavage sites and to design synthetic peptides that mimic their effects in the native protein. A peptide corresponding to the inhibitory domain was characterized by Hashimoto et al. *(10)*. The loss of calmodulin binding upon cleavage at residue 392 supported the earlier tentative identification of the calmodulin-binding domain based on its sequence similarity with other calmodulin-binding domains *(11)*. NMR studies of a calcineurin B-binding peptide identified by limited proteolysis were used to characterize its interaction with calcineurin B *(12)*.

The similar patterns of proteolysis with proteases with different specificities suggests that the selective cleavages are caused by the presence of flexible and exposed domains rather than of specific amino acid sequences. The presence of these hinge regions has been confirmed by the elucidation of the crystal structure of recombinant calcineurin expressed in *Escherichia coli (13)* and that of the complex of the 43-kDa calcineurin derivative complexed with FKBP/FK506 *(14)*. Thus, limited proteolysis can be used to confirm the similarity of the solution and crystal structures of the protein as well as to help design recombinant proteins suitable for structure–function relationships.

2. Materials
2.1. Proteins and Reagents

1. Bovine brain calcineurin purified as aforementioned *(15)*.
2. Ram testis calmodulin purified as aforementioned *(16)* or recombinant Xenopus calmodulin expressed in *E. coli* *(17,18)*.
3. Lactoperoxidase 200 U/mg (5 mg/mL) from Boehringer-Mannheim Corporation.
4. Glucose oxidase (grade I), from Boehringer-Mannheim Corporation. 15 mg/mL in 0.05 M potassium phosphate buffer, pH 7.0.
5. Na^{125}I (specific activity 16 mCi/µg of iodine) from Amersham Pharmacia Biotech.
6. Sulfosuccinimidyl-6-(biotinamido) hexanoate (Sulfo-NHS-LC-Biotin) 4 mg/mL in 2 mM sodium acetate, pH 6.0 from Pierce.
7. Vectastain reagents from Vector Laboratories (catalog no. PK–4001).
8. Enhanced chemiluminescence (ECL) Western blotting detection reagents from Amersham Pharmacia Biotech (catalog no. RPN 2209).
9. Calmodulin-Sepharose 4B from Amersham Pharmacia Biotech.

2.2. Proteases

1. Clostripain (from Boehringer Mannheim Corporation) activated by overnight incubation at 0–4°C of a 5 mg/mL solution in 0.05 M Na$_2$HPO$_4$, pH 7.9, containing 10 mM dithiothreitol (DTT) *(19)*. The activated enzyme is stored at –70°C for up to 2 mo without significant loss of activity.
2. Chymotrypsin (from Worthington Biochemical Corporation) 2 mg/mL in 0.001 N HCl is stored at –70°C up to 6 mo without significant loss of activity.
3. Trypsin treated with L-(tosylamido 2-phenyl) ethyl chloromethyl ketone (TPCK) (from Worthington Biochemical Corporation) 2 mg/mL in 0.001 N HCl is stored at –70°C up to 6 mo without significant change in activity.

2.3. Protease Inhibitors (from Sigma)

1. N-*p*-tosyl-L-lysine chloromethyl ketone (TLCK) 5 mg/mL in H$_2$O, stored at –20°C.
2. TPCK 4 mg/mL in dimethylsulfoxide (DMSO), stored at –20°C.
3. Soybean trypsin inhibitor 10 mg/mL in H$_2$O, stored at –20°C.

2.4. Protease Substrates

1. N-benzoyl-L-arginine-methyl ester (BAME): freshly prepared 0.25 mM solution in 0.1 M potassium phosphate buffer, pH 7.1 or in 10 mM Tris-HCl, pH 7.5, containing 0.3 mM CaCl$_2$ and 1.7 mM DTT.
2. N-acetyl-L-tyrosine ethyl ester (ATEE): freshly prepared 9.34 mM solution in 0.1 M potassium phosphate buffer, pH 7.1.

2.5. Limited Proteolysis

1. Ca^{2+} 0.2 M, stored at 4°C.
2. EGTA 0.2 M, pH 7.5, stored at room temperature.

3. DTT 0.5 M, stored at $-20°C$.
4. Buffer P1: 40 mM Tris-HCl, pH 7.5, 0.1 M NaCl, 1 mM MgCl$_2$, 0.1 mM EGTA, and 1 mM DTT.
5. Buffer P2: 40 mM Tris-HCl, pH 8.0, 0.1 M NaCl, 1 mM MgCl$_2$, 8 mM DTT, and 1 mM CaCl$_2$.
6. Acrylamide, *bis*-acrylamide are products of Polysciences Inc., and glycine, SDS, and Coomassie blue R–250 are from Bio-Rad.

2.6. Calmodulin Overlays

1. Buffer C1: 50 mM Tris-HCl, pH 7.5, 0.2 NaCl, and 1 mM CaCl$_2$,
2. Buffer C2: 50 mM Tris-HCl, pH 7.5, 5 mg/mL bovine serum albumin (BSA), 0.1% (v/v) Tween-20, and 1 mM CaCl$_2$.

2.7. Calcineurin Assays

1. Phosphopeptide: Synthetic peptide DLDVPIPGRFDRRVSVAAE (from Peninsula Laboratories, Inc.) corresponding to the phosphorylation site of bovine cardiac cyclic-AMP dependent protein kinase regulatory subunit type II [R$_{II}$peptide *(20)*] purified by high-performance liquid chromatography (HPLC) and phosphorylated with [γ^{32}P]ATP with cAMP-dependent protein kinase as described by Hubbard and Klee *(21)*.
2. *p*-Nitrophenyl phosphate: 0.1 M in H$_2$O.
3. Buffer A1: 20 mM Tris-HCl, pH 8.0, 0.1 M NaCl, 6 mM MgCl$_2$, 0.1 mM DTT, and 0.1 mg/mL BSA.
4. Buffer A2: 40 mM Tris-HCl, pH 8.0, 0.1 M NaCl, 0.5 mM DTT, and 0.1 mg/mL BSA.
5. Stop solution: 5% trichloroacetic acid containing 0.1 M potassium phosphate, pH 7.0.
6. Dowex AG 50W-X8 (200–400 mesh) from Bio-Rad.

2.8. Microsequencing

1. Protein sequenator (Model 477A from PE Applied Biosystems).
2. Polyvinylidene fluoride (PVDF) membranes supplied in sheets by Millipore Co. as Immobilon-P transfer membranes (catalog no. IPVH 15150).
3. Sequencing-grade reagents, glass fiber filters, and Biobrene from PE Applied Biosystems.
4. Transfer buffer: 10 mM CAPS (3-(cyclohexylamino)–1-propanesulfonic acid) from Sigma, pH 11, containing 10% methanol, kept at 4°C overnight.
5. Denaturing solution: 0.1 M Tris-HCl, pH 8.0, 1% sodium dodecyl sulfate (SDS), 8 M urea, and 0.8 M DTT.

2.9. Calmodulin-Sepharose Affinity Chromatography

1. Buffer S1: 40 mM Tris-HCl, pH 7.5, 0.05 M NaCl, 3 mM MgCl$_2$, 0.1 mM DTT, 1 µg/mL leupeptin, and 0.2 mM CaCl$_2$.
2. Buffer S2: 40 mM Tris-HCl, pH 7.5, 0.2 M NaCl, 3 mM MgCl$_2$, 0.1 mM DTT, 1 µg/mL leupeptin, and 0.2 mM CaCl$_2$.

3. Buffer S3: 40 mM Tris-HCl, pH 7.5, 0.2 M NaCl, 1 mM MgCl$_2$, 0.1 mM DTT, 1 µg/mL leupeptin, and 2 mM EGTA.

2.10. HPLC

1. C$_{18}$ µBondapak reverse-phase column (3.9 × 30 cm, 10-µm beads), from Waters Associates).
2. 0.45-µm nitrocellulose filters from Millipore Co.
3. Solvents: 0.1% trifluoroacetic acid (TFA) and CH$_3$CN made 0.1% (v/v) TFA.
4. Savant speed vac concentrator.

3. Methods
3.1. Protease Assays

The specific activities of chymotrypsin, trypsin and clostripain are determined under standard conditions as described by Schwert and Takenaka *(22)*. ATEE hydrolysis is monitored by the decreased absorbance at 237 nm (chymotrypsin) and BAME hydrolysis by the increased absorbance at 253 nm (trypsin and clostripain). One unit is the amount of chymotrypsin that catalyzes the hydrolysis of 0.1 µmol of ATEE (decreased absorbance of 0.0075) per min, and the amounts of trypsin and clostripain that catalyze the hydrolysis of 0.01 µmol of BAME (increased absorbance of 0.003) per min under the standard assay conditions. The rates of hydrolysis are also determined under the conditions used for the limited proteolysis experiments to correct for the effect of CaCl$_2$ and EGTA on the activities of trypsin and chymotrypsin.

3.1.1. Clostripain

1. In 1-cm pathlength quartz cuvets, add 2.8 mL of 0.25 mM BAME in 10 mM Tris-HCl, pH 7.5, containing 0.3 mM CaCl$_2$ and 1.7 mM DTT.
2. After 5 min preincubation at 25°C, the reaction is started by addition of 0.2 mL of a freshly diluted solution of clostripain (3–6 µg [35–70 U]/mL in 10 mM Tris-HCl, pH 7.5, containing 0.3 mM CaCl$_2$ and 1.7 mM DTT).

3.1.2. Chymotrypsin and Trypsin

1. In 1-cm pathlength quartz cuvets, pipet 2.8 mL of 9.34 mM ATEE or 0.25 mM BAME in phosphate buffer (standard conditions) or in Buffer P1 containing either 0.6 mM CaCl$_2$ or 2 mM EGTA (limited proteolysis conditions).
2. After 5 min equilibration at 25°C, the reaction is started by addition of 0.2 mL of a freshly diluted solution of chymotrypsin (15–30 µg [25–50 U]/mL in 0.001 M HCl) or trypsin (10–20 µg [30–60 U]/mL in 0.001 M HCl).
3. The specific activities of chymotrypsin and trypsin under standard conditions and in the presence of EGTA (1500 U/mg and 3300 U/mg, respectively) are 80% of those measured in the presence of Ca^{2+}.

3.2. Limited Proteolysis of Calcineurin

3.2.1. Buffer Exchange

1. Calcineurin (4 mg in 18 mL of 40 mM Tris-HCl, pH 7.5, containing 0.1 M NaCl, 1 mM MgCl$_2$, 0.1 mM EGTA, 0.2 mM DTT, 10% (v/v) glycerol, 1 μg/mL leupeptin, and 10 μg/mL soybean trypsin inhibitor) is concentrated three- to fourfold by overnight dialysis against 2 L of buffer P1 containing 50% (v/v) glycerol. The concentrated calcineurin (4 mg in 4.5 mL) is passed through a 1.5 × 50-cm column of Sephadex G-100 equilibrated with buffer P1 to remove protease inhibitors and glycerol.
2. Protein concentration is determined spectrophotometrically using a coefficient $\varepsilon\ ^{1\%}_{277nm} = 7.4$ (C. Klee and P. Stemmer, unpublished observations).

3.2.2. Proteolysis of Calcineurin with Clostripain, Chymotrypsin, and Trypsin

1. The time dependence of the proteolysis of calcineurin with proteases with different specificities is first analyzed to demonstrate the existence of selective cleavage sites and to identify optimal conditions to isolate and characterize the resulting proteolytic derivatives.
2. For clostripain digestion, calcineurin (0.12 mg in 1.2 mL of buffer P2, with or without 15 μM calmodulin) is preincubated for 1 min at 30°C and proteolysis is started by addition 17 μL or 12 μL of activated clostripain (5 mg/mL) to final concentrations of 70 μg (10^3 U) or 50 μg (700 U)/mL in the presence and absence of calmodulin, respectively *(8)*.
3. For chymotrypsin and trypsin digestions, calcineurin (0.12 mg in 1.2 mL of buffer P1 containing either 0.6 mM CaCl$_2$ or 2 mM EGTA) is preincubated for 1 min at 30°C and proteolysis is started by addition of 65 μL of a freshly diluted solution (20 μg/mL in 0.001 M HCl) of chymotrypsin or trypsin to final concentrations of 1.5 U/mL chymotrypsin or 3.5 U/mL trypsin.
4. At appropriate times, duplicate 0.1 mL aliquots are placed in Eppendorf tubes containing 20 μL of 0.02 mg/mL TLCK (clostripain digest), or 5 μL of 0.2 mg/mL TPCK (chymotryptic digest), or 5 μL of 0.2 mg/mL soybean trypsin inhibitor (tryptic digest).
5. One set of samples is placed in dry ice for SDS gel electrophoresis and the other set is stored at –70°C for enzymatic assays.

3.3. Separation of Proteolytic Fragments by SDS Gel Electrophoresis

3.3.1. SDS Gel Electrophoresis

Fifty microliters of freshly prepared denaturing solution are added to the 0.1-mL aliquots stored in dry ice and then immediately placed in a boiling bath. After standing in the boiling bath for 1 min, the samples are subjected to SDS gel electrophoresis in 7.5–15% gradients of acrylamide in 0.1% SDS *(23)*

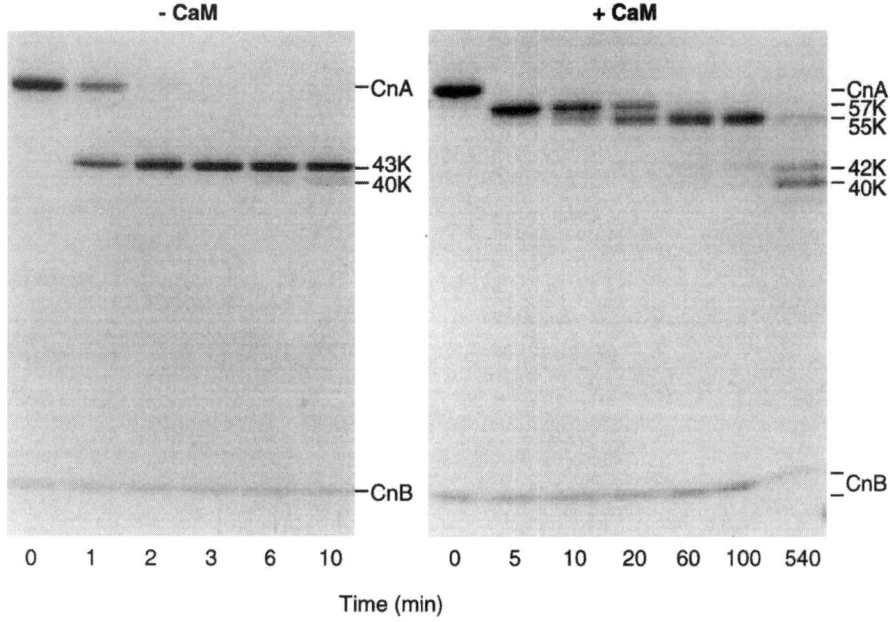

Fig. 2. Time course of proteolysis of calcineurin in the presence and absence of calmodulin. Calcineurin (0.1 mg/mL) was digested with clostripain 0.7×10^3 or 10^3 U/mL in the presence of Ca^{2+} or Ca^{2+}/calmodulin (CaM), respectively, as described in the text. At the indicated times aliquots, corresponding to 5 µg calcineurin, were subjected to SDS-PAGE and stained with Coomassie blue (Reproduced from **ref. 8**).

on two different sets of gels (one for Coomassie staining and one for calmodulin overlays). The samples are cooled on ice prior to loading onto the gels and adjusted to 3 mM EGTA to avoid Ca^{2+}-induced mobility shifts.

3.3.2. Identification of Calcineurin A Derivatives

One set of gels with 4–5 µg protein per lane is stained with Coomassie blue. The size of the proteolytic fragments is determined by comparison with the R_f values of protein standards (phosphorylase b, 97 kDa; BSA, 67 kDa; catalase, 58 kDa; fumarase, 48 kDa; actin, 42 kDa; lactate dehydrogenase, 36 kDa; and β-lactoglobulin, 17.5 kDa). Representative patterns of limited proteolysis of calcineurin in the presence and absence of calmodulin are shown in **Fig. 2** (clostripain digests) and in the presence of Ca^{2+} and EGTA in **Fig. 3** (tryptic and chymotryptic digests). The different patterns of the sequential degradation of calcineurin observed under these different conditions is evidence that Ca^{2+}-binding to calcineurin B and calmodulin binding to calcineurin A are accompanied by pronounced and distinct conformational changes of calcineurin A.

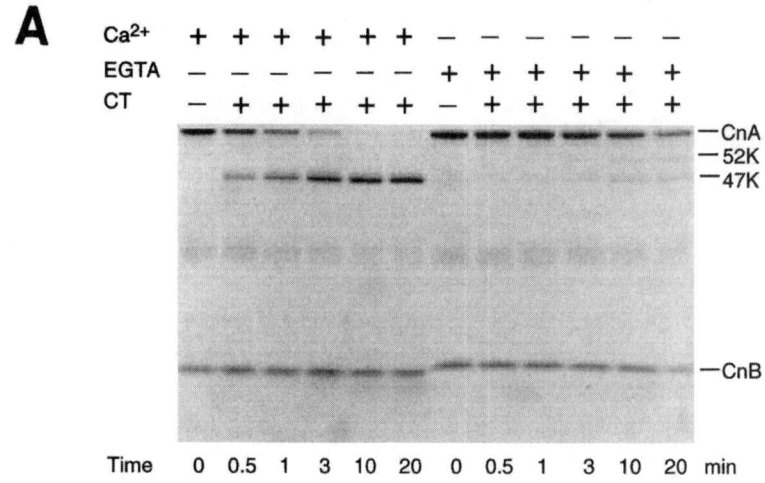

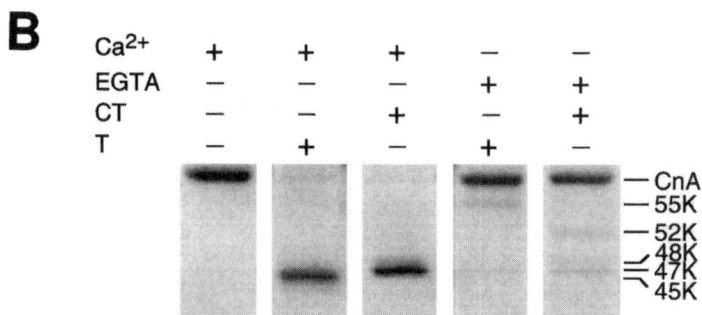

Fig. 3. Proteolysis of calcineurin in the presence and absence of Ca^{2+}. Calcineurin (0.1 mg/mL) was digested as described in the text with chymotrypsin (CT, 2 U/mL; T, 3.5 U/mL) in the presence of Ca^{2+} or EGTA for the indicated times (**A**). A comparison of the chymotryptic and tryptic digests (5 min) is shown in (**B**). Aliquots (4 µg calcineurin) were subjected to SDS-PAGE and stained with Coomassie blue.

The rates of proteolysis at the different sites are determined by quantitative densitometric analysis of the calcineurin fragments using the public domain NIH Image 1.55 program for Macintosh computers written by Wayne Rasband at the U.S. National Institutes of Health and available from the Internet by anonymous FTP from zippy.nimh.nih.gov (24). Calcineurin A and calcineurin B are used as standards assuming that the color responses of calcineurin derivatives are proportional to their estimated sizes. The rates of proteolysis by trypsin and chymotrypsin in the presence of EGTA are corrected for the 20% decrease of their specific activities measured as aforementioned.

3.4. Identification of Calmodulin-Binding Fragments

The second set of gels with 3.5 µg calcineurin/lane is used for [^{125}I]-calmodulin gel overlay *(15,25)* or with 0.6 µg calcineurin/lane, for biotinylated-calmodulin overlay *(26a)*.

3.4.1. [^{125}I]-Calmodulin Gel Overlay

Iodination of calmodulin. All operations are carried out in an iodination hood.

1. To a solution of calmodulin (1.4 mg in 900 µL of 0.05 M potassium phosphate buffer, pH 7.0) add 20 µL of 5 mM CaCl$_2$, 2 µL of 5 mg/mL lactoperoxidase, 2 µL of 15 mg/mL glucose oxidase, 20 µL of 0.01 M NaI, and 10 mCi of Na^{125}I. Start the reaction by the addition of 20 µL of 0.055 M glucose. Incubate at room temperature for 60 min and stop the reaction by the addition of 30 µL of 2-mercaptoethanol. Purify the iodinated calmodulin using a Pharmacia PD10 column (9 mL) equilibrated and eluted with 0.04 M Tris-HCl, pH 8.0, 0.1 M NaCl. Collect 0.5 mL fractions. Iodinated calmodulin, monitored by gamma counting, is eluted in tubes 6–9. The two peak fractions are pooled and stored at –70°C for up to 3 mo. The solution of iodinated calmodulin (2.2 mol I/mol, specific activity = 100 Ci/mmol) is about 6×10^{-5} M.
2. Gel overlay. After SDS-polyacrylamide gel electrophoresis (PAGE), fix the gel in 500 mL of 25% isopropanol, 10% acetic acid with three changes over 15 h. All procedures are conducted at room temperature. To document the Ca^{2+}-dependence of the labeling, duplicate gels are processed in an identical manner, except that 1 mM EGTA is substituted for CaCl$_2$ in buffer C1. After 10–15 min washing in 500 mL deionized water, the gel is soaked in 500 mL of buffer C1, with three changes over 15 h. Soak the gel in 500 mL buffer C1 containing 1 mg/mL serum albumin for 2 h. Transfer the gel to 100 mL of buffer C1 containing 10 nM [^{125}I]-calmodulin in a shallow plastic tray, which is gently agitated for 2–3 min to disperse [^{125}I]-calmodulin and the incubation is continued for 12 h. After washing with 500 mL of buffer C1, with at least four changes over 36 h, the gel is stained for 2 h in staining solution (Coomassie blue 0.075% in 20% isopropanol, 6.5% acetic acid), and destained in 10% isopropanol and 10% acetic acid. The calmodulin binding fragments, identified by autoradiography (2–24 h exposure) of the gel wrapped in Saran Wrap, are shown in **Fig. 4**.

    [^{125}I]-calmodulin binding is quantitated by gamma counting of the excised radiolabeled protein bands and calcineurin fragments by densitometric analysis of the Coomassie blue patterns. As illustrated in **Fig. 4** and **Table 1**, in the presence of calmodulin the calmodulin-binding domain is protected. A 14-kDa calmodulin-binding fragment, that is further degraded to an 8-kDa calmodulin-binding fragment upon prolonged incubation, is detectable after 20 min incubation *(8)*. In contrast, when proteolysis is carried out in the absence of calmodulin, calcineurin A is rapidly converted to a 43-kDa species that no longer binds calmodulin (data not shown).

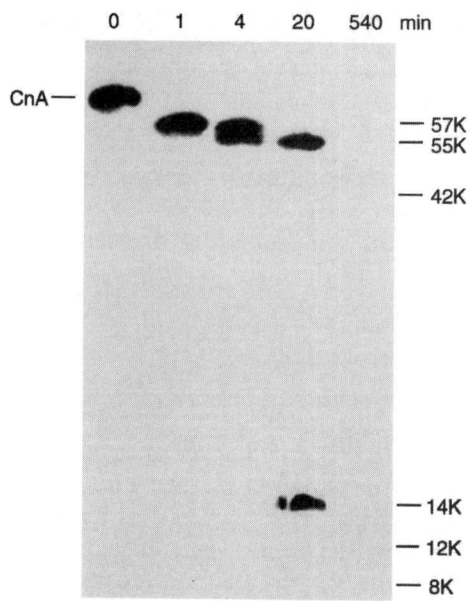

Fig. 4. ^{125}I-Calmodulin binding to calcineurin A fragments generated in the presence of calmodulin. Calcineurin (0.4 mg/mL) was digested with clostripain (10^3 U/mL) in the presence of Ca^{2+} and 13 μM calmodulin. At the indicated times aliquots, corresponding to 3.5 μg calcineurin, were subjected to the [^{125}I]-calmodulin gel overlay and autoradiography as described in the text. Upon prolonged incubation (540 min), low levels of radioactivity associated with 42-, 12-, and 8-K bands are not visible in the photograph of the autoradiogram (Reproduced from **ref. 8**).

3.4.2. Biotinylated-Calmodulin Overlay

1. Biotinylation of calmodulin *(26b)*. Dialyze calmodulin (3.5 mg in 1 mL of Tris-HCl, pH 8.0) against 1 L of 0.1 M sodium bicarbonate at 4°C for 2 d with 4 changes. Add Sulfo-NHS-LC-Biotin to the dialyzed protein to a final molar ratio (Sulfo-NHS-LC-Biotin: calmodulin) of 3.5:1. Incubate the mixture at room temperature for 3 h. Add 100 μL of 1 M ethanolamine, pH 9.0, to a final concentration of 0.1 mM to block excess reagents and allow the reaction to proceed for 2 h at room temperature. Separate the labeled protein from excess reagents by gel-filtration on a Sephadex G-50 column (0.6 × 28 cm), equilibrated and eluted with TBS. Collect 0.5-mL fractions. The protein-containing fractions monitored by absorbance at 276 nm are pooled (tube 10–17), dialyzed against 2 L of TBS at 4°C for 24 h with two changes, and stored at –70°C. The biotinylated calmodulin concentration is calculated using an extinction coefficient $\varepsilon_{276nm} = 3300$ *(27)*.
2. Biotinylated-calmodulin overlay. Electrotransfer the proteins to PVDF membranes at 0.3 amp for 45 min at 4°C using transfer buffer. Block the membrane with buffer C2 for 1 h at room temperature. Incubate the membrane in 10–20 mL

Table 1
[^{125}I]-Calmodulin Binding to Clostripain Fragments of Calcineurin[a]

Calcineurin A derivatives (M_r)	Protein[b] (pmol)	[^{125}I]-CaM[c]	
		U	U/mol
CnA	70	70	1.00
55000	15	21	0.40
42000	43	4	≤ 0.09
40000	17	0.6	≤ 0.03
14000	nd[d]	12.2	
8000	nd[d]	2.2	

[a]The Coomassie staining intensity and [^{125}I]-CaM binding capacity of calcineurin A and its derivatives were obtained after a 40 min clostripain digestion of calcineurin in the presence of calmodulin (Reproduced from **ref. 8**).
[b]Calcineurin fragments were quantitated as described in the text.
[c]One unit is arbitrarily defined as that amount of [^{125}I]-CaM bound to 1 pmol of calcineurin A.
[d]Not detectable.

buffer C2 containing 0.5 µg/mL biotinylated calmodulin for 25 min at room temperature. Wash three times for 5 min with 100 mL buffer C2. Incubate in 10 mL buffer C2 containing avidin DH and biotinylated horseradish peroxidase H (Vectastain reagents A and B) for 25 min at room temperature as described in the manufacturer's instructions. Incubate for 1 min in ECL reagent (8 mL reagent 1 [luminol/enhancer] mixed with 8 mL reagent 2 [stable hydrogen peroxide]) according to the Amersham Pharmacia Biotech instructions. Expose for 3–50 s to hyperfilm ECL for optimal sensitivity. Biotinylated calmodulin is quantitated by densitometric analysis of the autoradiograms and the amount bound/mol of calcineurin calculated on the basis of the quantitation of the Coomassie-stained bands (*see* **Subheading 3.3.2.**) assuming 100% transfer of the proteins during the electrotransfer. The biotinylated calmodulin overlay assay has the advantage to be rapid and avoids the use of radioactive reagents. The [^{125}I]-calmodulin gel overlay is less convenient, but yields a more reliable quantitation of the calmodulin binding. The same procedures can be used to analyze the binding of calcineurin B to calcineurin A and the isolation of the calcineurin B-binding domain of calcineurin A.

3.5. Enzymatic Assays

The protein phosphatase and *p*-nitrophenyl phosphatase activities of calcineurin digests are measured with the set aside frozen aliquots.

3.5.1. Protein Phosphatase Activity

1. Mix 20 µL of buffer A1 containing 3×10^{-8} M calcineurin with 20 µL buffer A1 containing 3×10^{-8} M calmodulin, or 20 µL buffer A1 containing 3×10^{-7} M

calcineurin with 20 µL buffer A1. Preincubate for 1 min at 30°C and start the reaction by addition of 20 µL buffer A1, containing 1–5 µM phosphorylated substrate, and either 2 mM $CaCl_2$ to measure Ca^{2+} and calmodulin-stimulated activity, or 3 mM EGTA to measure Ca^{2+}-independent activity.
2. Allow the reaction to proceed for 10 min and stop the reaction by the addition of 0.5 mL stop solution.
3. Isolate the released inorganic phosphate (^{32}P) by chromatography on Dowex–50 columns. Prepare 0.5 mL Dowex–50 columns converted to the H^+ form by sequential washing with 10 mL H_2O, 1 mL 1 M NaOH, 2 mL 1 N HCl, and 4 mL H_2O. Apply the reaction mixtures to the columns and wash with 0.5 mL H_2O. Collect the flowthrough and the 0.5-mL wash directly into scintillation vials, and quantitate the released ^{32}P by scintillation counting.
4. Rate constants, determined as aforementioned (15), are used to calculate the specific activity of calcineurin and of its derivatives (nmol/min/mg) normalized to 1 µM substrate.

3.5.2. p-Nitrophenyl Phosphatase Activity

The p-nitrophenyl phosphatase activity of calcineurin (28) is monitored with a spectrophotometer equipped with a microcell attachment.

1. 4×10^{-7} M calcineurin in 0.2 mL buffer A2 containing either 1 mM $CaCl_2$, with or without 18×10^{-7} M calmodulin, or 1 mM EGTA is preincubated for 5 min at 23°C and the reaction is started by addition of 10 µL 0.1 M p-nitrophenyl phosphate.
2. The reaction is allowed to proceed for 5 min and the initial rates are calculated using an extinction coefficient of ε_{400nm} = 15,400 at pH 8.0 (29).

As shown in **Fig. 5**, digestion of calcineurin in the absence of calmodulin, conditions that convert calcineurin A to the 43,000 M_r derivative, results in a rapid and complete loss of calmodulin dependence and an apparent partial loss of Ca^{2+} dependence for both protein phosphatase and p-nitrophenyl phosphatase activities. A low-residual Ca^{2+} concentration may be sufficient to activate this enzyme, which has a very high affinity for Ca^{2+}. Further degradation to 40,000 and 38,000 M_r species is accompanied by a progressive loss of both activities. In contrast, when digestion is carried out in the presence of calmodulin, the 57,000 M_r and, to a lower extent, the 55,000 M_r species have a greatly enhanced Ca^{2+}-independent p-nitrophenyl phosphatase activity, but their Ca^{2+}-independent protein phosphatase activity is stimulated only 1.5–2-fold. Ca^{2+}-binding to calmodulin present in the digest is likely responsible for the 40-fold stimulation observed upon addition of Ca^{2+}. The expression of recombinant calcineurin derivatives corresponding to the 40,000, 57000, and 55,000 M_r enzymes would help to establish the molecular basis for the different requirements of the two phosphatase activities of calcineurin and help to clarify

Calcineurin Structure

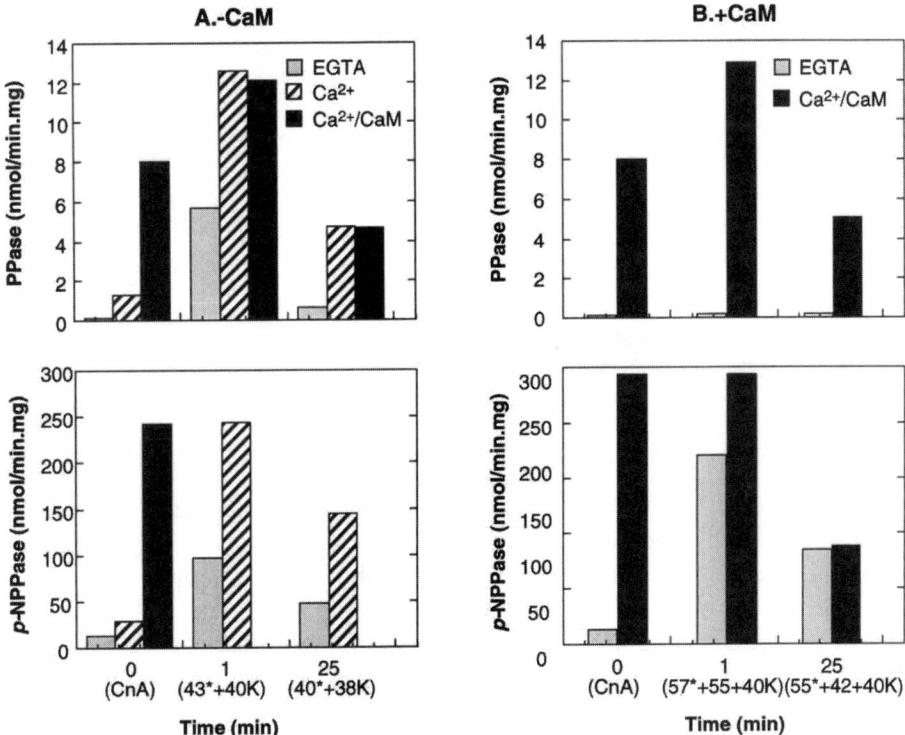

Fig. 5. Effects of proteolysis on protein phosphatase (PPase) and *p*-nitrophenyl phosphatase (*p*-NPPase) activity of calcineurin. Calcineurin (0.6 mg/mL) was digested with clostripain (530 U/mL) in the presence of Ca^{2+} (**A**) or Ca^{2+}/CaM (**B**) as described in the text. The phosphatase activities were measured in the presence of EGTA, Ca^{2+} and Ca^{2+}/CaM (**A**), and EGTA and Ca^{2+}/CaM (**B**) as indicated in the figure. The M_r of the calcineurin derivatives present in the digests are indicated in the figure (the asterisks indicate the predominant species).

why the p-nitrophenyl phosphatase is stimulated, whereas the protein phosphatase is inhibited by the immunosuppressive drugs, FK506 and cyclosporin A *(30)*.

3.6. Microsequencing of Calcineurin Fragments

For microsequencing and peptide isolation the concentration of calcineurin in the digests is increased to 0.4–1 mg/mL and that of clostripain decreased to 500 U/mL. A preliminary time course of the digestion, as aforementioned, is performed to identify the conditions yielding suitable amounts of the species to be tested. The proteolytic derivatives of calcineurin, separated by SDS-gel electrophoresis (10–20 µg/lane), are electrotransferred to PVDF membranes as aforementioned, and subjected to automated Edman degradation *(31)* to identify their amino-terminal sequences, as described by Matsudeira *(32)*. Care should be taken to use clean instruments and gloves for all manipulations.

3.6.1. Staining of Protein Bands

1. Wash the membrane with deionized water for 5 min.
2. Stain the membranes with a Coomassie blue solution (0.05% Coomassie blue and 50% methanol) for 5 min.
3. Destain with 50% methanol: 10% acetic acid for 5 min.
4. Rinse in deionized water for 5–10 min.
5. Photograph the membranes.
6. Cut the protein bands, and store them in individual Eppendorf tubes at –20°C.

3.6.2. Amino Terminal Sequence Determination

Protein bands transfered to the PVDF membranes are placed in the sequencing cartridge and covered with a precycled Biobrene-coated, trifluoroacetic acid-activated, glass fiber filter according to the manufacturer's instructions (Applied Biosystems User Bulletin issue 32, 1987).

As shown in **Table 2**, most of the calcineurin A proteolytic derivatives that have lost the calmodulin-binding and the inhibitory domain have retained the blocked amino-terminus of the native enzyme. Thus, the regulatory domain is located in the carboxyl-terminal third of the protein molecule. The identification of small fragments lost during the washing of the gels and the transfer to the PVDF membranes is needed to identify the location of the cleavage sites that result in the activation of the enzyme and its conversion to calmodulin independence.

3.7. Isolation of Small Calcineurin Fragments

The small proteolytic fragments resulting from the digestion of calcineurin in the presence of calmodulin can be separated from the large fragments by gel-filtration chromatography on a Sephadex-G50 column. The small fragments resulting from a short digestion in the absence of calmodulin are more easily purified by affinity chromatography on calmodulin-Sepharose.

3.7.1. Calmodulin-Sepharose Fractionation

1. Pack a 0.8 × 0.4 cm column with 0.6 mL of a one to two slurry of calmodulin-Sepharose equilibrated with buffer S1.
2. Place the column in a 13-mL tube and centrifuge for 1 min at 0°C in a Eppendorf centrifuge (5810R) to remove excess buffer.
3. Apply the calcineurin digest (0.2 mg in 0.5 mL buffer P1 made 3 mM $MgCl_2$ and enough $CaCl_2$ to insure a concentration of free $CaCl_2$ of 0.5 mM), place the col-

Table 2
Amino-Terminal Sequence of Calcineurin A and Its Proteolytic Derivatives

CalcineurinA derivatives (M_r)	Amount sequenced (pmol)	Amino-terminal sequence
CnA	180	None detected
57000[a]	180	None detected
55000[b]	40	V V K A V P
42000[b]	40	A V P
43000[c]	200	None detected
14000[b]	nd[d]	Q F N - S P

Calcineurin A was digested with clostripain (1[a] and 20[b] min digests with CaM and 0.5[c] min digest without CaM) in the presence of Ca^{2+} as indicated in the text. [d]Not detectable. Because of its comigration with calcineurin B the amount of 14-K fragment could not be measured, but it could be sequenced since calcineurin B has a blocked amino terminus *(33)*. The fourth residue of the 14K was not conclusively identified (Reproduced from **ref. 8**).

umn in a 13-mL test tube and collect the flowthrough fraction by gravity. The column is then centrifuged as aforementioned to collect the residual fluid trapped in gel matrix.
4. Wash the column with 0.2 mL of buffer S2 as aforementioned to ensure complete recovery of the small fragments not bound to the column. The combined eluates (flowthrough fraction) are then subjected to HPLC.
5. The calcineurin derivatives bound to calmodulin-Sepharose can be eluted, as aforementioned, by four successive washes with 0.1 mL of buffer S3 and analyzed by SDS gel electrophoresis.

All steps are carried out at 0°C in an ice bucket.

3.7.2. Peptide Separation by HPLC

1. Pass the flowthrough fraction from the calmodulin-Sepharose column (0.7 mL) through a 0.45-µm millipore filter.
2. Apply the filtered sample to a C-18 column equilibrated in 0.1% TFA.
3. Elute with a 70-min 0–50%, linear acetonitrile gradient in 0.1% TFA at a flow rate of 1.5 mL/min. The fractions are collected every 30 s.
4. Monitor the fractions containing peptide fragments for absorbance at 215 and 280 nm and pool the peak fractions.
5. The pooled fractions, flash evaporated in a speed vac concentrator, can be stored at –20°C or dissolved in 20 µL of 30% acetonitrile in 0.1% TFA for microsequencing.
6. The amino acid sequence of the peptides are determined as described in **Subheading 3.6.2.**, except that the samples are directly applied onto the precycled Biobrene-coated TFA treated filters according to the manufacturer's instructions.

4. Notes

1. This method, based on the selective proteolytic cleavage of poorly structured hinge regions between functional domains of proteins, is a powerful tool to identify these functional domains and to isolate truncated proteins that have preserved their catalytic activity but lost their regulatory properties. Protein motifs such as the binding domains for regulatory proteins (the calmodulin and calcineurin B-binding and the autoinhibitory domains in the case of calcineurin) identified by limited proteolysis can then be used to design synthetic peptides and to test their effects on enzyme activity and regulation. An even more attractive approach is to use the information obtained by limited proteolysis to design recombinant protein derivatives to study structure–function relationships or truncated proteins more suitable for crystallization and NMR studies.
2. Conformational changes accompanying ligand binding may affect the rate of cleavage at specific bonds and may thereby be identified by a quantitative analysis of the effect of ligand binding on these cleavage rates. The sensitivity of this method is enhanced by the irreversibility of the proteolytic cleavages that magnifies transient structural changes often difficult to observe by alternative methods.
3. It is important to keep in mind that optimal conditions for the digestion procedures should be determined by preliminary trials as aforementioned. The detection and isolation of transient intermediates require short incubation times and low-protease concentrations. The isolation of fragments particularly sensitive to proteolysis, such as the calmodulin-binding domain of calcineurin, require specific conditions to protect the fragments against proteolysis. Under these conditions, their isolation requires longer incubation times and higher protease concentrations. The limited number of bonds cleaved by proteases with strict specificities (clostripain or endoproteinase Arg-C) facilitates the analysis of the proteolytic patterns and enhances the probability to obtain homogenous peptides.
4. With recent developments in protein characterization by mass spectrometry that eliminate the need for the purification of individual peptides and greatly increase the sensitivity of microsequencing, limited proteolysis may become the method of choice for structural studies of very scarce proteins.

References

1. Linderstrom-Lang (1952) The initial stages in the breakdown of proteins by enzymes. *Lane Med. Lectures* **6,** 53–72.
2. Manalan, A. S. and Klee CB (1983) Activation of calcineurin by limited proteolysis. *Proc. Natl. Acad. Sci. USA* **80,** 4291–4295.
3. Blumenthal, D. K., Takio, K., Edelman, A. M., Charbonneau, H., Titani, K., Walsh, K. A., and Krebs, E. G. (1985) Identification of the calmodulin-binding domain of skeletal muscle myosin light chain kinase. *Proc. Natl. Acad. Sci. USA* **82,** 3187–3191.
4. James, P., Vorherr, T., Krebs, J., Morelli, A., Castello, G., McCormick, D. J., Penniston, J. T., De Flora, A., and Carafoli, E. (1989) Modulation of the erythro-

cyte Ca²⁺ ATPase by selective calpain cleavage of the calmodulin-binding domain. *J. Biol. Chem.* **264,** 8289–8296.
5. Ikebe, M., Stepinska, M., Kemp, B. E., Means, A. R., and Hartshorne, D. J. (1987) Proteolysis of smooth muscle myosin light chain kinase. Formation of inactive and calmodulin-independent fragments. *J. Biol. Chem.* **262,** 13,828–13,834.
6. Charbonneau, H., Kumer, S., Novack, J. P., Blumenthal, D. K., Griffin P. R., Shabanowitz, J., et al. (1991) Evidence for domain organization within the 61-kDa calmodulin-independent cycle nucleotide phosphodiesterase from bovine brain. *Biochemistry* **30,** 7931–7940.
7. Klee, C. B., Ren, H., and Wang, X. (1998) Regulation of the calmodulin-stimulated protein phosphatase, calcineurin. *J. Biol. Chem.* **273,** 13,367–13,370.
8. Hubbard, M. J. and Klee, C. B. (1989) Functional domain structure of calcineurin A: mapping by limited proteolysis. *Biochemistry* **28,** 1868–1874.
9. Stemmer, P. M. and Klee, C. B. (1994) Dual calcium ion regulation of calcineurin by calmodulin and calcineurin B. *Biochemistry* **33,** 6859–6866.
10. Hashimoto, Y., Perrino, B. A., and Soderling, T. R. (1990) Identification of an autoinhibitory domain in calcineurin. *J. Biol. Chem.* **265,** 1924–1927.
11. Kincaid, R. L, Nightingale, M. S., and Martin, B. M. (1988) Characterization of a cDNA clone encoding the calmodulin-binding domain of mouse brain calcineurin. *Proc. Natl. Acad. Sci. USA* **85,** 8983–8987.
12. Anglister, J., Ren, H., Klee, C. B., and Bax, A. (1995) NMR identification of calcineurin B residues affected by binding of a calcineurin A peptide. *FEBS Lett.* **375,** 108–112.
13. Kissinger, C. R., Parge, H. E., Knighton, D. R., Lewis, C. T., Pelletier, L. A., Tempczyk, A., et al. (1995) Crystal structure of human calcineurin and the human FKBP12-FK506-calcineurin complex. *Nature* **378,** 641–644.
14. Griffith, J. P., Kim, J. L., Kim, E. E., Sintchak, M. D., Thomson, J. A., Fitzgibbon, M. J., et al. (1995) X-ray structure of calcineurin inhibited by the immunophilin-immunosuppressant FKBP12-FK506 complex. *Cell* **82,** 507–522.
15. Klee, C. B., Krinks, M. H., Manalan, A. S., Cohen, P., and Stewart, A. A. (1983) Isolation and characterization of bovine brain calcineurin: a calmodulin-stimulated protein phosphatase. *Methods Enzymol.* **102,** 227–244.
16. Newton, D. L., Klinks, M. H., Kaufman, J. B., Shiloach, J., and Klee, C. B. (1988) Large scale preparation of calmodulin. *Prep. Biochem.* **18,** 247–259.
17. Chien, Y. H. and Dawid, I. B. (1984) Isolation and characterization of calmodulin genes from Xenopus laevis. *Mol. Cell Biol.* **4,** 507–513.
18. Kuboniwa, H., Tjandra, N., Grzesiek, S., Ren, H., Klee, C. B., and Bax, A. (1995) Solution structure of calcium-free calmodulin. *Nat. Struct. Biol.* 2. 768–776.
19. Porter, W. H., Cunningham, L. W., and Mitchell, W. M. (1971) Studies on the active site of clostripain. The specific inactivation by chloromethyl ketone derived from N-tosil-L-lysine. *J. Biol. Chem.* **246,** 7675–7682.
20. Blumenthal, D. R., Takio, K., Edelman, A. M., Charbonneau, H., Titani, K., Walsh, K. A., and Krebs, E. G. (1986) Dephosphorylation of cAMP-dependent protein

kinase regulatory subunit (type II) by calmodulin-dependent protein phosphate. Determinants of substrate specificity. *J. Biol. Chem.* **261,** 8140–8145.
21. Hubbard, M. J. and Klee, C. B. (1991) Exogenous kinase and phosphatase as probes of intracellular modulation, in Molecular Neurobiology, A practical approach (Chad, J. and Wheal, H., eds.), IRL, Oxford, pp. 135–157.
22. Schwert, G. W. and Takenaka Y. (1955) A spectrophotometric determination of trypsin and chymotrypsin. *Biochim. Biophys. Acta* **16,** 570–578.
23. Laemmli, U. K. (1970) Cleavage of structural proteins during the assembly of the head of bacteriophage T4. *Nature (London)* **227,** 680–685.
24. Rasband, W. S. and Bright, D. S. (1995) NIH image: A public domain image processing program for the macintosh. *Microb. Anal.* **4,** 137–149.
25. Glenney, J. R., Jr. and Weber, K. (1980) Calmodulin-binding proteins of the microfilaments present in isolated brush borders and microvilli of intestinal epithelial cells. *J. Biol. Chem.* **255,** 10,551–10,554.
26a. Smith, G. P. and Scott, J. K. (1993) Libraries of peptides and proteins displayed on filamentous phage. *Methods Enzymol.* **217,** 228–257.
26b. Gao, Z. H., and Zhong, G. (1999) Calcineurin B- and calmodulin-binding prefernces identified with phage-displayed peptide libraries. *Gene* **228,** 51–59.
27. Klee, C. B. and Vanaman, T. C. (1982) Calmodulin. *Adv. Protein Chem.* **35,** 213–321.
28. Pallen, C. H. and Wang, J. H. (1983) Calmodulin-stimulated dephosphorylation of p-nitrophenyl phosphate and free phosphotyrosine by calcineurin. *J. Biol. Chem.* **258,** 8550–8553.
29. Bessey, O. A. and Love, R. H. (1952) Preparation and measurement of the purity of the phosphatase reagent, disodium p-nitrophenyl phosphate. *J. Biol. Chem.* **196,** 175–178.
30. Liu, J., Farmer, J. D. Jr., Lane W, S, Friedman J., Weissman, I., and Schreiber, S. L. (1991) Calcineurin is a common target of cyclophilin-cyclosporin A and FKBP-FK506 complexes. *Cell* **66,** 807–815.
31. Edman, P. and Begg, G. (1967) A protein sequenator. *Eur. J. Biochem.* **1,** 80–91.
32. Matsudaira, P. (1987) Sequence from picomole quantities of proteins electroblotted onto polyvinylidene difluoride membranes. *J. Biol. Chem.* **262,** 10,035–10,038.
33. Aitken, A., Klee, C. B., and Cohen, P. (1984) The structure of the B subunit of calcineurin. *Eur. J. Biochem.* **139,** 663–671.

Index

A

α-Lactalbumin
  calcium-binding properties, 211, 217–221
  calcium stoichiometry determination, 214–216
  structure, 211–212
ALG-2, *see also* Calpains and EF-hand proteins
  bacterial expression, 236–237
  chemical cross-linking, 236, 238–239
  purification, 236–238
  role in apoptosis, 235–236
Aluminum, 16–18
Analytical Ultracentrifugation, caseins, 104–107
Annexins, 6, 35, 225–227, *see also* S100 proteins
  bacterial expression, 227–229
  purification, 227, 229–230
  recombinant annexin II tetramer, 230
Apoptosis linked gene-2, *see* ALG-2

B, C

Bacterial Transformation, 162, 164
C2 domain proteins, 6, 35,
  calcium-binding mode, 305–306
    monitoring by NMR, 308–315
  membrane association, 295–296
Cadherins, 199
  bacterial expression, 200, 203
  calcium-binding, 205–207
  cell adhesion assay, 199, 201–202
  inhibition, 202
  purification, 203–205
Calbindin, *see also* EF-hand proteins
  calcium buffering role, 39, 175
  calcium removal, 183
  purification, 177–183
Calcineurin,
  assays, 320, 327–329
  calmodulin binding fragments, 320, 325–327
  function and regulation, 317–318
  proteolyis, 320–325
Calcium-binding proteins,
  extracellular, 4, 35
  intracellular, 6, 35–36
Calcium,
  binding to proteins,
    binding constant determination, 217–219, 299–300, 314
    co-ordination modes, 219–221
    detection by calcium-selective electrode, 205–207, 209
    detection by NMR spectroscopy, 308–315
    detection by PVDF radioisotope blotting, 205, 207–208
    stoichiometry determination, 214–216, 299–300, 301, 308
  cellular diffusion kinetics, 39–41

335

chelation and decontamination, 28–30, 298
complex equilibrium, 32–33
cytoplasmic concentration, 32
in evolution, 42–44
indicator dyes, 38–39, 217–219
isotope distribution, 41–44
membrane interaction, 33–34
metabolism, pumps and buffering, 5, 36–38, 44–48
physical properties, 22–23, 32
removal from proteins, 183
solubility, 6, 23–28
Calcium/calmodulin-dependent kinase I *see* CaMKI
Calcium/calmodulin-dependent kinase IIα *see* CaMKIIα
Calcium pump,
  calmodulin-binding domain, 14,15
  physiological role, 44
Calmodulin, *see also* EF-hand proteins
  affinity chromotography, 320, 330–331
  hydrophobic patch, 12
  overlays, 320
  plant isoforms, 143–144
    concentration determination, 148
    expression and purification, 144–147
  structure, 7–10
  targets, 8, 11–15
Calpain, *see also* EF-hand proteins and ALG-2
  activity and regulation, 51–53, 57–58, 60–61, 243–244
  calcium dependence, 55–57
  crystallization, 62–63, 244–246
  expression, purification, and storage 61–62
  structure 53–55, 247–256
    C2-like domain, 58–59
    calcium induced rearrangements, 251–254
    EF-hands, 247–251, 255–256
    heterodimerization, 255
    refinement, 246–247
    target/inhibitor binding, 254
Calsequestrin,
  calcium-binding models, 284–285
  folding, 284
  function, 282–283, 291–292
  purification, 285–286
  structure,
    crystallization, 286
    refinement, 286–287
    thioredoxin fold, 287–288
CaMKI,
  calmodulin-binding domain, 12, 13
CaMKIIα,
  calmodulin-binding domain, 12, 14
Caseins,
  aggregation, 102–106, 125–135
  micelles of, 98–102
  molecular modeling, 104, 116–124, 132–133
  radius of gyration, 122
  solubility,
    phosphate effects, 111
    self-association model, 111–115
    thermodynamic linkage, 102–103
  temperature effects, 107–111
Cellular adhesion proteins, *see* Cadherins
Chemical cross-linking,
  DSG (lysine specific), 238–239

Index

Chromotography, *see also* HPLC and Protein purification
  anion-exchange, 154–155, 164, 169, 176, 179–182, 203–204, 229, 237
  calmodulin-affinity, 320, 330–331
  gel filtration, 164, 171, 179, 180, 204, 229, 237
  heparin-Sepharose affinity, 229, 232
  hydroxyapatite, 229, 232
  phenyl sepharose,
    calcium-dependent hydrophobic affinity, 144–147, 164, 169–171, 191, 285–286
    hydrophobic interaction, 157–158
Circular dichroism spectroscopy, peptide aggregation, 125–129, 130–131
Coagulation factors, *see* γ-Carboxyglutamic acid proteins

D

ΔG,
  of solubility, 27, 28,
DANSYL, see Fluorescence, FRET
Desalting, *see* Chromotography, gel filtration

E

EDTA, *see* Calcium, chelation
EF-hand proteins,
  calcium co-ordination, 10, 11, 35, 249–251
  conformational changes on calcium-binding, 11, 251–254, 256–257, 265–267
  superfamily, 6

EGF domains,
  homology in γ-Carboxyglutamic acid proteins, 82
EGTA, *see* Calcium, chelation
Epidermal growth factor domains, *see* EGF domains
Eukaryotic protein expression, 191–196, 274–275

F

Fluorescence,
  calcium-binding dyes, 217–219
  FRET, 295–301
Fluorescence resonance energy transfer, see Fluorescence, FRET
Free energy *see* ΔG
FT-IR,
  protein calcium-binding mode, 219–221

G

γ-Carboxyglutamic acid proteins,
  calcium binding,
    catalytic domain, 89
    EGF-like domain, 87–88
    Gla residues, 82–86
  domain architecture, 82
  in blood, 81–82
  phospholipid binding, 83–84, 85–86
GCAP, 261–265, *see also* ROS-GC
Gel Filtration chromatography, *see* Chromotography, gel filtration
Gla residue proteins, *see* γ-Carboxyglutamic acid proteins
Guanylate cyclase activating protein, see GCAP
Guanylate cyclase assay, 269–272, 275

H

High performance liquid chromotography, *see* HPLC
High pressure liquid chromotography, *see* HPLC
Hill coefficient, *see* Calcium, binding to proteins, stoichiometry determination
HPLC,
   calcium-binding stoichiometry determination, 214–216
   DEAE-HPLC, 156–157
   reverse-phase, 321, 331
Hydroxyapatite, *see* Chromotography, hydroxyapatite

L

Lactalbumin, *see* α-Lactalbumin
Lysozyme (calcium-binding), 212–214, *see also* α-Lactalbumin

M

Magnesium, 23–26, 29–32
Microsequencing, 320, 329–330
Milk proteins, *see* Caseins, α-Lactalbumin and Lysozyme (calcium-binding)
MLCK,
   calmodulin-binding domain, 12, 14
Myosin Light Chain Kinase *see* MLCK
Myristoyl-calcium switches, *see* Neurocalcin, regulation and function

N

Neurocalcin,
   bacterial expression, 272
   guanylate cyclase activation, 269–272, 275
   purification, 272
   regulation and function, 264, 267–269
   structure,
      EF-hands, 265–267, 271,
      refinement, 273–274
NMR spectroscopy,
   ^{1}H-^{15}N correlation (HSQC), 308, 313
   manganese broadening, 310, 312, 315
   monitoring calcium-binding, 308–315
   structure determination, 308, 311–312
Nuclear magnetic resonance spectroscopy, *see* NMR spectroscopy

P

PCR, 187–188
Phospholipid vesicles, 297–298
Phenyl-sepharose chromatography, calcium-dependent, *see* Chromotography, calcium-dependent hydrophobic affinity
   hydrophobic interaction, 157–158
Photoreceptor proteins, *see* GCAP, Neurocalcin, and ROS-GC
Polymerase chain reaction, *see* PCR
Proteases, 319, 321
   calpain, *see* Calpain
   inhibitors, 319
   substrates, 319
Protein C, *see* γ-Carboxyglutamic acid proteins

Protein purification, *see also* Chromotography and HPLC
  acetone powder and extraction of bacterial cell pellet, 163, 165–169
  ammonium sulfate precipitation, 152–153, 155–156, 190–191
  heat treatment, 148, 177–179
Protein sequencing, *see* Microsequencing
Proteolysis, 317–332
Prothrombin, *see* γ-Carboxyglutamic acid proteins

R

Retinal guanylyl cyclase, *see* ROS-GC
Rod outer-segment guanylyl cyclase, *see* ROS-GC
ROS-GC, 261–265, *see also* GCAP and Neurocalcin

S

S100 proteins, *see also* EF-hand proteins, Annexins and GCAP
  bacterial expression, 186–190
  biochemical properties, 72–74, 185
  chromosomal organization, 69–72
  eukaryotic expression, 191–196
  functions, 74–76
  in disease, 76–77, 185–186
  purification, 187–191
Size-exclusion chromatography, *see* Chromotography, gel filtration

Synaptotagmin, 305–307, *see also* C2 domain proteins

T

Thrombin, *see* γ-Carboxyglutamic acid proteins
Transformation, *see* Bacterial transformation
Troponin C, *see also* EF-hand proteins
  cardiac, 151–152
    bacterial expression, 153–154
    purification, 154–158
    storage, 158
    UV absorbance, 158
  skeletal, 161–162
    bacterial expression, 162–165
    C/N domain purification, 163–169
    intact protein purification, 163–171

V

Vesicles, *see* Phospholipid vesicles

X

X-ray crystal structures,
  α-Lactalbumin, 211–212
  calmodulin, 7–10
  calpain, 53–61, 247–256
  calsequestrin, 286–292
  lysozyme (calcium-binding), 212
  neurocalcinin, 265–269, 273